Mechatronics and Robotics

Mechatronics and Robotics

New Trends and Challenges

Edited by

Marina Indri and Roberto Oboe

CRC Press is an imprint of the
Taylor & Francis Group, an **informa** business

First edition published 2020
by CRC Press
6000 Broken Sound Parkway NW, Suite 300, Boca Raton, FL 33487-2742

and by CRC Press
2 Park Square, Milton Park, Abingdon, Oxon, OX14 4RN

CRC Press is an imprint of Taylor & Francis Group, an Informa business

Library of Congress Cataloging-in-Publication Data

Names: Indri, Marina, editor. | Oboe, Roberto, editor.
Title: Mechatronics and robotics : new trends and challenges / edited by
Marina Indri, Roberto Oboe.
Description: Boca Raton : CRC Press, 2020. | Includes bibliographical
references and index.
Identifiers: LCCN 2020030663 (print) | LCCN 2020030664 (ebook) | ISBN
9780367366582 (hardback) | ISBN 9780429347474 (ebook)
Subjects: LCSH: Mechatronics. | Robotics.
Classification: LCC TJ163.12 .M43327 2020 (print) | LCC TJ163.12 (ebook)
| DDC 621--dc23
LC record available at https://lccn.loc.gov/2020030663
LC ebook record available at https://lccn.loc.gov/2020030664

Visit the Taylor & Francis Web site at
http://www.taylorandfrancis.com

and the CRC Press Web site at
http://www.crcpress.com

ISBN: 978-0-367-56204-5 (pbk)
ISBN: 978-0-367-36658-2 (hbk)
ISBN: 978-0-429-34747-4 (ebk)

DOI: 10.1201/9780429347474

Typeset in Times
by Deanta Global Publishing Services, Chennai, India

Dedication

To our families

Contents

Preface

Why another book on Mechatronics and Robotics? Aren't these disciplines mature? In other words, what's new? Indeed, Mechatronics and Robotics are no longer young disciplines, but they are still rapidly growing, thanks to new technologies that are spreading just now, and new developments are expected for them in the very next years.

In fact, even if Mechatronics (officially born in 1969) has now a well-defined and fundamental role in strict relation with Robotics, it is now undergoing a deep revolution in the industrial context, pushed by the Industry 4.0 and Internet of Things concepts. A similar fast-paced evolution is involving the everyday life as well, in scenarios characterized by a growing collaboration between humans and robots.

Several books on Mechatronics and Robotics can be found, but they often come from a single research group or they deal with some specific aspects of these disciplines.

The idea underlying this book is to take a different approach and to offer a wide overview of new research trends and challenges for both Mechatronics and Robotics, through the contributions of top researchers around the world. This book is not intended to be a textbook. Instead, it should lead the reader on a scientific journey from advanced actuators and sensors to human-robot interaction, through robot control, navigation, planning and programming issues, with representative experts of their research areas as accompanying guides. Hot topics in both Mechatronics and Robotics are presented, like, for example, multiple-stage actuation to cope with conflicting specification of large motion-spans and ultra-high accuracy, model-based control for high-tech mechatronic systems, modern approaches of software systems engineering to robotics, humanoids for human assistance, etc. For each topic, the reader has the possibility to enhance his/her competences from the specific contributions provided by the authors, but also from the wide, accurate analysis of the state-of-the-art included in each chapter. New techniques are investigated in approaching design of mechatronic systems, and both industrial and service robotics scenarios are considered, addressing also some examples of interaction between humans and mechanisms in contexts like medical, nursing and surgery applications.

We do believe that the readers will enjoy this special journey through the new trends and challenges in Mechatronics and Robotics, at the end of which they will have earned a deeper knowledge on new technologies, applications and methods, to be later used in their own academic or industrial settings.

Marina Indri and Roberto Oboe

About the Editors

Marina Indri received her M.Sc. degree in Electronic Engineering in 1991 and the Ph.D. in Control and Computer Engineering in 1995, both from Politecnico di Torino, Italy. She is currently associate professor at the Department of Electronics and Telecommunications of Politecnico di Torino, teaching Robotics and Automatic Control.
She coauthored about 100 papers in international journals, books and conference proceedings, in the fields of robotics and automatic control. Her current research interests are in the industrial and mobile robotics areas. She participated in various research projects, and was involved also in research joint activities with industrial partners. She received the best paper award in Factory Automation at ETFA 2014, the 2nd prize of the euRobotics Technology Transfer Award in 2014 and was among the finalists of the same Award in 2017 for joint works with COMAU S.p.A.

She is a member of the IEEE Industrial Electronics Society (IEEE-IES) and chair of the IEEE IES Technical Committee on Factory Automation. She serves as an associate editor of the IEEE Transactions on Industrial Informatics, and as a technical editor of the IEEE/ASME Transactions on Mechatronics.

Roberto Oboe obtained a Laurea degree in Electrical Engineering in 1988 from the University of Padova, Padova, Italy. During 1990 he was a visiting researcher at the Department of Electrical Engineering of the Keio University - Yokohama - Japan. He obtained a Doctorate degree from the University of Padova in 1992. From 1993 to 2003, he has been with the Dipartimento di Elettronica e Informatica of the University of Padova. From 2003 to 2008 he was associate professor in Automatic Control at the Department of Mechanical and Structural Engineering, University of Trento. Since 2008, he has been an associate professor in Automatic Control at the University of Padova, Dep.t of Management and Engineering. In 1996 visited the Biorobotics Laboratory of the Univ. of Washington, Seattle, where he participated to the realization a system for the manipulation of objects in shared virtual environments. In 1997, ked at the Jet Propulsion Laboratory (JPL), where he has been working on Internet-based telerobotics with haptic feedback. In 2000, he was visiting scholar at the Center for Magnetic Recording Research, UCSD. In 2002, he as a visiting faculty member at the Department of Mechanical Engineering, Univ. of California, Berkeley. His interests are in the fields of data storage, parametric identification of mechanical systems, control and applications of MEMS devices, applied digital control, telerobotics, virtual mechanism, haptic devices, and biomedical equipment. He is author/coauthor of almost 200 contributions, plus 4 international patents. He is senior member of the IEEE. For the IEEE Industrial Electronics Society (IES), he served as vice president for Planning and Development, chair of the Technical Committee on Motion Control and senior AdCom member. He is currently vice president for Technical activities, chair of the Management Committee of the IEEE/ASME Transactions on Mechatronics, and associate editor of the IEEE Transactions on Industrial Electronics.

Contributors

Arash Ajoudani
Istituto Italiano di Tecnologia
Genova, Italy

Darwin G Caldwell
Istituto Italiano di Tecnologia
Genova, Italy

Henry Carrillo
Pontificia Universidad Javeriana
Bogotà, Colombia

José A. Castellanos
Universidad de Zaragoza
Zaragoza, Spain

Yasutaka Fujimoto
Yokohama National University
Yokohama, Japan

Hubert Gattringer
Johannes Kepler University Linz
Linz, Austria

Van-Dung Hoang
Quang Binh University
Đ`ông Hô'i, Vietnam

Marina Indri
Politecnico di Torino
Torino, Italy

Takahiro Ishikawa
Keio University
Tokyo, Japan

Makoto Iwasaki
Nagoya Institute of Technology
Nagoya, Japan

Kang-Hyun Jo
University of Ulsan
Ulsan, Korea

Yasir Latif
University of Adelaide
Adelaide, Australia

Jinoh Lee
Istituto Italiano di Tecnologia
Genova, Italy

Alex Lotz
Technische Hochschule Ulm
Ulm, Germany

Matthias Lutz
Technische Hochschule Ulm
Ulm, Germany

Claudio Melchiorri
University of Bologna
Bologna, Italy

Yasue Mitsukura
Keio University
Tokyo, Japan

Andreas Müller
Johannes Kepler University Linz
Linz, Austria

Toshiyuki Murakami
Keio University
Tokyo, Japan

Roberto Oboe
Università degli Studi di Padova
Padova, Italy

Naoki Oda
ChitoseInstitute of Science and Technology
Hokkaido, Japan

Kouhei Ohnishi
Keio University
Tokyo, Japan

Tom Oomen
Technische Universiteit Eindhoven
Eindhoven, The Netherlands

Lorenzo Sabattini
University of Modena and Reggio Emilia
Italy

Christian Schlegel
Technische Hochschule Ulm
Ulm, Germany

Cristian Secchi
University of Modena and Reggio Emilia
Italy

Bruno Siciliano
Università di Napoli Federico II
Napoli, Italy

Klemens Springer
Engel Austria GmbH
Schwertberg, Austria

Dennis Stampfer
Technische Hochschule Ulm
Ulm, Germany

Maarten Steinbuch
Technische Universiteit Eindhoven
Eindhoven, The Netherlands

Satoshi Suzuki
Tokyo Denki University
Tokyo, Japan

Yuta Tawaki
Keio University
Tokyo, Japan

Nikos Tsagarakis
Istituto Italiano di Tecnologia
Genova, Italy

Toshiaki Tsuji
Keio University
Tokyo, Japan

Luigi Villani
Università di Napoli Federico II
Napoli, Italy

1 Mechatronics versus Robotics

Marina Indri and Roberto Oboe

CONTENTS

1.1 MECHATRONICS: DEFINITIONS AND EVOLUTION

The name *mechatronics* was coined in 1969 (40 years ago!) by Ko Kikuchi, who subsequently became president of Yaskawa Electric Corporation [10,32]. The word is composed of "mecha" from mechanism, i.e., machines that move (or mechanics), and "tronics" from electronics, and reflects the original idea at the basis of this discipline, i.e., the integration of electrical and mechanical systems into a single device.

The spread of this term has been growing over the years, and different definitions have been given, each time adding something and/or underlining some aspect not previously highlighted. The analysis of the various definitions throughout the years can help us to understand how mechatronics was considered at the beginning and what it represents nowadays.

In Bolton [6], mechatronics is defined as the "integration of electronics, control engineering, and mechanical engineering", thus recognizing the fundamental role of control in joining electronics and mechanics. An *official* definition was given in the mid-1990s by a technical committee of the International Federation for the Theory of Machines and Mechanism: "Mechatronics is the synergistic combination of precision mechanical engineering, electronic control and systems thinking in the design of products and manufacturing processes" [10]. The key point in such a definition was the concept of *synergistic combination*, distinguishing mechatronics from the classical *concurrent engineering* approach, in which groups of the same project team work separately, sharing the overall obtained results but only partially sharing the specific design decisions during the project development.

Subsequent definitions in the late 1990s put in evidence the goal of designing "*improved* products and processes" [5], thanks to the synergistic use of the different mechatronics components, and of achieving an "*optimal design* of electromechanical products" [48], recognizing at the same time the role of other disciplines beyond electronics, control engineering, and mechanical engineering, such as computer engineering and communication/information engineering.

At the beginning of the 21st century, mechatronics already had a well-defined and marked identity as an ever-growing engineering and science discipline, thanks to the continuous advancement in enabling technologies, such as multi-sensor fusion, motion devices, very large integrated circuits, microprocessors and microcontrollers, and system and computational intelligence software and techniques [29]. The importance of sensors and of communication capabilities for mechatronic systems has grown steadily. Currently, an intelligent mechatronic system is supported by various sensing devices, and it can be of micro/nano size as well as highly integrated in a multi-system overall architecture.

1.2 MECHATRONICS VERSUS ROBOTICS

What is the difference between mechatronics and robotics? A robot is commonly considered as a typical mechatronic system, which integrates software, control, electronics, and mechanical designs in a synergistic manner. Robotics can be considered as a part of mechatronics; i.e., all robots are mechatronic systems, but not all mechatronic systems are robots. Fukuda and Arakawa [15] highlighted how autonomy and self-decision making represent the key points to classify systems in the fields of mechatronics and robotics. All machines that do not have any kind of autonomy in their behavior, because they simply *automatically* act according to the inputs they receive (directly or indirectly) from humans, are strictly pure mechatronic systems. All robots (manipulators, mobile robots, etc.) instead have a certain degree of autonomy, and hence, they can be considered as special mechatronic systems, which can be classified both in the mechatronics and in the robotics fields. If a robot has also a proper self-decision-making function, allowing it to autonomously determine its behavior, it is then classified into the fields of robotics only: according to Fukuda and Arakawa [15], it is then something more than a pure mechatronic system.

Such a classification is not (and cannot be) so rigid, because the level of autonomy of a mechatronic system and a robot could be very different. In Novianto [33], it is said that "the main difference is inputs are *provided* to mechatronics systems whereas robotics systems *acquire* inputs by their own".

The capability of a robot of acquiring inputs on its own is strongly related to its *awareness* of what is happening around it. Sensors and information from the environment can be used by a robot not only to automatically act according to the way it has been programmed but also to vary its behavior in an autonomous manner.

The industrial robots have generally a quite limited (or null) awareness of the events occurring in the surrounding environment, so that they can be considered as standard examples of good mechatronic systems. On the contrary, the last generations of robots, devoted to various, not strictly industrial, applications, are characterized by an ever-growing level of autonomy and awareness of the environment. Examples are given by service robots, underwater and aerial robots, and biologically inspired robots; all these kinds of robots exploit various technologies beyond mechatronics, e.g., automotive issues, smart machine technology, and software and communication structures.

1.3 MECHATRONICS AND ROBOTICS: NEW RESEARCH TRENDS AND CHALLENGES

As mentioned in the previous sections, drawing a sharp border between mechatronics and robotics is impossible, as they share many technologies and objectives. A mechatronic system makes use of sensors, actuators, and controllers, like a basic robot's equipment. They usually have a specific task to accomplish, and this is performed in a known and fixed scenario. Advanced robots, on the other hand, usually plan their actions by combining an assigned functional task with the knowledge about the environment in which they operate. By using a simplified approach, advanced robots could be defined as mechatronic devices governed by a "smart brain", placed at a higher hierarchical level. This definition, however, does not cover some new mechatronic devices, used in tight interaction with humans, such as an active rehabilitation orthosis. For these types of devices, user safety is the primary issue. Clearly, for the achievement of intrinsically safe behavior, the use of smart control strategy (the above-mentioned "smart brain") is not enough, and the use of actuators that will never harm the user (even in case of control failure) is mandatory. This simple example tells how it is difficult to characterize new research trends and challenges exclusively pertaining to mechatronics or robotics. Having this in mind, in the following, we will present what we consider the "hot topics" in both fields, with attention to new fields of application, new challenges to the research communities, and new technologies available. The next subsections will follow the reminder of the book, in which researchers from different institutions will provide their view on specific subjects.

1.3.1 ADVANCED ACTUATORS FOR MECHATRONICS

Actuators are building blocks of any mechatronic system. Such systems, however, have a huge application span, ranging from low-cost consumer applications to high-end, high-precision industrial manufacturing equipment. Actuators have to provide the driving force to a mechanism, which in turn, has to perform some actions or movements. The technologies available to produce such driving forces are quite varied. We can have traditional electromagnetic actuators, piezoelectric, capacitive, pneumatic, hydraulic, etc. Each of them has different peculiarities in terms of range of force generated, speed of response, and accuracy. In addition to the actuator itself, reduction gears, recirculating ball screws, etc. are often used to convert the actuator output in order to be coupled to the mechanical load. Such fixtures are usually introducing friction and backlash, thus reducing the level of precision achievable. The solution to this problem has been addressed by either designing new actuators [14] or introducing sophisticated modeling and compensation of the non-idealities introduced by the motion transmission device [41]. The design of new actuators can follow different paths, driven by the specific needs of the application. In high-precision applications, the use of a secondary actuator, placed in series to a primary one, is an effective way to obtain at the same time large motion spans and high accuracy. This solution, at first introduced in consumer products like Hard Disk Drives [31], is now being utilized in manufacturing plants when a single actuator cannot achieve at the same time the prescribed range of motion and accuracy [9]. The use of multiple extra actuators, however, requires the development of new control strategies to cope with the diverse interaction that may arise compared with standard systems with a single actuator [56]. Two–degrees of freedom (d.o.f) actuators can also be designed in order to combine two motions, like the roto-translational direct drive motor presented in Tanaka et al. [51]. Actuators can also be tailored to the specific application, and they can be designed either to move along a specific path in space (see the half-circle-shaped tubular permanent magnet motor presented in Omura et al. [35]) or to provide a variable compliance in order to increase the safety of all mechatronic devices that directly interact with the user [44]. Finally, among all possible actuator technologies available for a mechatronic application, the designer must account for the cost/performance trade-off. For some inexpensive, ultra-low-cost applications, such as those for consumer devices, a new type of actuator makes use of shape memory alloy (SMA) wires, which contract and relax according to their operating temperature [30]. Their response time is in the order of tens to hundreds of milliseconds, but they can be profitably used in many mechatronic applications, thanks also to their self-sensing capabilities [4].

Among all the possible directions mentioned, we will examine in greater depth the main issues arising in the realization of a high-end industrial manufacturing system. For the achievement of the highest performance, accurate modeling and compensation of the non-idealities of the transmission gears is mandatory. Additionally, multiple-stage actuation is becoming the "mantra" for the high-end designer, as it makes it possible to cope with the usual conflicting specifications of large motion spans and ultra-high accuracy. Both concepts will be developed by Prof. Iwasaki of the Nagoya Institute of Technology (Japan).

1.3.2 ADVANCED SENSORS FOR MECHATRONICS

Like actuators, sensors are needed in all mechatronic systems in order to guarantee the achievement of the desired performance in spite of all disturbances and uncertainties. Indeed, there are numerous references on traditional sensors for mechatronic systems (see for instance Regtien [39]), mainly aimed at the description of standard sensor technologies and their use in measuring some quantity to be controlled. The technology advancements and the manufacturing cost reduction, however, have opened new perspectives in the use of sensors in mechatronic systems. This is the case, for instance, of micro-electromechanical system (MEMS) sensors [47]. Thanks to the push coming from the consumer electronics world, the once expensive accelerometers and gyroscopes have evolved into ultra-small ($2 \times 2 \times 1$ mm) and low-cost (less than 1 USD) commodities, which can be placed virtually everywhere in a mechatronic system to better monitor and/or control it. Of course, the use of additional sensors to achieve higher control performances requires smarter designs and better understanding of the underlying system dynamics. Applications of MEMS sensors in the control of mechatronic devices can be found in the area of vibration control, where they can be used to sense the oscillations of the mechanical load due to flexible couplings between actuator and load [3]. The same sensors can be also used to measure and compensate torque/force ripples caused by the actuator non-idealities, thus allowing a better overall performance [2].

In addition to classical sensors capable of measuring a single process variable, we see a growing number of applications in which the state of the process to be controlled and possibly, its colocation with respect to the operating scenario are determined by properly combining measurements from different sensors. This is not only the case for autonomous vehicles [23] but refers also to floor-cleaning robots, smartphones, inertially stabilized cameras, etc. In this scenario, humans make their appearance as sources of disturbance (e.g., obstacles to be avoided) or as partners to collaborate with (e.g. in assistive mechatronic devices). This deep interaction between mechatronic devices and humans requires specific sensors, and this subject is being tackled by Prof. Murakami of the Keio University (Japan) and his collaborators, who will further explore some aspects on the use of vision systems and other sensors (tactile, brain-to-computer interfaces, etc.) for the implementation of safe and reliable interaction of humans with mechatronic devices, like an active wheelchair.

1.3.3 MODEL-BASED CONTROL TECHNIQUES FOR MECHATRONICS

Control is what ties together mechanisms, actuators, and sensors in order to perform an assigned task with a prescribed degree of accuracy, speed, and robustness, all in spite of the possible disturbances. Usual designs rely on feedback to synthesize the proper command for the actuator, and this per se requires an accurate tailoring of the control around the nominal plant, its possible variations, and the disturbances acting on the system. The simple mathematical modeling of the plant is no longer sufficient in this scenario, and new techniques and procedures have been developed through the years, aimed at identifying not only a linear approximation of the process to be controlled but also all disturbances and non-linearities affecting it [21,45,50]. But feedback control of a mechatronic device relies on a measurement of errors between target and actual motion, which unfortunately, tells us that with this approach, it is almost impossible to achieve a null error by relying exclusively on feedback.

The usual practice, in this scenario, is to add some feedforward action, as it is typically done in computer numerical control (CNC) when implementing the tracking of a trajectory. In advanced, ultra-high-precision control, the use of feedforward is gaining more and more support, thanks to the development of very accurate models for all system components. A notable area in which the demanded performance is at the highest level is the photolithography in integrated circuit manufacturing. Here, the most advanced control techniques have been developed, and the most relevant results (applicable in many other highly demanding mechatronic applications) will be presented by Prof. Omen and Prof. Steinbuch of the Eindhoven University of Technology (the Netherlands).

1.3.4 CONTROL AND MANIPULATION

The traditional definition of robotics deals with articulated, multi-d.o.f. mechanisms, capable of manipulating objects and tools along a desired path, in terms of position and orientation. On this subject, numerous publications and books are available, dealing with standard motion control, force control, visual feedback, etc. [11,20,46]. The most recent trends and challenges are oriented toward new robotics applications, requiring advanced manipulation capabilities and possible collaborations with the human operator, so that the most classical approaches in the fields of task space control, robot compliance behavior, and force and interaction control need to be revised, e.g., including joint trajectory generation to create compliant motion for manipulators [24], developing new solutions for task space control guaranteeing a compliant behavior for possible redundant d.o.f. [42], or exploiting force control and optimized motion planning to achieve autonomous manipulation capabilities [40]. In the first chapter entirely devoted to robotics, Prof. Siciliano and Prof. Villani of the University of Napoli (Italy) will survey the motion control problem as the basis for more advanced algorithms. Particular attention will be devoted to manipulators having non-negligible joint elasticity, as well as to redundant robotic systems, exploiting a large number of degrees of freedom to simultaneously execute multiple tasks. Several applications, in both industrial and service robotics, require the physical interaction of the robot manipulator with the environment, or possibly with the human operator. Since the pure motion control strategies are not suitable for handling such situations, the main interaction control approaches will be also addressed in the same chapter, which will be completed by a discussion about future directions in robot control, including a list of recommended reading for a deeper analysis.

1.3.5 NAVIGATION, ENVIRONMENT DESCRIPTION, AND MAP BUILDING

Mobile robots, introduced several decades ago [17], have evolved from simple "devices on wheels", with very simple reaction control (e.g., aimed at implementing simple collision avoidance), to very complex systems, with a rich set of sensors and actuators, under the control of sophisticated software, which allows the robot to autonomously move in an unknown environment. Halfway between fully autonomous mobile robots and simple reacting mechanisms, we can find automatic guided vehicles (AGVs), which are nowadays profitably deployed in many industrial scenarios [52]. Most of them navigate the environment by following some magnetic or optical reference placed under or over the floor. This, however, makes any modification of the paths somewhat hard, and there are also problems in dealing with turnarounds in the case of unexpected obstacles on the path. So, even in industry, there is a lot of interest in those technologies developed in the field of totally autonomous mobile robots, which do not make use of special guiding infrastructure to reach the final target position. In this regard, critical for the completion of the tasks assigned to a mobile robot is its ability to detect its current position with respect to the environment in which it moves and to the final goal. Additionally, the capability of finding a way to reach the final goal is necessary [49]. Prof. Castellanos of the University of Saragozza (Spain) and his collaborators will illustrate the most recent advancements in the area of mobile robot navigation, starting from the above-mentioned basic issues and getting to the so-called SLAM (Simultaneous-localization-and-mapping).

1.3.6 PATH PLANNING AND COLLISION AVOIDANCE

In the previous subsection, we mention the need for modern mobile robots to localize themselves in an unknown environment and to find their way to the final goal. This goal-reaching feature is also present in standard industrial robotics, where, however, the operating scenario is slightly different compared with that of mobile robots. For instance, in an industrial robotic cell, the manipulator is an articulated mechanism, for which the computation of possible collisions with the obstacles in the workspace is far more complicated than in the case of box-shaped mobile robots. Additionally, path planning is no longer merely done by constructing a path from starting to end position, composed by lines and circular arcs [27], but it takes advantage of the availability of high computational power to use more complicated curves definitions, like high order polynomials and NURBS [37]. Planning can be done also by taking into account additional objectives, like the minimization of time or energy consumption [53], and even by exploiting the redundancy of the manipulator, when available [16]. Eventually, the trajectory planning can also consider dynamic aspects such as the structural flexibility of the manipulator and in turn, design a trajectory that does not excite the structural modes or produce a null residual vibration of the end effector at final target position (or while following a desired path) [55]. Prof. Müller of the Johannes Kepler University (Austria) and his collaborators will present the most recent advancements in the field of optimal path planning and collision avoidance, both dealing with standard and redundant robots and investigating methods to incorporate the existence of obstacles in the robot workspace in the path optimization problem.

1.3.7 ROBOT PROGRAMMING

During the early age of industrial robotics, programming was restricted to the construction of a list of actions to be performed by a single manipulator, possibly interfaced with some sensors and coordinated with other manipulators placed in the same production line [36]. In all robotics textbooks, a little portion was devoted to programming, and usually, the subject was treated as something handled by robot manufacturers, who developed proprietary programming languages (e.g., VAL by PUMA). Each manufacturer also developed closed ecosystems, in which only the proprietary peripherals (like vision systems) could be seamlessly integrated into the robotic cell. Programming, in such a scenario, was an art, mastered by a few experts with a deep understanding of the architecture of the robotic system (i.e., the mechanical device, its sensors, the controlling unit, and all the peripherals), and adding new features (e.g., by using the readings of new sensors) was extremely difficult. Since early stage, robot programming has evolved in many directions in order to cope with the requests for creating new robotic equipment, possibly composed of multiple robotic devices and sensors, manufactured by different companies or developed ad hoc. With the emerging need for new and complex robotic equipment, its programming is becoming more and more a team exercise with parallel development of different functionalities. Both academia and industry have tried to respond to such requests for a unique digital industrial platform for robotics. The present scenario is vast; the chapter authored by Prof. Schlegel and his collaborators of the Technische Hochschule Ulm (Germany) will investigate the most advanced solutions nowadays available for robot programming, explaining the step change from framework-specific programming to technology-agnostic modeling, separation of roles, and composition. Eclipse-based open-source tooling with repositories of software components and robotic applications is accessible to make the first steps.

1.3.8 NETWORK ROBOTICS

Recently, there has been growing interest in and research activity on cooperative control and motion coordination of multiple robots. Such interest is mainly due to the growing possibilities enabled by robotic networks, not only in the monitoring of natural phenomena and the enhancement of human capabilities in hazardous and unknown environments but also in the industrial scenario, where a team of networked robots can be used to flexibly implement a production cycle [8]. In this scenario, the coordination between robots becomes a key issue to exploit as much as possible the potentialities of a team cooperatively carrying out a common task. Multi-robot coordination addresses several issues, e.g., centralized and decentralized control, formation control [1], consensus networks [34], coordinated trajectory tracking, and communication infrastructures and resources (heavily exploited by the most recent *cloud robotics* solutions [25]). Prof. Melchiorri of the University of Bologna (Italy) and his collaborators from University of Reggio Emilia and Modena (Italy) will consider a team of mobile robots, equipped with general-purpose tools and coordinated along complex trajectories, to be employed as an automated solution for highly flexible and variable production scenarios. In particular, the multi-robot system is partitioned into two groups: one of independent robots (acting as supervisors and defining the production cycle) and the other one composed of dependent robots, actually provided by tools and acting as workers.

1.3.9 INTELLIGENT, ADAPTIVE HUMANOIDS FOR HUMAN ASSISTANCE

Humanoid robots, for long time present only in sci-fi movies and novels and more recently, as expensive demonstrators of technological achievements [19], are now making their actual appearance in real-world applications, where they are promising to bring a new form of bilateral and assistive interaction with humans. The envisioned applications of humanoid robots are countless, and they all have in common the fact that this type of robot will closely interact with humans in the same places where they live and work. Such environments, indeed, have stairs, doors, windows, etc. and are full of different objects to be grasped, manipulated, and moved around. Humanoid robots represent a great challenge for both science and technology, as their realization requires a deep understanding of essential aspects of biomechanics (at least those needed for the replication of some human ability, like walking, stair climbing, object grasping, perception, etc. [22]) and the deployment (and sometimes the development) of many different technologies for sensing and actuation [12]. Additionally, as humans, it is expected that humanoid robots will interact with their world in an adaptive way, learning by experience. The state of the art and the most recent achievements in this research area of robotics will be investigated by Prof. Caldwell and his collaborators of the Italian Institute of Technology, with particular focus on the COMAN humanoid robot developed at IIT.

1.3.10 ADVANCED SENSORS AND VISION SYSTEMS

The most recent research results in machine vision and advanced sensors are leading to significant improvements in various fields, in particular in autonomous vehicle navigation, with several applications from robotics to space exploration [28]. Autonomous navigation can be achieved by giving the robot the capability of planning a global path toward the target position [54], [13], of locating itself with respect to certain benchmarks (or with respect to a given map), and of recognizing obstacles that must be avoided [49]. Such capabilities require the combination of data coming from different types of sensors (e.g. vision sensors, omnidirectional cameras, laser range finders, and geographical positioning systems [GPS]), whose characteristics and performances must be exploited in a robust manner, taking into account the environmental conditions in which the autonomous agent has to move [38,43]. For example, the environment could be indoor or outdoor, static or dynamic, structured or unstructured, described by means of landmarks or by an occupancy grid map.

Prof. Kang-Hyun Jo from University of Ulsan (Korea) and his collaborator from Ho Chi Minh City University of Technology and Education (Vietnam) will survey the main issues of visual odometry (such as the extraction and matching of feature descriptors and the estimation of the robot rotation by using an omnidirectional vision system) as well as some advanced sensors successfully employed for autonomous navigation, like laser range finders and GPS, providing general guidelines for sensor combination.

1.3.11 HUMAN–ROBOT INTERACTION

A strong human–robot interaction (HRI) requires that humans and robots share similar sensing capabilities. The haptic sense in particular allows humans to recognize various physical characteristics of an object by simply touching it. The "real-haptics" technology allows the reconstruction of the haptic sense and can lead to a new generation of robots, ready for a more complete interaction with humans in various fields, from daily life to medical and rehabilitation robotics [7,18]. Moreover, a smooth and compliant manipulation capability can be exploited in the Robot Learning from Demonstration framework for the development of teaching interfaces that allow the robot stiffness to be changed by physically interacting with it [26].

The potentialities of real haptics as a new way to interact with robots are discussed by Prof. Kohuei Ohnishi from Keio University (Japan) in the last chapter, which provides an introduction to the basic principles of the feedback of the tactile sensation between the robot and the human together with some experimental examples.

REFERENCES

1. G. Antonelli, F. Arrichiello, F. Caccavale, and A. Marino. Decentralized time-varying formation control for multi-robot systems. *The International Journal of Robotics Research*, 33(7), pages 1029–1043, 2014.
2. R. Antonello, A. Cenedese, and R. Oboe. Torque ripple minimization in hybrid stepper motors using acceleration measurements. In *Proceedings 18th World Congress of the International Federation of Automatic Control (IFAC)*, 44(1), pages 10349–10354, Jan 2011.
3. R. Antonello, A. Cenedese, and R. Oboe. Use of MEMS gyroscopes in active vibration damping for HSM-driven positioning systems. In *IECON 2011 - 37th Annual Conference on IEEE Industrial Electronics Society*, pages 2176–2181, Nov 2011.
4. R. Antonello, S. Pagani, R. Oboe, M. Branciforte, and M.C. Virzi. Use of antagonistic shape memory alloy wires in load positioning applications. In *Industrial Electronics (ISIE), 2014 IEEE 23rd International Symposium on*, pages 287–292, Jun 2014.
5. S. Ashley. Getting a hold on mechatronics. *Mechanical Engineering*, 119(5), pages 60–63, 1997.
6. W. Bolton. *Mechatronics: Electronic Control Systems in Mechanical Engineering*. Longman, 1995.
7. S.A. Bowyer, B.L. Davies, and F. Rodriguez y Baena. Active constraints/virtual fixtures: A survey. *IEEE Transactions on Robotics*, 30(1), pages 138–157, 2014.
8. F. Bullo, J. Cortés, and S. Martínez. *Distributed Control of Robotic Networks: A Mathematical Approach to Motion Coordination Algorithms: A Mathematical Approach to Motion Coordination Algorithms*. Princeton Series in Applied Mathematics. Princeton University Press, 2009.
9. Y.-M. Choi and D.-G. Gweon. A high-precision dual-servo stage using halbach linear active magnetic bearings. *Mechatronics, IEEE/ASME Transactions on*, 16(5), pages 925–931, Oct 2011.
10. R. Comerford. Mecha... what? *IEEE Spectrum*, 31(8), pages 46–49, 1994.
11. J.J. Craig. *Introduction to Robotics: Pearson New International Edition: Mechanics and Control*. Pearson Education Limited, 2013.

12. P. Dario, C. Laschi, and E. Guglielmelli. Sensors and actuators for 'humanoid' robots. *Advanced Robotics*, 11(6), pages 567–584, 1996.
13. N.E. Du Toit and J.W. Burdick. Robot motion planning in dynamic, uncertain environments. *IEEE Transactions on Robotics*, 28(1), 2012.
14. Y. Fujimoto and K. Ohishi. Newest developments and recent trends in sensors and actuators; a survey. In *Industrial Electronics Society, IECON 2013 - 39th Annual Conference of the IEEE*, pages 80–87, Nov 2013.
15. T. Fukuda and T. Arakawa. Intelligent systems: Robotics versus mechatronics. *Annual Reviews in Control*, 22, pages 13–22, 1998.
16. M. Galicki. Time-optimal controls of kinematically redundant manipulators with geometric constraints. *Robotics and Automation, IEEE Transactions on*, 16(1), pages 89–93, Feb 2000.
17. W. Grey Walter. An electromechanical animal. *Dialectica*, 4, pages 42–49, 1950.
18. R. Groten, D. Feth, R.L. Klatzky, and A. Peer. The role of haptic feedback for the integration of intentions in shared task execution. *IEEE Transactions on Haptics*, 6(1), pages 94–105, 2013.
19. M. Hirose and K. Ogawa. Honda humanoid robots development. *Philosophical Transactions of the Royal Society of London A: Mathematical, Physical and Engineering Sciences*, 365(1850), pages 11–19, 2007.
20. S. Hutchinson, G.D. Hager, and P.I. Corke. A tutorial on visual servo control. *IEEE Transactions on Robotics and Automation*, 12(5), pages 651–670, Oct 1996.
21. M. Iwasaki, M. Yamamoto, H. Hirai, Y. Okitsu, K. Sasaki, and T. Yajima. Modeling and compensation for angular transmission error of harmonic drive gearings in high precision positioning. In *Advanced Intelligent Mechatronics, 2009. AIM 2009. IEEE/ASME International Conference on*, pages 662–667, Jul 2009.
22. B. Jakimovski. *Biologically Inspired Approaches for Locomotion, Anomaly Detection and Reconfiguration for Walking Robots*. Cognitive Systems Monographs. Springer Berlin Heidelberg, 2011.
23. J. Jurado, K. Fisher, and M. Veth. Inertial and imaging sensor fusion for image-aided navigation with affine distortion prediction. In *Position Location and Navigation Symposium (PLANS), 2012 IEEE/ION*, pages 518–526, Apr 2012.
24. M. Kazemi, J.S. Valois, J.A. Bagnell, and N. Pollard. Robust object grasping using force compliant motion primitives. In *Proceedings of Robotics: Science and Systems*, 2012.
25. B. Kehoe, S. Patil, P. Abbeel, and K. Goldberg. A survey of reasearch on cloud robotics and automation. *IEEE Transactions on Automation Science and Engineering*, 12(2), pages 398–409, 2015.
26. K. Kronander and A. Billard. Learning compliant manipulation through kinesthetic and tactile human-robot interaction. *IEEE Transactions on Haptics*, 7(3), pages 367–380, 2014.
27. J.C. Latombe. *Robot Motion Planning*. The Springer International Series in Engineering and Computer Science. Springer, 1991.
28. T. Luettel, M. Himmelsbach, and H.-J. Wuensche. Autonomous ground vehicles–concepts and a path to the future. *Proceedings of the IEEE*, 100(Special Centennial Issue), pages 1831–1839, 2012.
29. R.C. Luo and Y.W. Perng. Advances of mechatronics and robotics. *IEEE Industrial Electronics Magazine*, 5(3), pages 27–34, Sep 2011.
30. M. Moallem and V.A. Tabrizi. Tracking control of an antagonistic shape memory alloy actuator pair. *Control Systems Technology, IEEE Transactions on*, 17(1), pages 184–190, Jan 2009.

31. K. Mori, T. Munemoto, H. Otsuki, Y. Yamaguchi, and K. Akagi. A dual-stage magnetic disk drive actuator using a piezoelectric device for a high track density. *Magnetics, IEEE Transactions on*, 27(6), pages 5298–5300, Nov 1991.
32. T. Mori. Yasakawa internal trademark application memo. Technical Report 21.131.01, Yasakawa Electric Corporation, July 12 1969.
33. R. Novianto. *Website*, 2015. http://www.ronynovianto.com.
34. R. Olfati–Saber, J. A. Fax, and R. M. Murray. Consensus and cooperation in networked multi–agent systems. *Proceedings of the IEEE*, 95(1), pages 215–233, 2007.
35. M. Omura, T. Shimono, and Y. Fujimoto. Development of a half-circle-shaped tubular permanent magnet machine. In *Industrial Electronics Society, IECON 2013 - 39th Annual Conference of the IEEE*, pages 6114–6119, Nov 2013.
36. R.P. Paul. *Robot Manipulators: Mathematics, Programming, and Control : The Computer Control of Robot Manipulators*. Artificial Intelligence Series. MIT Press, 1981.
37. L. Piegl and W. Tiller. *The NURBS Book*. Springer Verlag, 2nd edition, 1997.
38. Li Q., L. Chen, M. Li, S.-L. Shaw, and A. Nuchter. A sensor-fusion drivable-region and lane-detection system for autonomous vehicle navigation in challenging road scenarios. *IEEE Transactions on Vehicular Technology*, 63(2), pages 540–555, 2014.
39. P.P.L. Regtien. *Sensors for Mechatronics*. Elsevier, 2012.
40. L. Righetti, M. Kalakrishnan, P. Pastor, J. Binney, J. Kelly, R.C. Voorhies, G.S. Sukhatme, and S. Schaal. An autonomous manipulation system based on force control and optimization. *Autonomous Robots*, 36, pages 11–30, 2014.
41. M. Ruderman, T. Bertram, and M. Iwasaki. Modeling, observation, and control of hysteresis torsion in elastic robot joints. *Mechatronics*, 24(5), pages 407–415, 2014.
42. H. Sadeghian, L. Villani, M. Keshmiri, and B. Siciliano. Task-space control of robot manipulators with null-space compliance. *IEEE Transactions on Robotics*, 30(2), pages 493–506, Apr 2014.
43. D.O. Sales, D.O. Correa, L.C. Fernandes, D.F. Wolf, and F.S. Osório. Adaptive finite state machine based visual autonomous navigation system. *Engineering Applications of Artificial Intelligence*, 29, pages 152–162, 2014.
44. R. Schiavi, G. Grioli, S. Sen, and A. Bicchi. VSA-II: A novel prototype of variable stiffness actuator for safe and performing robots interacting with humans. In *Robotics and Automation, 2008. ICRA 2008. IEEE International Conference on*, pages 2171–2176, May 2008.
45. J. Schoukens, A. Marconato, R. Pintelon, Y. Rolain, M. Schoukens, K. Tiels, L. Vanbeylen, G. Vandersteen, and A. Van Mulders. System identification in a real world. In *Advanced Motion Control (AMC),2014 IEEE 13th International Workshop on*, pages 1–9, Mar 2014.
46. L. Sciavicco and B. Siciliano. *Modelling and Control of Robot Manipulators*. Advanced Textbooks in Control and Signal Processing. Springer London, 2012.
47. D.K. Shaeffer. MEMS inertial sensors: A tutorial overview. *Communications Magazine, IEEE*, 51(4), pages 100–109, Apr 2013.
48. D. Shetty and R.A. Kolk. *Mechatronis Systems Design*. PWS Pub. Co., 1997.
49. R. Siegwart, I.R. Nourbakhsh, and D. Scaramuzza. *Introduction to Autonomous Mobile Robots*. Intelligent robotics and autonomous agents. MIT Press, 2011.
50. J. Swevers, F. Al-Bender, C.G. Ganseman, and T. Projogo. An integrated friction model structure with improved presliding behavior for accurate friction compensation. *Automatic Control, IEEE Transactions on*, 45(4), pages 675–686, Apr 2000.
51. S. Tanaka, T. Shimono, and Y. Fujimoto. Development of a cross-coupled 2dof direct drive motor. In *Industrial Electronics Society, IECON 2014 - 40th Annual Conference of the IEEE*, pages 508–513, Oct 2014.

52. G. Ullrich and P.A. Kachur. *Automated Guided Vehicle Systems: A Primer with Practical Applications*. Springer Berlin Heidelberg, 2015.
53. D. Verscheure, B. Demeulenaere, J. Swevers, J. De Schutter, and M. Diehl. Time-optimal path tracking for robots: A convex optimization approach. *Automatic Control, IEEE Transactions on*, 54(10), pages 2318–2327, Oct 2009.
54. V. Vonasek, M. Saska, K. Kosnar, and L. Preucil. Global motion planning for modular robots with local motion primitives. In *Robotics and Automation (ICRA), 2013 IEEE International Conference on*, pages 2465–2470, 2013.
55. D.G. Wilson, R.D. Robinett, and G.R. Eisler. Discrete dynamic programming for optimized path planning of flexible robots. In *Intelligent Robots and Systems, 2004. (IROS 2004). Proceedings. 2004 IEEE/RSJ International Conference on*, volume 3, pages 2918–2923, vol.3, Sep 2004.
56. Y. Yazaki, H. Fujimoto, K. Sakata, A. Hara, and K. Saiki. Settling time shortening method using final state control for high-precision stage with decouplable structure of fine and coarse parts. In *Industrial Electronics Society, IECON 2014 – 40th Annual Conference of the IEEE*, pages 2859–2865, Oct 2014.

2 Advanced Actuators for Mechatronics

Makoto Iwasaki

CONTENTS

2.1 INTRODUCTION

As actuators for mechatronic systems, e.g., electromagnetic/electrostatic machines, hydraulic/pneumatic machinery, piezoelectric/microelectromechanical system (MEMS) devices, etc., a wide variety of power sources are applied to actuate the mechanisms to achieve the required specifications, such as high accuracy, fast response, and mechanical vibration suppression, together with sensor techniques and control algorithms. In the actuator design for mechatronics, the selection of actuator as well as the controller design for the whole mechatronic system should be indispensable: that is, it is quite important to know "how the actuators should be specifically designed for the actual applications and should be practically controlled to provide the required performance".

In the subsection, promising advanced actuators are introduced as an example state of the art in mechatronic applications, e.g., industrial robots, precision positioning devices in various manufacturing machines, mass-storage devices, etc. The examples presented here are actuators including strain wave gearings and piezoelectric devices, which are especially designed for specified applications and driven by the suitable control schemes for them. The actuators with strain wave gearings allow mechanical elements to be compact and light-weight. This is especially useful in applications to industrial robots, where the precision control for static position accuracy can be achieved by model-based control approaches. The piezoelectric actuators drive precision stages to achieve the high-precision positioning performance as well as to suppress the mechanical resonant vibration by applying the self-sensing technique to the position control as a practical control approach.

2.2 ACTUATORS INCLUDING STRAIN WAVE GEARINGS AND APPLICATIONS TO PRECISION CONTROL

Strain wave gearings, such as harmonic drive gearings (HDGs), are widely used in a variety of industrial applications, e.g., industrial and/or humanoid robots, precision positioning devices, etc., because of their unique kinematics and high-performance attributes such as simple and compact mechanism, high gear ratios, high torque, and zero backlash. Figure 2.1 shows a schematic configuration of the laboratory prototype as an experimental positioning device, which is comprised of an actuator (alternating current [ac] motor) with an encoder, a harmonic drive gearing, an inertial load, and a load side encoder. The actuator with gearing is manufactured with compactness, like a conventional motor, which allows the mechanical component to be small-size and light-weight. The actuator including the gearing is generally controlled in a semi-closed control manner: actuator states (displacement and/or velocity) can be directly detected and controlled by sensors on the actuator shaft, while load states are indirectly controlled with the apparent sensor resolution due to the gear ratio.

However, since the transmission system inherently possesses nonlinear attributes known as "angular transmission errors (ATEs)" due to structural errors and

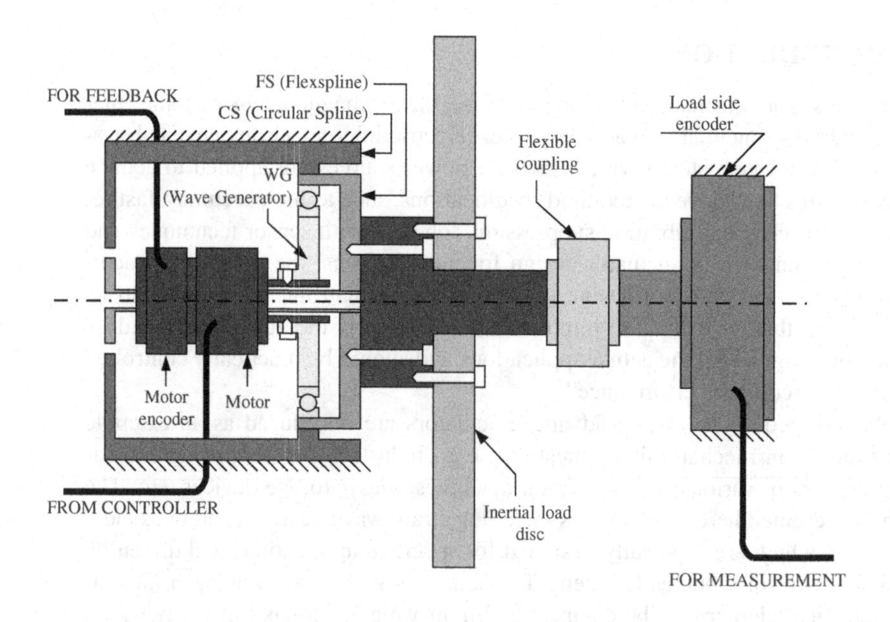

Figure 2.1 Experimental setup with actuator including harmonic drive gearing.

flexibility, the ideal control accuracy corresponding to the apparent resolution cannot be attained at the output of gear in devices. Although the gearing possesses a zero backlash property in the mechanism, the ATEs, due to kinematic errors and nonlinear elastic deformations in the gearing still remain, resulting in deterioration in the control performance [1]. Here, ATEs can be classified into types. One is the synchronous component, which is synchronous to the rotation of the gearing. The other is the "nonlinear elastic component", which behaves as a nonlinear phenomenon with hysteresis due to the nonlinear elastic deformations. The nonlinear elastic attribute may cause the static control accuracy to deteriorate due to the essential transmission error with hysteresis, especially in the micro-displacement region, e.g., several tens to hundreds of arc-sec (1 arc-sec = 1/3600 degrees) in the load angle. On the basis of these two components of ATEs, a precise modeling methodology for ATEs is provided to improve the static positioning accuracy, where the nonlinear elastic component in the micro-displacement region is mathematically modeled by applying a modeling framework for the rolling friction with hysteresis attributes [2] as well as the conventional modeling for the synchronous component by spectrum analyses for rotation angle [1]. This transmission error model is then adapted to the positioning system as in model-based feedforward (FF) compensation manner in order to improve the settling accuracy.

2.2.1 MODELING OF ANGULAR TRANSMISSION ERRORS

The angular transmission error θ_{TE} is generally defined by a motor angular displacement θ_M, a load angular displacement θ_L, and a gear ratio N as follows [1].

$$\theta_{TE} = \theta_L - \frac{\theta_M}{N} \tag{2.1}$$

In the following, in order to improve the static positioning accuracy at the settling, θ_{TE} is handled as the sum of the synchronous component θ_{Sync} and the nonlinear elastic component θ_{Hys}.

The synchronous component θ_{Sync} is basically caused by kinematic errors in the teeth of the flex spline (FS) and circular spline (CS) and assembling errors in the shaft of gearing and load, which is synchronous to the relative rotation of the wave generator (WG), FS, and CS [1]. Here, θ_{Sync} synchronizing to the motor angle is mathematically formulated as the following periodical pulsation for θ_M:

$$\theta_{Sync}(\theta_M) = \sum_{i=1}^{n_M} A_M(i)\cos\left(i\theta_M + \phi_M(i)\right) \tag{2.2}$$

where

i: harmonic order for motor angle,
A_M: amplitude of harmonics, and
ϕ_M: phase of harmonics.

Figure 2.2 shows comparative waveforms of the synchronous component in the experiment (solid line) and the model (broken line), where the horizontal axis of the

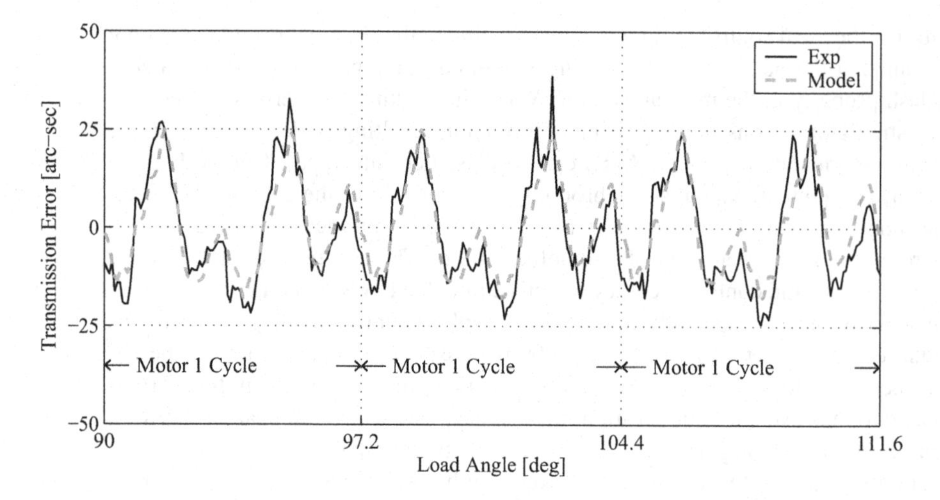

Figure 2.2 Comparative waveforms of synchronous component in experiment and model.

load angle is magnified to scale three revolutions of the motor. From the figure, the model can precisely reproduce the actual synchronous component of transmission error.

In order to analyze and model the nonlinear elastic component, on the other hand, sinusoidal angular motions with amplitude of 180, 45, and 5 degrees on the motor shaft have been driven, where an extra-low frequency of 0.05 Hz was given to eliminate effects of inertia force and dynamic torsional vibration on the responses. Figure 2.3 shows waveforms of motor angle (upper) and transmission error (lower) for three sinusoidal angular references ((a): 180, (c): 45, and (f): 5 degrees, respectively). From the figure, the level of transmission error iterates between about 0 and 80 arc-sec at every velocity reversal point, i.e., at 10, 20, and 30 s, while vibratory components due to the synchronous component are superimposed for the reference amplitude of 180 and 45 degrees. Figure 2.4, on the other hand, indicates Lissajous's figures of transmission errors for the sinusoidal motions in Figure 2.3 in order to clarify the level fluctuation in the error as a hysteresis behavior. The figures obviously suggest that 1) the nonlinear elastic component of transmission error shows a hysteresis characteristic with the reproducibility, superimposing the periodical synchronous component for motor angle in large-amplitude motions, and 2) the inclination of the hysteresis curve at the velocity reversal is constant and independent of the sinusoidal amplitude.

Here, this nonlinear characteristic appears to be in common with the rolling friction behavior by elastic deformations and hysteresis attributes in micro-displacement regions of various motion devices. The nonlinear elastic component θ_{Hys}, therefore, is mathematically formulated by a hysteresis model with the nonlinear elastic deformation in FS, referring to a modeling framework of the conventional rolling friction [2].

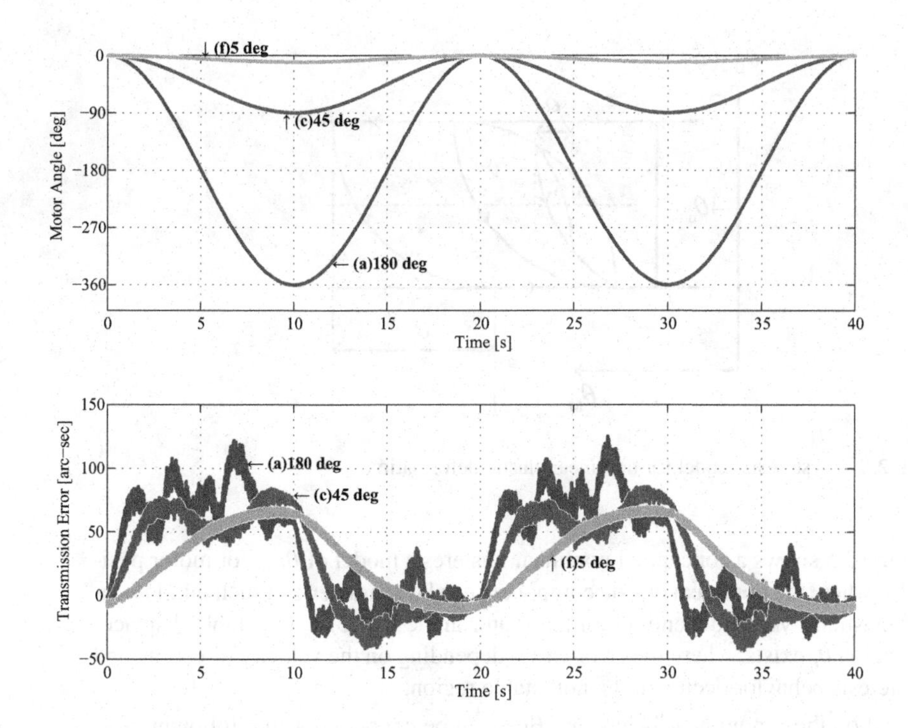

Figure 2.3 Waveforms of angular transmission error for sinusoidal angular input.

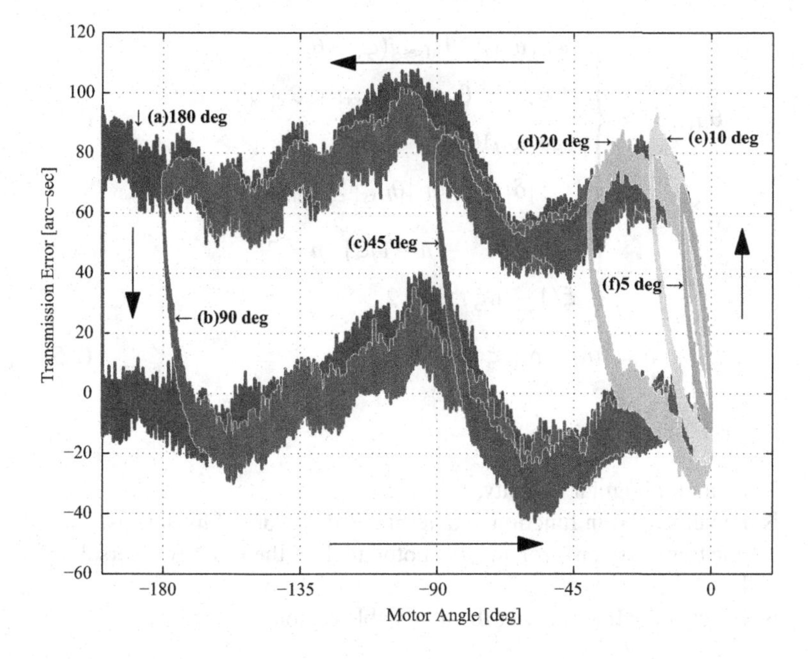

Figure 2.4 Hysteresis characteristics of angular transmission error.

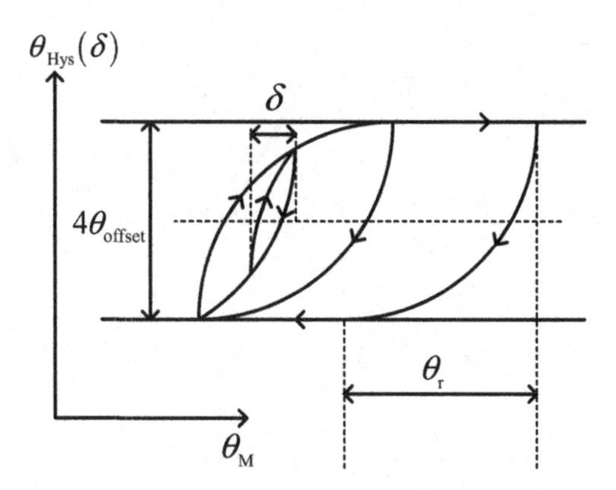

Figure 2.5 Hysteresis model for nonlinear elastic deformation.

Figure 2.5 shows a conceptual nonlinear hysteresis model of θ_{Hys} for motor position θ_M, where the nonlinear elastic component can be characterized such that 1) nonlinear elasticity varies depending on the motor angle θ_M, 2) a non-stable displacement region θ_r exists, where the error varies depending on the velocity direction, and 3) hysteresis behavior occurs in the non-stable region.

Based on these nonlinear behaviors, θ_{Hys} can be expressed as the following formula for the displacement δ after velocity reversal:

$$\theta_{Hys}(\delta) = \begin{cases} \operatorname{sgn}(\omega_M)(2\theta_{\text{offset}} g(\xi) - \theta'_{Hys}) \\ \quad : |\delta| < \theta_r \text{ and } \left|\theta_{Hys}\right| < \theta_{\text{offset}} \\ \operatorname{sgn}(\omega_M)\theta_{\text{offset}} \\ \quad : |\delta| \geq \theta_r \text{ or } \left|\theta_{Hys}\right| \geq \theta_{\text{offset}} \end{cases} \tag{2.3}$$

$$g(\xi) = \begin{cases} \frac{1}{2-n}\left(\xi^{n-1} - (n-1)\xi\right) : n \neq 2 \\ \xi(1 - \ln \xi) : n = 2 \end{cases} \tag{2.4}$$

$$\delta = |\theta_M - \delta_0|, \xi = \delta/\theta_r \tag{2.5}$$

where

ω_M is the motor angular velocity,
$\operatorname{sgn}(\omega_M)$ is the velocity sign function, n determines the width of hysteresis,
θ'_{Hys} and δ_0 are the transmission error and the motor angle at the velocity reversal, and
$2\theta_{\text{offset}}$ is the level fluctuation after the non-stable region.

2.2.2 FEEDFORWARD COMPENSATION IN PRECISION POSITIONING

The angular transmission error model has been adopted to the positioning system as a model-based FF compensation manner in order to improve the steady state angular error due to the transmission error. Figure 2.6 shows a block diagram of the proposed angular transmission error compensation in a conventional positioning system, where the compensation angular input $N\theta^*_{TE} = N(\theta_{Sync} + \theta_{Hys})$ is subtracted from the original motor angular reference θ^*_M.

In the following experimental verifications, inching motions of 240 steps (amplitude of 43.56 degrees in load side and an interval period of 2 s for every step) have been performed. Notice here that the inching step amplitude of 43.56 degrees was intentionally given in order to provide a different transmission error profile of the synchronous component in each inching motion. FB controller $C(s)$ denotes a P-PI compensator (proportional for position, and proportional and integral for velocity).

Figures 2.7 and 2.8 show waveforms of the settling performance in motor and load angle responses, where 240 inching trials are superimposed, and the target angle is indicated as 0 arc-sec in each figure. In Figure 2.8, responses denote: (a) without transmission error compensation, (b) with compensation for only synchronous component θ_{Sync}, (c) with compensation for only nonlinear elastic component θ_{Hys}, and (d) with compensation for both θ_{Sync} and θ_{Hys}, respectively. From Figures 2.7a and 2.8a, the transmission error obviously causes steady state errors in load angle, while the motor can be settled at the exact target angle.

Notice here that the scattered responses are observed during the transient in motor and load angles (overshoot and undershoot in responses) and at the settling in load angle (steady settling level), since the angular transmission error varies depending on the angle. In Figure 2.8d with the compensation on the other hand, the proposed compensation can suppress the scatter at the settling in load angle, with the result that the motor settling in Figure 2.7d is shifting to a shorter target angle due to the angular

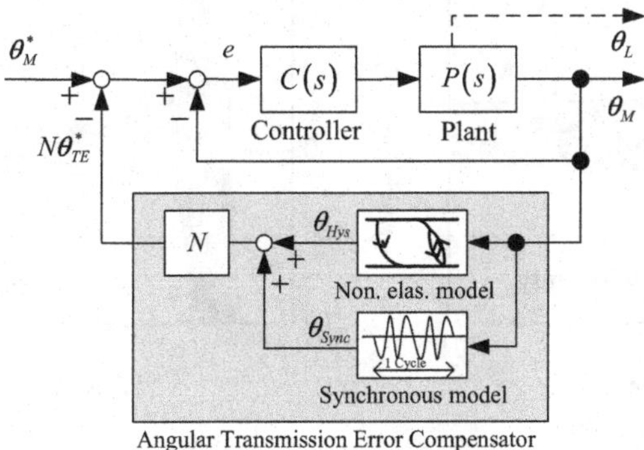

Figure 2.6 Block diagram of angular transmission error compensator.

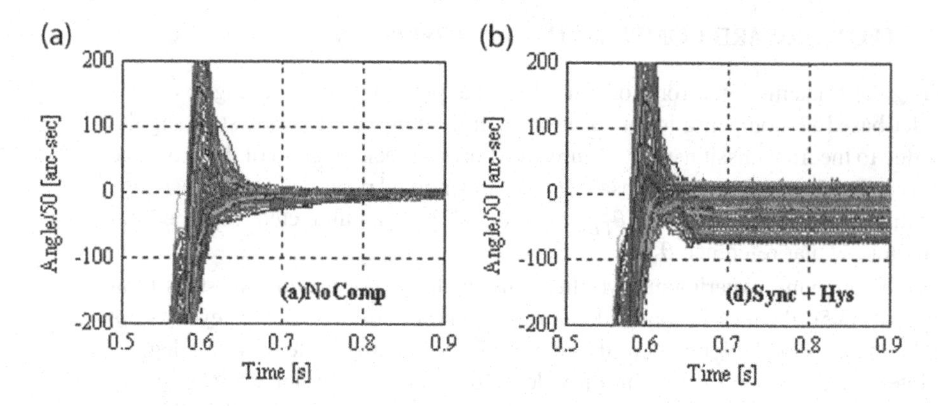

Figure 2.7 Magnified response waveforms of motor angle at settling.

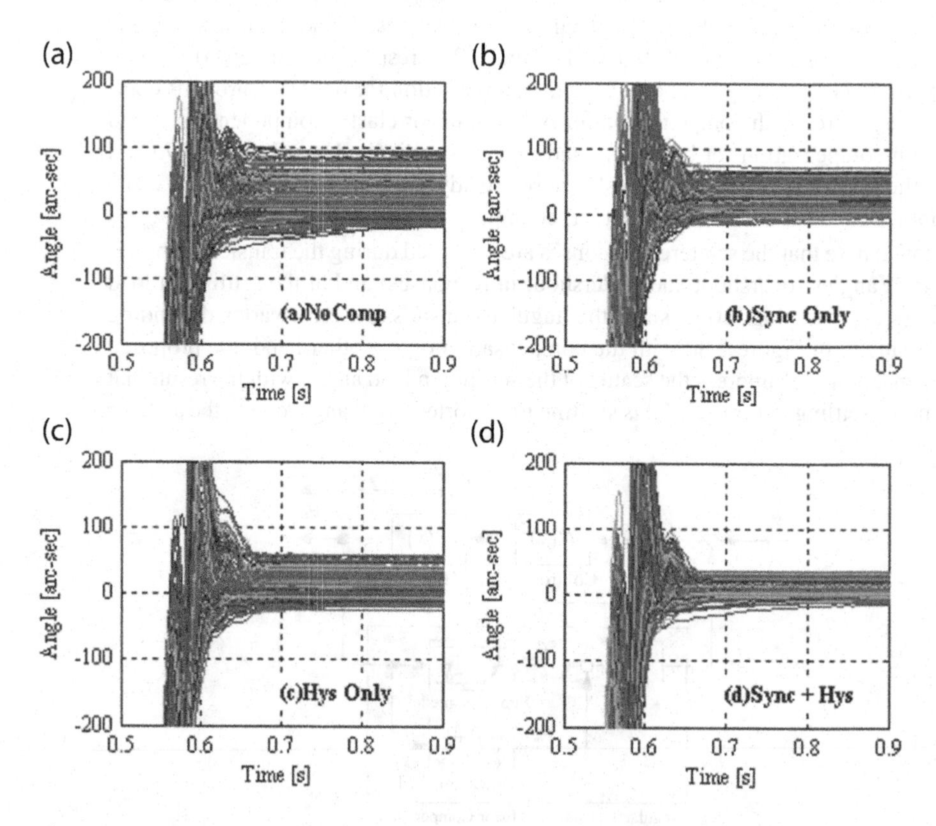

Figure 2.8 Magnified response waveforms of load angle at settling.

reference compensation as shown in Figure 2.6. The applications of each component alone in Figure 2.8b and c show insufficiencies of the compensation compared with Figure 2.8d.

2.3 PIEZOELECTRIC ACTUATORS WITH SELF-SENSING TECHNIQUES FOR PRECISION STAGE

Piezoelectric actuators have been widely applied to high-precision positioning and tracking applications, such as nano-positioning stage systems, scanning tunneling microscopes, atomic force microscopes, nano-robotic manipulators, etc. [3–6]. The piezo-driven systems generally include uncertain components, such as nonlinearities due to hysteresis and creep phenomena and resonant vibrations in mechanisms, which lead to deterioration in the positioning and/or tracking accuracy. In order to provide higher-precision control performance, therefore, the nonlinear characteristics as well as resonant vibrations should be compensated. Hysteresis is one of the typical nonlinear properties and appears between the applied voltage and output displacement of the piezoelectric actuator.

This section presents a vibration suppression approach for resonant vibration modes considering the hysteresis property without any additional sensor in the piezo-driven stage. Since the piezoelectric element has the function of both sensor and actuator, the detection and suppression of the mechanical resonant vibration can be simultaneously achieved by self-sensing techniques [5,7]. In the self-sensing technique, a simple bridge circuit is only added to the system to separate the intermingled sensor and actuator signals. A minor feedback loop can then be designed to suppress the resonant vibration modes on the basis of the detected signal through the bridge circuit, where the reduction of sensitivity characteristics around the vibration frequencies is especially considered to attenuate the residual vibration for the reference and disturbance. A major loop for augmented plant including the minor loop, on the other hand, can be designed by considering the expansion in the servo bandwidth and the system stability.

2.3.1 SYSTEM CONFIGURATION OF PIEZO-DRIVEN STAGE SYSTEM

Figure 2.9 shows an overview of a target piezo-driven stage with a high-resolution capacitive position sensor (P-622.1, Physik Instrumente GmbH & Co. KG) as a laboratory setup.

The stage (25 $\times$ 30mm) is supported in the single direction by elastic hinges, where the maximum stroke of the stage is 300 μm through the integrated displacement amplifier mechanisms. Figure 2.10, on the other hand, shows the control system configuration, where the stage position y is detected by the capacitive sensor and is transferred in a controller board (DS1103, dSPACE GmbH) through an sensor amplifier (E-610, Physik Instrumente GmbH & Co. KG) with the sampling period of 0.1 ms. The piezoelectric actuator is driven by a power amplifier (HSA 4014, NF

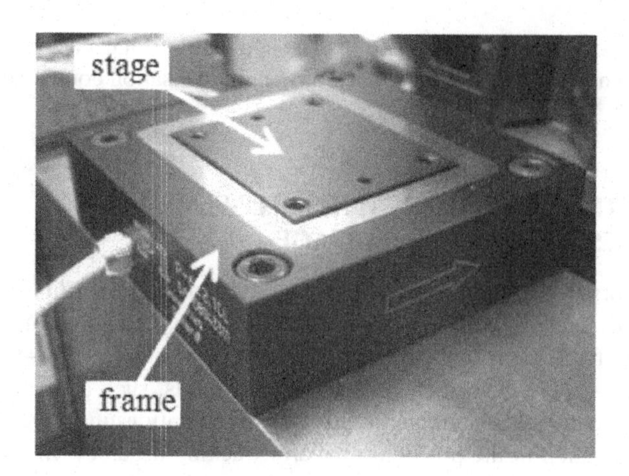

Figure 2.9 Overview of piezo-driven stage.

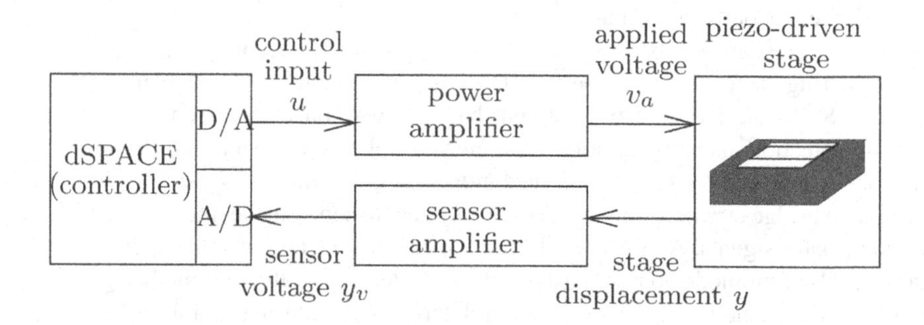

Figure 2.10 Control system configuration of piezo-driven stage.

Corporation) with the control input u generated by the feedback controller. Solid lines in Figure 2.11 show a frequency characteristic of the sensor voltage y_v for the control input u.

From the figure, since the mechanism includes mechanical vibration modes due to the elastic hinges in the high-frequency range, a mechanical part as the linear dynamics can be formulated as a plant mathematical model $P(s)$, that consists of two vibration modes and a dead time component as presented in the following equations:

$$P(s) = \frac{y_v}{u} = K_a P_m(s) P_s(s) \tag{2.6}$$

$$P_m(s) = \frac{y}{v_a} = K_g \left(\sum_{n=1}^{2} \frac{k_n}{s^2 + 2\zeta_n \omega_n s + \omega_n^2} \right) \tag{2.7}$$

$$P_s(s) = \frac{y_v}{y} = K_s \frac{\omega_s}{s + \omega_s} e^{-Ls} \tag{2.8}$$

where

K_a: gain of power amplifier,
K_g: linear plant gain, K_s: gain of sensor amplifier,
ω_n: natural angular frequency of nth vibration mode,
ζ_n: damping coefficient of nth vibration mode,
k_n: modal constant of nth vibration mode,
ω_s: angular frequency of lowpass filter in the sensor amplifier, and
L: equivalent dead time, respectively.

Since the cut-off frequency of the power amplifier is over 10 kHz, the dead time is caused by the capacitive sensor and the sensor amplifier. Broken lines in Figure 2.11, on the other hand, show a frequency characteristic of the mathematical model $P(s)$.

On the basis of the identified mathematical model, the following are controller design procedures for the piezo-driven stage system.

2.3.2 DESIGN OF PLANT SYSTEM INCLUDING BRIDGE CIRCUIT

In the self-sensing technique, since the piezoelectric element acts as both the sensor and the actuator, the voltage of the piezoelectric element includes both sensing and actuating states. The voltage, therefore, should be isolated by using a bridge circuit [5,7] as shown in Figure 2.12.

In the figure, the piezoelectric element can be expressed by an output voltage v_p due to piezoelectric effect and an equivalent capacitance C_p, while v_c denotes

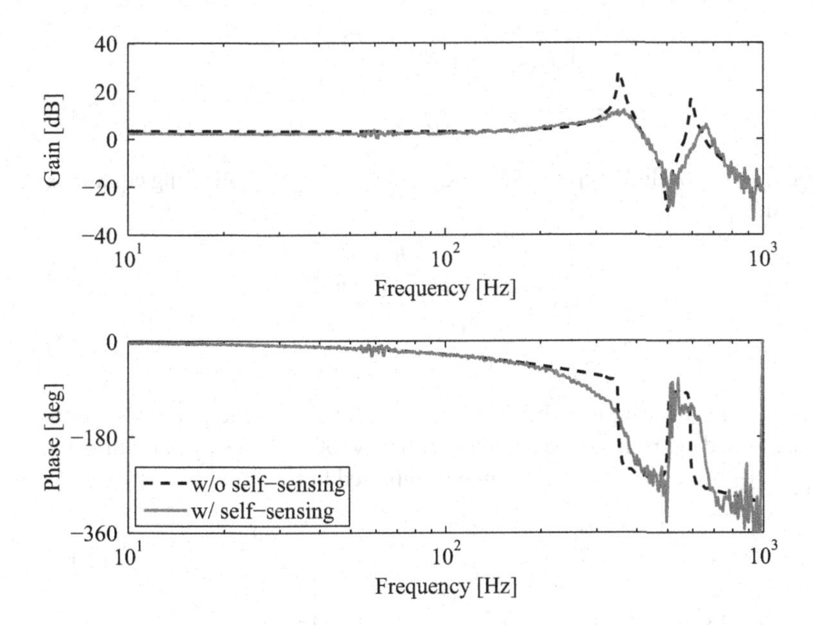

Figure 2.11 Frequency characteristics of sensor voltage y_v for control input u.

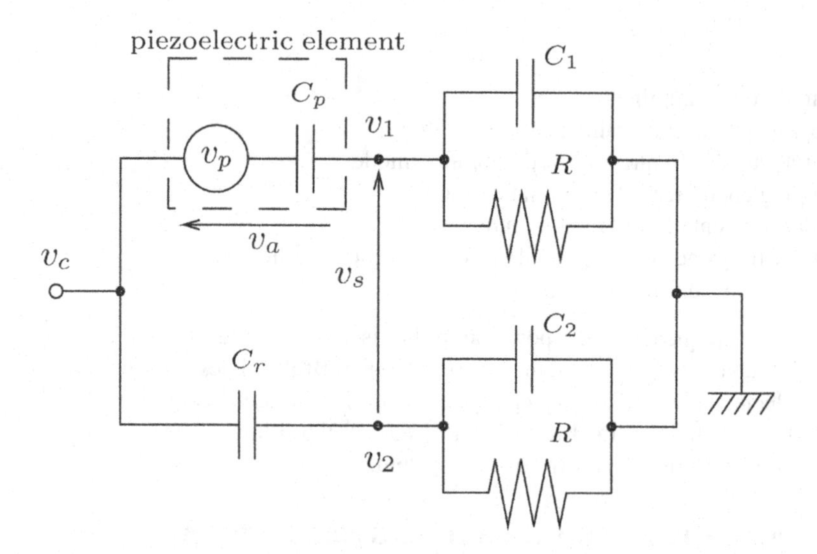

Figure 2.12 CC bridge circuit for isolation of signals in voltage.

the supplied voltage to bridge circuit, v_s denotes the sensor voltage, v_a denotes the applied voltage to the piezoelectric element (actuation voltage), C_r, C_1, C_2 denote the capacitances, and R denotes the resistance to remove the offset of signal. From the figure, voltage equations of the bridge circuit are given as follows.

$$v_1 = \frac{C_pRs}{1+(C_p+C_1)Rs}\{v_c+v_p\} \tag{2.9}$$

$$v_2 = \frac{C_rRs}{1+(C_r+C_2)Rs}v_c \tag{2.10}$$

In order to satisfy the bridge balance of $C_p = C_r$ and $C_1 = C_2$, the following equations can be derived.

$$v_s = v_1 - v_2 = \frac{C_pRs}{1+(C_p+C_1)Rs}v_p \tag{2.11}$$

$$v_a = v_c - v_1 = \frac{C_1}{C_p+C_1}v_c \tag{2.12}$$

From Equation 2.11, the output voltage v_p can be extracted by the potential difference between v_1 and v_2. Here, since angular frequency ω_b of the bridge circuit is set as $\omega_b \gg \frac{1}{(C_p+C_1)R}$ by R, Equation 2.11 can be simplified by:

$$v_s \simeq \frac{C_p}{C_p+C_1}v_p \tag{2.13}$$

From Equation 2.13, the sensor voltage v_s is proportional to the output voltage v_p. From Equation 2.12, on the other hand, the supplied voltage v_c to the bridge circuit

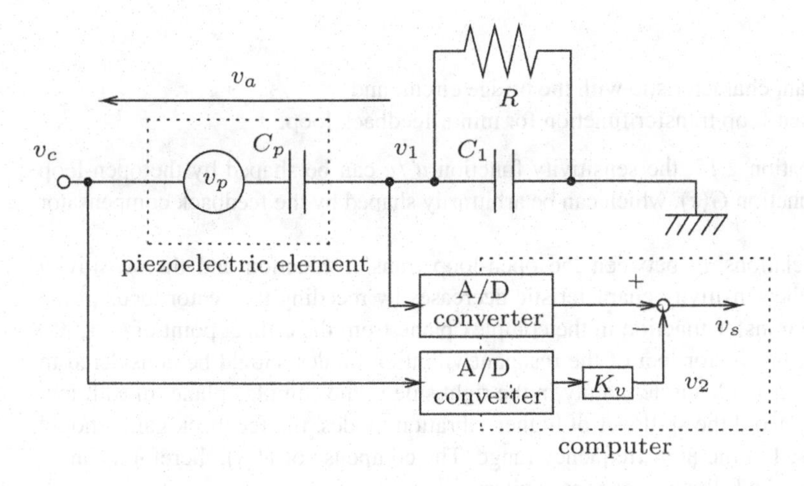

Figure 2.13 Virtual bridge circuit for isolation of signals in voltage.

is proportional to the actuation voltage v_a. However, it is difficult to exactly satisfy the bridge balance of $C_p = C_r$ and $C_1 = C_2$. Especially, C_r completely corresponding to the value of C_p cannot be always implemented by a circuit element. In order to solve the problem, a part of the CC bridge circuit in Figure 2.12 can be constructed by software as a virtual bridge circuit. From Equation 2.10, since v_2 is determined by supplied voltage v_c regardless of v_p, v_2 may not need to be directly measured by the analog circuit. Therefore, Figure 2.12 can be modified as Figure 2.13 by constructing the lower part of the circuit as the software.

In Figure 2.13, K_v can be given as follows based on Equation 2.10.

$$K_v = \frac{C_r Rs}{1 + (C_r + C_2)Rs} \tag{2.14}$$

As a result, the sensor voltage v_s in Equation 2.11 can be easily calculated only by R and C_1 as the circuit elements.

The dotted frame in Figure 2.14 shows a block diagram of plant with the bridge circuit, where K_ϕ is a coefficient of transformation from mechanical energy to electrical energy.

2.3.3 MINOR LOOP CONTROLLER DESIGN CONSIDERING VIBRATION SUPPRESSION

Based on the plant characteristic with the bridge circuit, a minor loop feedback controller $H(s)$ in Figure 2.14 can be designed to reduce the gain peaks at the resonant frequencies. From Figure 2.14, the sensitivity function is given as follows:

$$\frac{u'}{u} = \frac{1}{1 + K_a P_b(s) H(s)} = \frac{1}{1 + G(s)} \tag{2.15}$$

where

$P_b(s)$: plant characteristic with the bridge circuit and
$G(s)$: open-loop transfer function for minor feedback loop.

From Equation 2.15, the sensitivity function u'/u can be shaped by the open-loop transfer function $G(s)$, which can be arbitrarily shaped by the feedback compensator $H(s)$.

As a relationship between the open-loop transfer function and the sensitivity function, the sensitivity characteristic decreases by receding the vector locus of the open-loop transfer function in the complex plane from the critical point of $(-1, j0)$. Therefore, the vector loci of the resonant vibration modes should be considered to draw a larger mode circle mainly in the right side in the complex plane. In addition, in order to avoid the spillover of higher vibration modes, the feedback gains should be decreased in the high-frequency range. The compensator $H(s)$, therefore, can be designed as the following bandpass filter:

$$H(s) = K_h \cdot \frac{s^2}{s^2 + 2\zeta_h \omega_h s + \omega_h^2} \cdot \frac{\omega_l^2}{s^2 + 2\zeta_l \omega_l s + \omega_l^2} \tag{2.16}$$

where the designed parameters are listed in Table 2.1.

Figure 2.15 shows plant characteristics with and without the self-sensing minor loop.

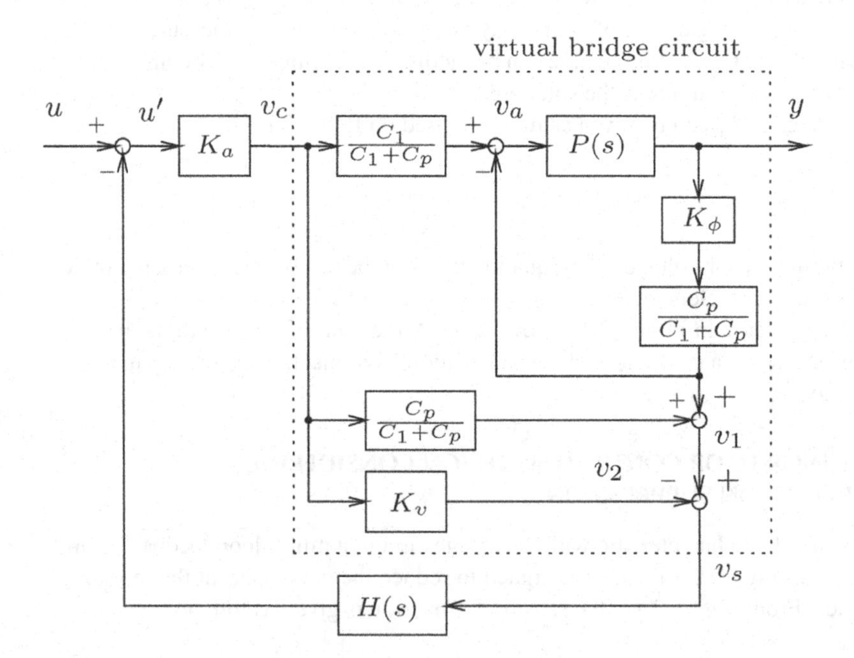

Figure 2.14 Block diagram of plant with self-sensing minor loop.

Table 2.1
Parameters of Minor Loop Feedback Compensator $H(s)$

K_h	**15**		
ω_h [rad/s]	$2\pi\times 700$	ζ_h	0.9
ω_l [rad/s]	$2\pi\times 2000$	ζ_l	0.2

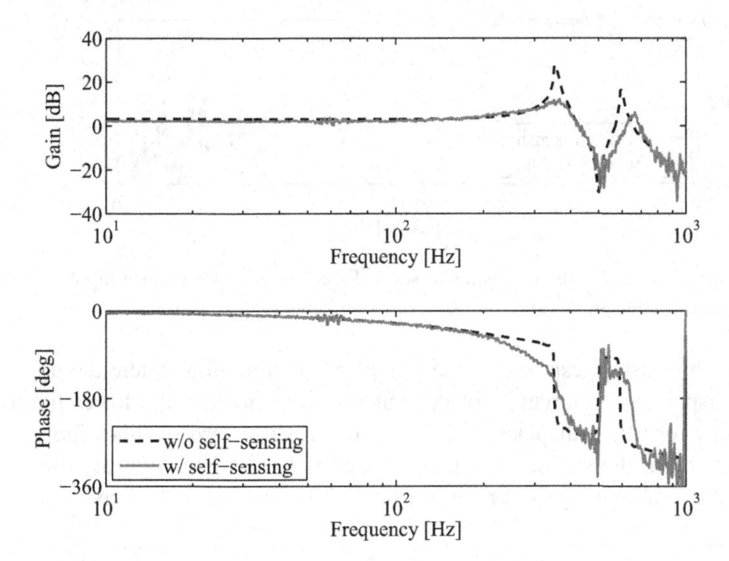

Figure 2.15 Plant frequency characteristics of stage displacement y for control input u with and without self-sensing minor loop.

Figure 2.16, on the other hand, shows plant characteristics with frequency variation of vibration modes due to the load mass variation, while $H(s)$ can be represented by the same structure and parameters as in Figure 2.16.

Gain peaks at resonant frequencies can be reduced by applying the self-sensing minor loop shown in solid lines of Figures 2.15 and 2.16 due to the direct detection of vibration modes. From these results, residual vibrations for the unknown disturbances can be suppressed by the improvement of minor loop sensitivity characteristic, while the servo bandwidth can be expanded by applying the self-sensing minor loop.

2.3.4 EXPERIMENTAL VERIFICATIONS

The effectiveness of the proposed control scheme has been verified by experiments using the piezo-driven stage shown in Figure 2.9. The controllers are implemented as discrete transfer functions by a bilinear transformation with the sampling time T_s = 0.1 ms.

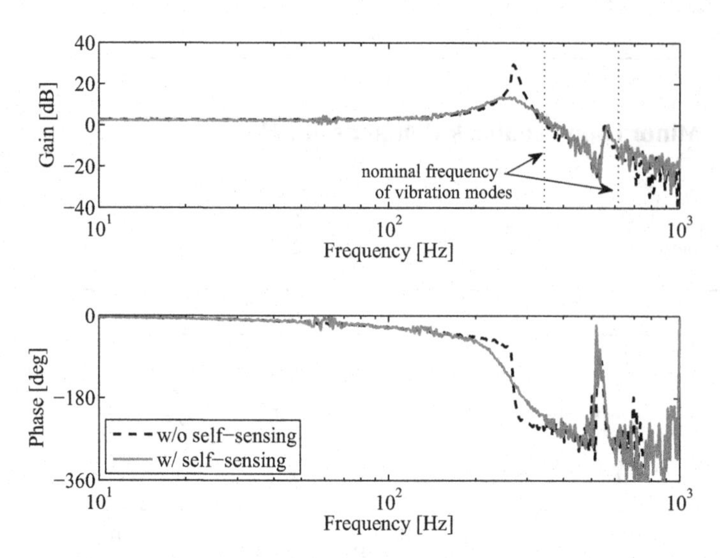

Figure 2.16 Plant frequency characteristics of stage displacement y for control input u with frequency variation in vibration modes.

Figure 2.17 shows step responses under the nominal condition, where the upper figure shows responses for target displacement stroke of 50 μm, the lower figure shows responses for target displacement stroke of 150 μm, broken lines indicate the responses without the self-sensing minor loop, and solid lines indicate the responses with the self-sensing minor loop. Figure 2.18, on the other hand, shows

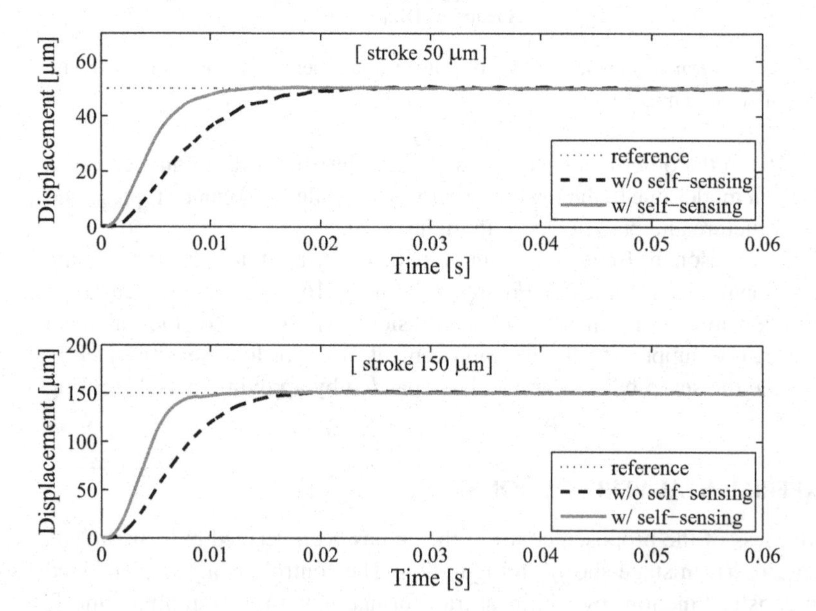

Figure 2.17 Step responses under nominal condition.

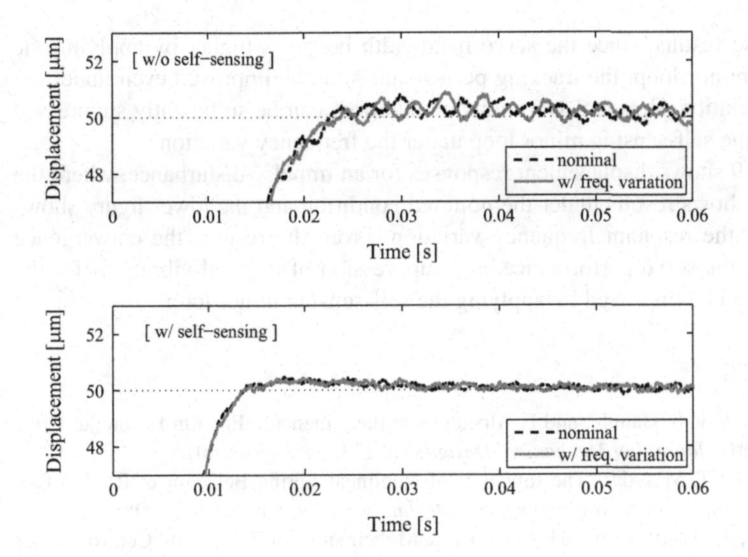

Figure 2.18 Step responses for target stroke of 50 μm under variation in resonant frequency.

magnified step responses at around the target displacement under the frequency variation in vibration modes corresponding to Figure 2.16, where the upper figure shows the responses without the self-sensing minor loop, the lower figure shows the responses with self-sensing the minor loop, broken lines indicate the results under nominal condition, and solid lines indicate the results under the resonant frequency variation.

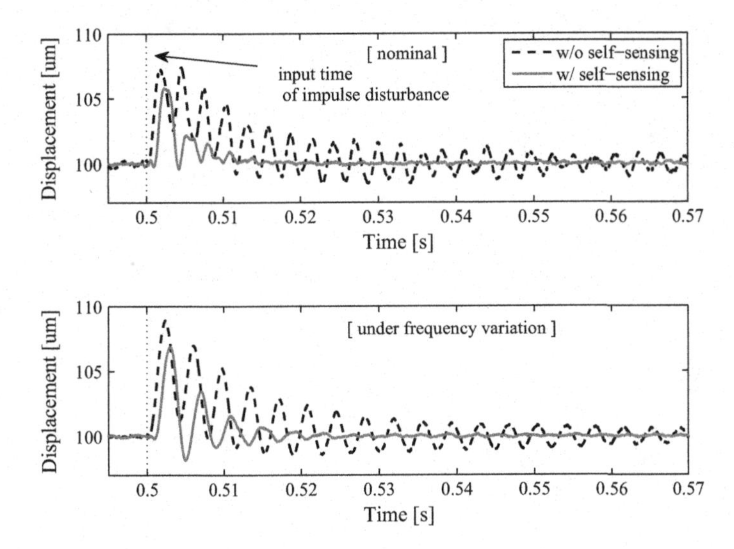

Figure 2.19 Responses for impulse disturbance.

From these results, since the servo bandwidth becomes higher by applying the self-sensing minor loop, the tracking performance can be improved even under the frequency variation. Residual vibrations, in addition, can be sufficiently suppressed by applying the self-sensing minor loop under the frequency variation.

Figure 2.19 shows displacement responses for an impulse disturbance, where the upper figure shows results under the nominal condition and the lower figure shows results under the resonant frequency variation. From the results, the convergence performance, the servo performance, and suppression of residual vibrations for the disturbance can be improved by applying the self-sensing minor loop.

REFERENCES

1. F. H. Ghorbel, P. S. Gandhi, and F. Alpeter, "On the Kinematic Error in Harmonic Drive Gears", *ASME Journal of Mechanical Design*, Vol.123, pp.90–97, 2001
2. J. Otsuka and T. Masuda, "The Influence of Nonlinear Spring Behavior of Rolling Elements on Ultraprecision Positioning", *Nanotechnology*, Vol.9, pp.85–92, 1998
3. S. Devasia, E. Eleftheriou and S. O. Reza Moheimani, "A Survey of Control Issues in Nanopositioning", *IEEE Transactions on Control Systems Technology*, Vol.15, No.5, pp.802–823, 2007.
4. S. O. Reza Moheimani and A. J. Fleming, "Piezoelectric Transducers for Vibration Control and Damping", *Springer-Verlag*, London, 2006.
5. N. Jalili, "Piezoelectric-Based Vibration Control", *Springer Science+Business Media*, 2010.
6. A. J. Fleming and K. K. Leang, "Design, Modeling and Control of Nanopositioning Systems", *Springer*, Cham, Switzerland, 2014.
7. J. J. Dosch, D. J. Inman and E. Garcia: "A Self-Sensing Piezoelectric Actuator for Colocated Control", *Journal of Intelligent Material Systems and Structures*, Vol.3, pp.166–185, 1992.

3 Advanced Sensors for Mechatronics

Naoki Oda, Toshiaki Tsuji, Yasue Mitsukura, Takahiro Ishikawa, Yuta Tawaki, Toshiyuki Murakami, Satoshi Suzuki, and Yasutaka Fujimoto

CONTENTS

3.1 STATE OF ART

As is well known, sensors and actuators are key technologies to develop industrial products and to achieve social innovation. In particular, intelligent sensor applications are important to improve system reliability and safety. The related technologies of intelligent transport systems (ITS) are given as a typical example of a sensor based intelligent system [1]. Recently, smart grids and smart communities have been taken up as one of the important topics of Green/Life innovation. In such a system, a LAN (local area network)–based sensors system is the key to the integration of several kinds of facilities, called a network node [2,3]. From a viewpoint of system integration, sophisticated methods are required to design and integrate sensor and actuator networks with hundreds of thousands of nodes. Needless to say, they bring a sustainable, safe, and comfortable life from several aspects. Figure 3.1 shows a conceptual diagram of system integration for human, machine, and environment in the network nodes. In such a system, vision-based environment recognition is one of the

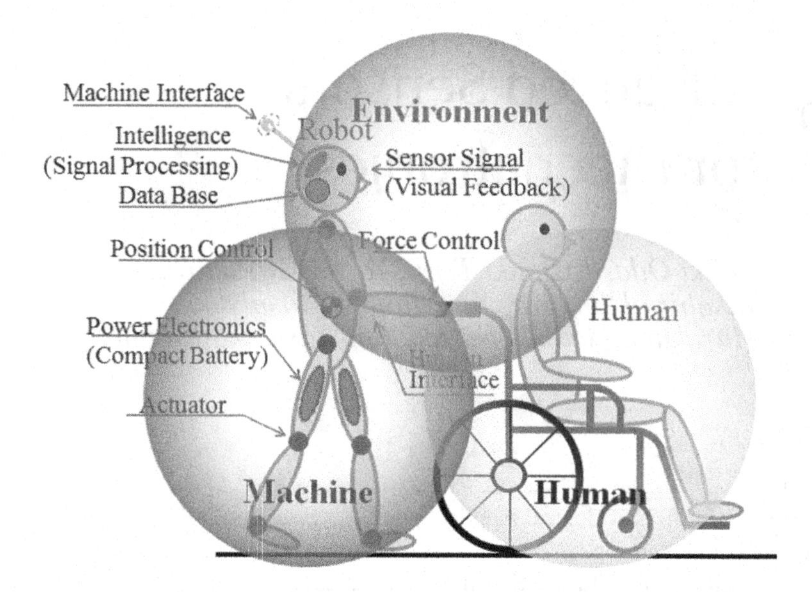

Figure 3.1 Sensors and actuators system

indispensable functions to bring sophisticated control of the system. Furthermore, the fusion system of the flexible control based on sensor information and the motion control has come to be shown in our real life as a human–machine interactive system, though the system construction has become very complicated. Then, the recognition of the environment, including human, by multiple sensors is the key issue for reliable interaction among environment, human, and machine.

To achieve the sophisticated and advanced system shown in the above-mentioned examples, because the compromise among the information processing and the regulating system of each node through the network of sensors and actuators becomes so important, it is necessary to develop a new and emerging research area of sensors and actuators technology. The key issues are summarized as:

- Advanced Vision-based Control Application
- Advanced Tactile Sensing
- Advanced Electroencephalography
- Advanced Human Motion Sensing
- Advanced Approaches for Human Motion Detection

In this section, the above-mentioned topics are briefly described.

3.2 ADVANCED VISION-BASED CONTROL APPLICATIONS

Vision sensors can provide environmental information for mechatronic systems, and the obtained relationship between the target system and its environments enables sophisticated functions in various control applications such as mobile systems, robotics, teleoperated systems, and so on. That is because the visual feedback control loop is very effective for improving the dexterity and flexibility of various tasks. With the improvements of vision sensors and computer technologies, the "Look and Move" cycle has been drastically shortened in recent developments, and it enables real-time and dynamic control using visual sensing information.

In this section, the visual feedback techniques are overviewed from the viewpoint of a control scheme. The visual sensing will play an important role in environmental interactions among the target system, humans, and environments. For example, the mobile robot can autonomously travel in the workspace while interacting with humans or avoiding obstacles. Moreover, the building map is also possible while identifying its own position in the map. This technology is so-called SLAM (simultaneous localization and mapping), which is remarkably developed in the area of mobile robot control. LIDAR sensors (LIDAR: light detection and ranging), such as laser scanners, are popular devices for detecting three-dimensional (3-D) information in the field of view. In the field of such active vision, the vision-based navigation is widely employed for autonomous mobile robots and unmanned vehicles.

The camera is also one of the frequently used sensors for vision sensing. The captured image includes versatile information for motion control or intelligent motion planning. In recent advances with respect to computing speed, the image processing time can be remarkably reduced, and it is possible to design visual servo control based on recognized data from the camera image. Figure 3.2 shows the typical controller structure. One is an image-based visual servo (IBVS), and the other is a

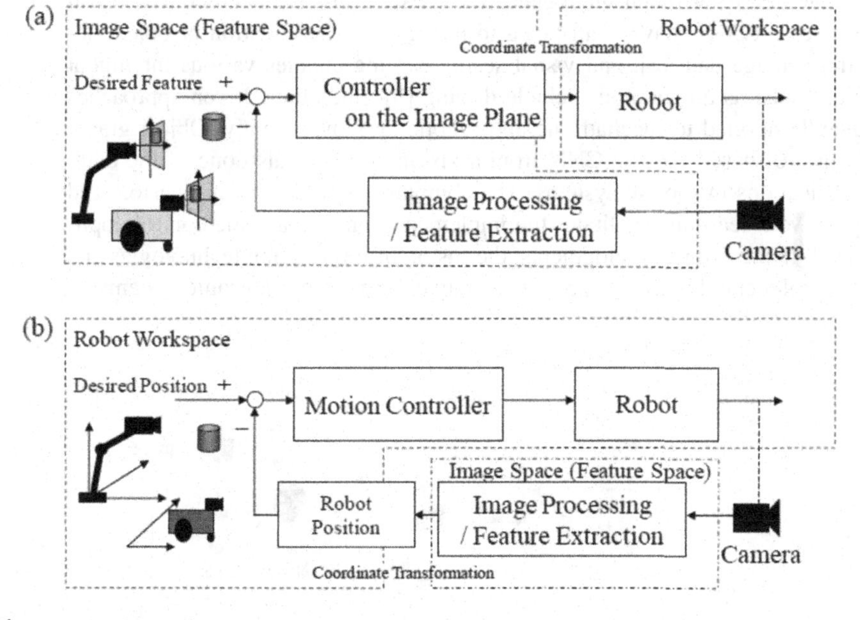

Figure 3.2 Controller structure in terms of control space. (a) Primary Space: Image Space (IBVS). (b) Primary Space: Workspace (PBVS).

position-based visual servo (PBVS). The detailed comparisons are well-described in tutorial literatures [4–6]. The main difference between them is a primary space for controller design as shown in Figure 3.2. In IBVS, the controller is designed directly on image space, so the 3-D reconstruction from image features is not required for motion control. On the other hand, the feedback position is obtained through 3-D reconstruction in PBVS, so generic motion controller design can be considered independent of vision data processing. However, the precise robot model and intrinsic/extrinsic camera parameters are required in PBVS.

The recent advancements in vision-based control also expand the system functions. In general, a faster inner control loop is required than the outer one in terms of controller stability. Due to the dead-time of image processing, the visual feedback loop is conventionally difficult to apply into the inner loop. Recently, several high-speed camera systems and computing environments are commercially available, so the flexible controller structure can be considered. Figure 3.3 shows an example controller structure with force feedback as outer control loop. The change of ROI (region of interest) in the image leads to various motion functions according to visual situations while considering the force contact with human, or environments. Since a wide view can be acquired by the vision sensor, the information selection by ROI contributes to relative and absolute motion capability, such as ordinary traveling, following humans, and avoiding obstacles, in accordance with surrounding changes. In the literature [7], a vision-based power assist function for wheelchair control is introduced as an example application.

In the past decade, much progress and innovative advancement has been made in the technical area of artificial intelligence (AI). Especially, the deep convolutional neural networks (CNN) have contributed to image-data classification or object detection from image data [8], and visual scene learning enables various intelligent applications such as autonomous vehicle driving [9]. Such data-driven approaches are frequently applied to mechatronic and robotic systems recently. Object grasping/picking [10] based on 3-D CNN from a vision sensor is also one of the data-driven applications in robotic systems. The computer architecture/chip customized for deep network learning facilitates the implementation of real-time control applications [11]. In the future developments, the fusion of information technologies and motion control technologies will be increasingly evolved in mechatronic systems.

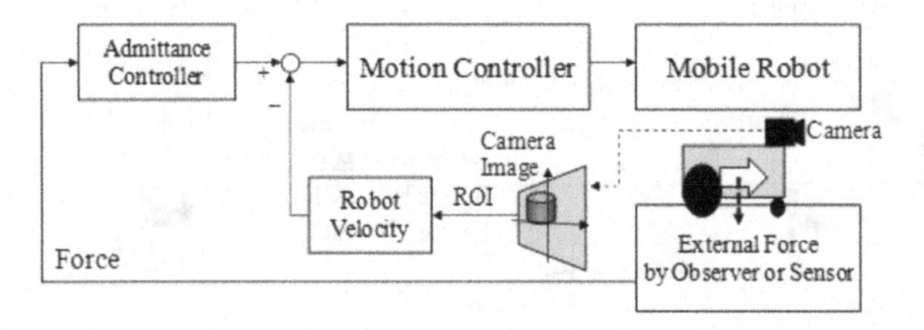

Figure 3.3 An example controller with vision-based inner loop.

3.3 ADVANCED TACTILE SENSING

3.3.1 CLASSIFICATION OF TACTILE SENSING

Tactile sensing techniques have been a popular subject, especially in robotics. This section describes the techniques for tactile sensors. First, the classification of tactile sensing techniques is described, and it is followed by descriptions of techniques in each classification.

Based on the location of the tactile sensors, tactile sensing can be categorized as intrinsic and extrinsic tactile sensing [12]. Intrinsic sensing is based on a force/torque sensor placed within the mechanical structure of the sensing system. The system with intrinsic sensing derives the contact information, like magnitude of force or contact location, using the force/torque sensor. Extrinsic sensing is based on sensors, which are often arranged in arrays that are mounted at or near the contact interface.

Extrinsic sensing is widely studied recently, and applying array sensors is a good solution to gather rich information on tactile sensing. On the other hand, intrinsic tactile sensing is useful as a simple solution in practical use, while it has a limitation that the force distribution is undetectable in return for the simplicity. Following are the descriptions of each classification.

3.3.2 INTRINSIC TACTILE SENSING

Intrinsic tactile sensing in a two-dimensional (2-D) plane has been widely applied in practice. The Wii balance board from Nintendo Co., Ltd. is the most typical and well-known application. It consists of four strain gauges detecting the vertical force and derives both the vertical force and the position of center of pressure. In 2014, Aisin Co., Ltd. applied a similar technique to a sleep monitoring system for a bed. These examples imply that tactile sensing is a good solution for detecting movements in daily life and that it can be applied as an interface for man–machine interaction.

Compared with the examples for 2-D tactile sensing, there are fewer examples of intrinsic tactile sensing in a 3-D system. The basic principle of intrinsic tactile sensing was studied in the 1980s. Salisbury proposed a tactile sensing method that identifies the contact point and the force vector on an insensitive end-effector using a six-axis force sensor [13]. Bicchi proposed the concept of intrinsic contact sensing that involves the identification of the contact point and force vector [14]. These studies estimate the contact force and the contact position under the assumptions of single contact and convex-shaped end-effector with large radii of curvature. External torque observer, an observer estimating external torque or external force from the input current and joint velocity information, is a good solution for sensorless force detection [15]. The method has the feature of intrinsic force sensing, while the contact position has to be given in the method. Iwata and Sugano developed an interface for human symbiotic robots. They realized whole-body tactile sensation by molding the end-effector of the force sensor into a shell shape with touch sensors [16]. Tsuji et al. have proposed "haptic armor", a calculation technique of intrinsic contact sensing, which simplifies the method by ignoring the spin resisting torque [17]. The mechanism has the advantage of reduced cost and less wiring while allowing six-axis resultant force detection. This method has been generalized for non-convex-shaped end-effectors, and it has been also applied for end-effectors with soft material [18]. The Fuwafuwa sensor module, a round, handsize, wireless device for measuring the shape deformations of soft objects such as cushions and plush toys, is one of the good solutions for external force detection by an end-effector with soft material [19]. Although it can detect only the scalar force information, the merit of the method is that it can be applied to an end-effector without any mechanical frames. Intrinsic tactile sensing has been applied as a technique that utilizes force sensors for a desk-type tactile interface [20]. A desk is a typical example of a tool for efficient work, and this idea can also be extended to common consumer electronics and furniture into our interfaces. The study indicates that intrinsic tactile sensing has an advantage in having fewer restrictions on the shape and constructions of the end-effector. Ready-made furniture or electrical appliances can be turned into a tactile interface, while extrinsic tactile sensors needs wiring and particular material on the surface.

3.3.3 EXTRINSIC TACTILE SENSING

Tactile sensor arrays, detecting the distribution of external force and therefore naturally extrinsic, have many kinds of application. Tactile sensor arrays are often introduced to beds and wheelchairs for prevention of pressure sores. The feature of measuring pressure distribution is also applied in the assessment of many machines. A robot hand and skin are also typical applications of tactile sensors. Most of the commercialized sensor arrays simply detect the pressure on each tactel, while some sensors detecting 3-D force on each tactel are also becoming popular.

Many transduction modes have been exploited for tactile sensors by now. Piezoresistors whose resistance changes with force have been utilized for tactile sensors [21,22]. One of the good examples is a tactile sensor with standing piezoresistive cantilevers embedded in an elastic material, which enables detection of shear stress [23]. Capacitive sensors, which consist of two conductive plates with dielectric material in between them, have an advantage of their small size [24]. This advantage allows a sensor array with higher position resolution, and therefore, many commercial touch sensors have been developed based on this method. Stray capacity and severe hysteresis are the major disadvantages. The optical mode is also one of the popular transductions using the change in light intensity, at media of different refractive indices, to measure the pressure [25]. Some commercial products have also been developed. Since the sensors are often bulky, and scalability is often required, a modular skin of pressure-sensitive elements, which are able to be folded and cut, has been developed [26]. Its sensor coverage is adjustable via the addition/deletion of modules, and sensor density is adjustable by folding the bandlike bendable substrate on which the sensor sits.

Soft materials are often introduced for tactile sensors because they increase the contact friction coefficient and protect the distributed embedded sensor device [27]. Wettels et al. have developed a biomimetic tactile sensor array consisting of a rigid core surrounded by a weakly conductive fluid contained within an elastic skin [28,29].

3.4 ADVANCED ELECTROENCEPHALOGRAPHY

In this section, human preference detection by simple electroencephalography (EEG) is illustrated. Many services that take personal preference into consideration have been provided in recent times. Hence, the process of the determination of personal preference, which was pioneered in Japan and is locally known as KANSEI, has been actively studied. Whereas sensitivity is generally inborn, the focus of KANSEI is considered to be a postnatal attribute. There are subjective and objective indexes in the method for determining personal preference. A subjective index is obtained by a questionnaire, whereas an objective index is determined by a bio-signal [30,31]. In addition, an objective index can be quantified, which enables an objective and engineered approach [32]. Incidentally, there have been many propositions regarding the relationship between an EEG signal and the preference determined by KANSEI in the analysis of a bio-signal [33–37]. The propositions are based on the idea that "the state of the brain should change if the state of the person changes because the brain governs the mind, consciousness, recognition, and senses" as well as other ideas [38,39]. EEG is one of the bio-signals used as indexes for determining preference in the present study. Based on our study, we propose various preference measurement systems for the olfactory sense, acoustic sense, haptic sense, taste sense, and visual sense (generally referred to as the five senses) as well as for a combination of the acoustic and visual senses. In this section, we introduce the procedure for measuring the EEG and describe the analysis method. We also describe a sample application of the developed preference measurement systems, namely, "Visualization of the Mind Status Using EEG". The EEG devices in Figure 3.4 are of the conventional type, which requires a long time to wear. The device in Figure 3.4a requires 45 min. to wear, and that in the Figure 3.4b requires about 30 min. In contrast, our simple EEG device (Figure 3.5) requires only 30 s to wear. It has only one electrode, which makes it easy to wear. The advantages of our device are as follows:
1. Reduced number of electrodes. 2. Does not require a gel to be worn. 3. The subject can easily wear it by himself or herself at anytime and anywhere.

Although this simplifies an understanding of preference, it cannot be easily used to obtain sequential measurements [40]. However, sequential measurements were re-

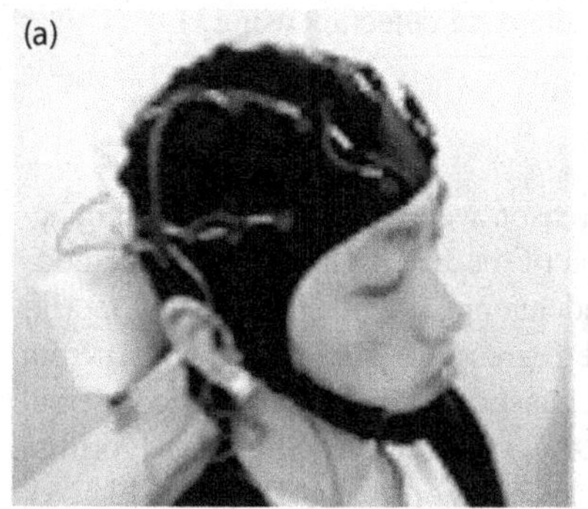

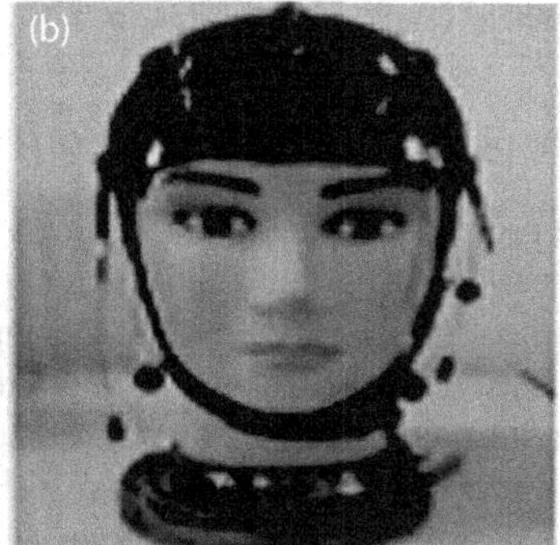

Figure 3.4 Conventional EEG devices.

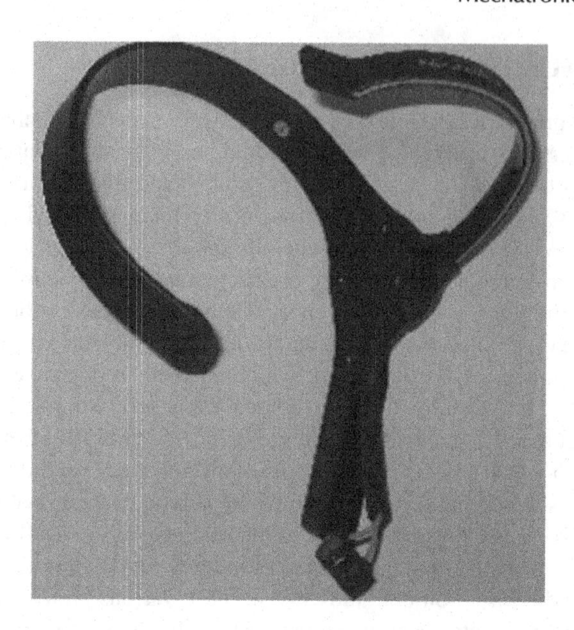

Figure 3.5 Our proposed EEG device.

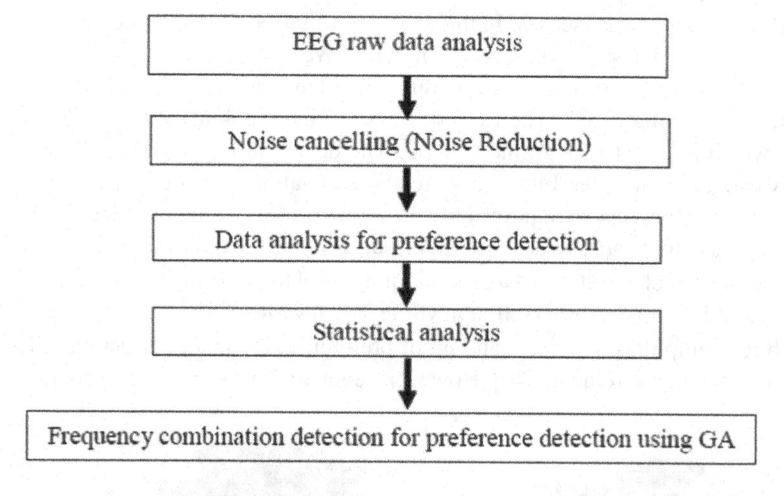

Figure 3.6 The procedure of the proposed analysis.

quired for the present investigation, for which reason we also observed the brain activity. Pre-processing is used to enable detection of the EEG. The available methods include Functional Analysis (FA) [41], independent component analysis (ICA), principal component analysis (PCA), and multiple regression. The data obtained from the pre-processing are then mined using a stochastic method such as incremental PCA, Fisher linear discriminant analysis (FLDA), sparse PCA (SPCA), Algorithm for Multiple Unknown Signals Extraction (AMUSE), or sparse FLDA (SFLDA). Furthermore, Genetic Algorithm (GA) is used for searching suitable combination of frequencies for preference detection (Figure 3.6).

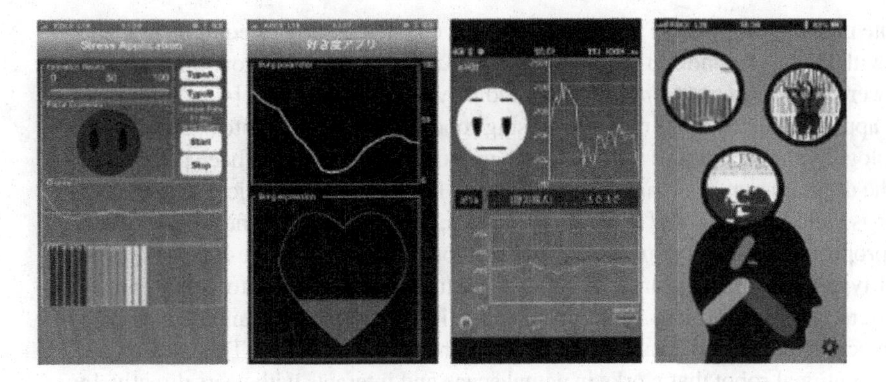

Figure 3.7 Real application using the proposed algorithm.

The data was then processed by our system, after which the results of the degree of human preference were sequentially obtained. By using this processing, the combinations of frequencies (not frequency band) are decided. The combination of 6 and 10 Hz was the favorite for the haptic sense, whereas the combination of 13 and 18 Hz was undesirable. With regard to the olfactory sense, the favorite combination was 4 and 7 Hz, whereas dislike was observed for a combination of 9 and 14 Hz. Furthermore, dozing was observed to be related to 5, 6, 9, and 16 Hz. Based on the observations, we developed different preference measurement systems (Figure 3.7).

3.5 ADVANCED HUMAN MOTION SENSING

3.5.1 OVERVIEW

Human motion sensing is one of the base technologies to support our civilization in many fields, including security, disaster defense, medical services, nursing care, marketing, and service activity. Recently, various sensors, devices, human–machine interfaces, and communication robots have been developed for it. Observation methods and information processing for human motion sensing are becoming increasingly complicated and sophisticated. Especially, since the target of the monitoring is human, not only engineering of sensor devices but also wide-ranging specialities such as mechatronics, robotics, cognitive science, psychology, sociology, bioinstrumentation, computer science, and ergonomics are required. Therefore, the human motion sensing is an integrated application mutually combined with the above-mentioned different specialities.

Approaches of human motion sensing are generally categorized into two types: passive remote sensing and active interactional sensing. The former is an observation of circumstances, and its representative examples are security systems and safety administration. On the other hand, the latter considers interaction between humans and machines and corresponds to man–machine interfaces like an intelligent mechatronics in a broad sense. A pioneering study of the latter seems to be *intelligent room* [42] in 1990s at the artificial intelligence (AI) laboratory in Massachusetts

Institute of Technology (MIT), where the system enabled control of equipment in the room with the human motion sensors for gesture and voice recognition. Similar concepts were found in *DreamSpace* (IBM) and *Oxygen Project* (MIT) [43]. After that, a new approach of physical interaction using robots was imported into such types of technologies, and techniques for the human motion monitoring have been progressed with the development of communication robots. In the field of robotics, such a robot system is called a *networked robot*. In the 2000s, three types of the networked robot were proposed in Japan: virtual-type, unconscious-type, and visible-type [44]. The virtual type is an agent or an icon robot in virtual space in order to communicate with users. The unconscious type is embedded in an environment and detects both human action and circumstance using remote sensing technologies. The visible type is a conventional robot that works in actual space and interacts with users directly. In short, passive and active approaches for the human motion sensing were re-realized as the unconscious type and visible type, respectively. Moreover, in Euro-America, several field tests like the URUS project (Ubiquitous Networking Robotics in Urban Settings Project) [45] are proceeding.

3.5.2 RECENT TRENDS

In the 2010s, applications using human motion sensing spread to individuals' health care, because people are taking a growing interest in health and medical services, and information and communication technology (ICT) is becoming widely used in our human society. In Europe and the United States, several social implementations based on eHealth and mHealth concepts to support public health by mobile devices are promoted as a government policy. Thanks to the rapid popularization of smartphones that are equipped with sensors and communication functions, various intelligent gadgets relating to the human monitoring for individuals are sold now, for instance, Jawbone UP (Jawbone Co., USA) [46] and Fitbit (Fitbit Inc., USA) [47]. These monitoring devices basically classify sensor signals to discriminate kinds of human motion or status. The accuracy of the discrimination depends on the following factors: 1) sensor and its position, 2) computation of feature information, 3) selection of classifier, and 4) sufficient amounts of data. In the following, a design step considering these factors is introduced through one case of activity recognition (AR).

3.5.3 ACTIVITY RECOGNITION AS CHILD'S MOTION SENSING

Activity Recognition is a generic term for technologies to identify human activities and actions from time-series data measured by various types of sensors such as acceleration sensor, camera, ultrasonic sensor, laser-range finder, depth sensor, and Kinect. Most basic AR is for acceleration sensors attached to the human body, and such AR began in the 2000s [48]. Since then, various types have been studied, such as a simple detection of ambulation [49], discrimination of several activities [50], and a precise action discrimination using a smart tag system [51]. Here, AR for children is introduced. This AR is a part of the functions of *Kinder-GuaRdian* system which is a child–parents–childminder support system in kindergarten, and it consists of a sensor node attached to the child, interactive measurement robots, and child-care data analyzers [52]. Recording the child's activity and finding his/her life rhythm, their information is utilized for his/her health care and growth management. Recently, in advanced countries, the rete of child obesity has been increasing [53], and it is said that childhood obesity causes lifestyle diseases when they become adults [54]; hence, such child AR may become important to national healthcare.

The following are the details of conputation for the child AR. As a constraint condition of the child AR, only one sensor is attached to the child's upper arm. The acceleration signal of the arm motion is measured by Bluetooth wireless three-axis acceleration sensor with 100 Hz sampling rate. In the case of our application, the following six kinds of features were computed: mean, standard deviation, energy, correlation, frequency domain entropy (FDE) [55], and cosine similarity. These features were computed against piecewise data separated by moving windows with 50% overlap from raw acceleration data. For later explanation, acceleration data of each axis are denoted as $a_x(t), a_y(t)$, and $a_z(t)$, where the notation of (t) $(t = 1, \cdots, T)$ means a sampling counter and T is the final time count. Energy feature is the sum of the squared discrete Fast Fourier Transform (FFT) components, and it is effective for accurate recognition of calm and active motions [56]. Describing an index of feature points as k, energy feature $\eta[k]$ is computed as $\eta_*[k] = \frac{1}{N/2-1}\sum_{j=1}^{N/2-1} |f_{*j}[k]|^2$, where $* = \{x,y,z\}$, N is an integer of power-of-two to specify the range of the moving window, and $f_{*j}[k] \in \mathscr{C}$ is a coefficient of discrete Fourier transformation computed by $f_{*j}[k] = \sum_{t=1+N(k-1)/2}^{N(k+1)/2} a_*(t) \cdot e^{-\frac{2\pi i}{N}jt}$, $j = 1, \cdots, \frac{N}{2} - 1$. Here, i is an imaginary unit, and note that f_0 is not computed to eliminate the Direct Current (DC) component. FDE is the normalized information entropy of the discrete FFT component magnitude of the signal. FDE supports discrimination of activities with similar features having similar energy values [57]. FDE $s_*[k]$ is calculated by $s_*[k] = -\sum_{j=1}^{N/2-1} p_{*j}[k] \cdot \log p_{*j}[k]$, where p_j is a probability distribution of each frequency and is computed by $p_j[k] = |f_j[k]|^2 / \sum_{l=1}^{N/2-1} |f_l[k]|^2$ for $j = 1, \cdots, N/2-1$.

In the case of our child AR, we tried to classify the following six types of activities of 16 3–5-years-old children: Sleeping, Hand motion, Sitting, Walking, Running, and Playing (in the room). Using the WEKA machine learning algorithms toolkit [58], the following representative classifiers, which do not depend on structure and properties of data, were selected: C4.5 Tree (J48), Naive Bayes, RandomForest, RandomTree, and REPTree [59]. C4.5 is the most popular classifier of a decision tree, although this is not as precise. Naive Bayes was selected as a representative probabilistic classifier. RandomForest is one of the meta classifier algorithms, combining several decision trees that are generated by bootstrapped sub-classes of data [60]. This algorithm was selected as a representative bootstrap method that was a popular and famous technique. RandomTree is an other type of decision tree, and this method chooses variables for bifurcation at random. REPTree generates a tree structure using information of gain and variance. Although the algorithm of REPTree resembles C4.5, this differs from C4.5 on the pruning method. For enhancement of

Table 3.1
Classification Accuracy of Childrens' Motion

Motion type	Sleeping	Hand motion	Sitting	Walking	Running	Playing
Best accuracies [%]	100[a]	84.8(RF)	88.1(NB)	83.3(NB)	98.4(NB)	86.6(NB)
Second accuracies [%]	96.9(CT)	81.2(NB)	84.1(RE)	72.9(RE)	96.6(RF)	83.9(CT)
Third accuracies [%]	-	75.0(RE)	80.4(RF)	70.8(RF)	90.8(RE)	75.5(RE)

Source: CT:C4.5 Tree, NB:Naive Bayes, RF:Random Forest, RT:RandomTree, RE:REPTree.
[a]NB, RF, RT, and RE.

the discrimination accuracies of the child's activities, a two-phase classification was performed first to separate six activities into two groups of calm and active motions, and then, a second classification was applied to each group separately. Finally, the accuracies were verified using 10-fold cross validation. The results are summarized in Table 3.1 according to each activity, with abbreviation of the best classifiers that lead the top three. In this case of child AR, it is found that the child's activity was well discriminated, as high as more than 80%, by Naive Bayes algorithm.

3.5.4 FOR FURTHER CONSIDERATION

In order to harness the benefits offered by human motion sensing, other aspects will be required in addition to adequate integration of technical factors as mentioned earlier: a) permeation into our daily life, b) sustainable effectiveness, and c) adaptation to individual differences. It is required to select monitoring methods depending on each person's life style after sufficient field survey and market need analysis for aspect a. For aspect b, various time span variations of persons, that is, short-term variation (fatigue and mental stress), medium-term (habituation and learning), and long-term (growth and aging), have to be coped with, and user experiences of them also should be taken into consideration. Nowadays, differences of personality such as liking and feeling can be guessed from big data analysis, and they may be utilized for aspect c. Both industry and academia desire to establish systematization involving the above-mentioned aspects and factors for human motion sensing.

3.6 ADVANCED APPROACHES FOR HUMAN MOTION DETECTION

In current years, many types of sensors have been installed throughout daily life thanks to the development of a semiconductor technology and decrease of the sensor prices [61]. They have been used not only for limited environments, such as factory automation, but also for open environments, such as human activities tracking. It is possible to analyze characteristics related to the motion performance and reveal the tacit knowledge of the motion by measuring human motion. Research that

Table 3.2
Sensor Examples for Measuring Human Motion

Sensor type	Application example
Acceleration sensor	Posture estimation
Gyro sensor	Posture estimation
Magnetic field sensor	Posture estimation
Infrared sensor	Posture estimation / Height estimation
Electromyography (EMG) sensor	Muscle motion estimation
Pressure sensor	Plantar pressure estimation
Camera	Entire body analysis

utilizes measurement of human motion has already advanced in several fields, such as sports [62] and rehabilitation [63].

There are various types of sensors for sensing human motion. Examples of the sensors are shown in Table 3.2. These sensors can be used in different ways depending on the features to be measured. A motion capture system using multiple cameras provides a comprehensive understanding of the entire body motion in the limited measurement environment [64]. On the other hand, the other mounted sensors can measure the motion in an external environment such as outdoors. An inertia navigation system (INS) is an algorithm using a foot mounted inertia measurement unit (IMU: an acceleration sensor, a gyro sensor, and a magnetic field sensor) to estimate 3-D position and orientation during human gait motion [65]. Besides, infrared sensors are applied to the foot as shown in Figure 3.8 for accurate state estimation [66]. In addition to measuring mechanical quantities directly, such as acceleration and angular velocity, there is another type of research that estimates the output torque indirectly by introducing a musculoskeletal model with electromyogram (EMG) sensors [67]. The sensing techniques can be applied to various research fields. Gait analysis is one of the most developed fields, in which different types of sensors are combined to estimate useful parameters during the gait cycle. For example, it is possible to compare differences in walking characteristics between a Parkinson's patient and a healthy person by using pressure sensors, acceleration sensors, and gyro

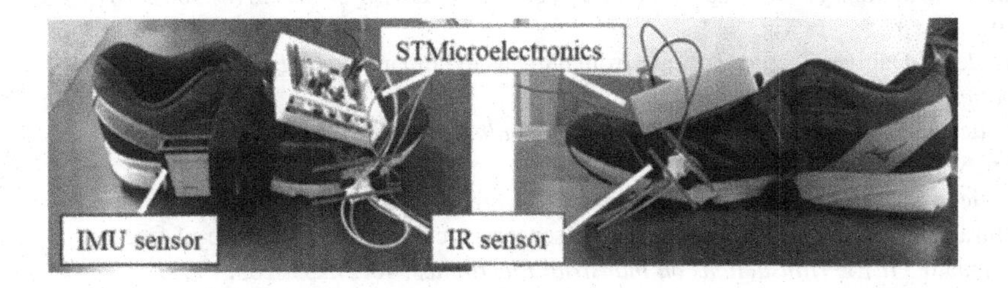

Figure 3.8 Foot mounted IMU-IR sensor system.

sensors [68,69]. Robotic assistance to human motion is also an applicable field of the sensing techniques. The assist robot is controlled based on the human state measured by the IMUs to follow the human. The robot applies the assist force to the human to prevent falling down when an unstable state is detected [70,71]. The human motion data can be captured easily with the development of the sensing techniques. In the future, the amount of data can keep increasing, and it is expected that further applications will be developed.

REFERENCES

1. R. Daily and D.M. Bevly. The use of gps for vehicle stability control systems. *Transactions on Industrial Electronics*, 51(2):270–277, 2004.
2. IECON2014. *Construction of HEMS in Japanese Cold District for Reduction of Carbon Dioxide Emissions*, 2014.
3. IECON2014. *Cost-effective Air Conditioning Control Considering Comfort Level and User Location*, 2014.
4. F. Chaumette and S. Hutchinson. Visual servo control parti: Basic approaches. *IEEE Robotics and Automation Magazine*, 13(4):83–90, 2006.
5. F. Chaumette and S. Hutchinson. Visual servo control partii: Advanced approaches. *IEEE Robotics and Automation Magazine*, 14(1):109–118, 2007.
6. S. Hutchinson, G. D. Hager, and P. I. Corke. A tutorial on visual servo control. *IEEE Transactions on Robotics and Automation*, 12(5):651–670, 1996.
7. N. Oda, M. Ito and M. Shibata. Vision-based motion control for robotic systems. *IEEJ Transactions on Electrical and Electronic Engineering*, 4(2):176–183, 2009.
8. Geoffrey E. Hinton Alex Krizhevsky and Ilya Sutskever. Imagenet classification with deep convolutional neural networks. *Advances in Neural Information Processing Systems 25 (NIPS2012)*, pages 1–9, 2012.
9. Jan Ben, Ayse Erkan, Marco Scoffier. Koray Kavukcuoglu, Urs Muller, Raia Hadsell, Pierre Sermanet and Yann LeCun. Learning long-range vision for autonomous off-road driving. *Journal of Field Robotics*, 26(2):120–144, 2009.
10. Joseph Del Preto Changhyun Choi, Wilko Schwarting and Daniela Rus. Learning object grasping for soft robot hands. *IEEE Robotics and Automation Letters*, 3(3):2370–2377, 2018.
11. Hyun Kim Duy Thanh Nguyen, Tuan Nghia Nguyen and Hyuk-Jae Lee. A high-throughput and power-efficient fpga implementation of yolo cnn for object detection. *IEEE Transactions on Very Large Scale Integration (VLSI) Systems*, 27(8):1861–1873, 2019.
12. Ravinder S. Dahiya. Tactile sensing - from humans to humanoids. *IEEE Transactions on Robotics*, 26(1):1–20, 2010.
13. Robotics and Automation. *Interpretation of Contact Geometries from Force Measurements*, volume 1, 1984.
14. Robotics and Automation. *Intrinsic Contact Sensing for Soft Fingers*, 1990.
15. T. Murakami, F. Yu, and K. Ohnishi. Torque sensorless control in multidegree-of-freedom manipulator. *IEEE Transactions on Industrial Electronics*, 40(2):259–265, 1993.
16. Robotics and Automation. *Whole-body Covering Tactile Interface for Human Robot Coordination*, volume 4, 2002.

17. T. Tsuji, Y. Kaneko, and S. Abe. Whole-body force sensation by force sensor with shell-shaped end-effector. *IEEE Transactions on Industrial Electronics*, 56(5):1375–1382, 2009.
18. T. Tsuji, N. Kurita, and S. Sakaino. Whole-body tactile sensing through a force sensor using soft materials in contact areas. *ROBOMECH Journal*, 1(1):1–11, 2014.
19. ACM Symposium on User Interface Software and Technology. *Detecting Shape Deformation of Soft Objects Using Directional Photoreflectivity Measurement*, 2011.
20. IECON2014. *Development of a Desk-type Tactile Interface Using Force Sensors*, 2014.
21. A.S. Fiorillo. A piezoresistive tactile sensor. *IEEE Transactions on Instrumentation and Measurement*, 46(1):15–17, 1997.
22. S. Stassi. Flexible tactile sensing based on piezoresistive composites: A review. *Sensors*, 14(3):5296–5332, 2014.
23. K. Noda. A shear stress sensor for tactile sensing with the piezoresistive cantilever standing in elastic material. *Sensors and Actuators A: Physical*, 127(2):295–301, 2006.
24. MFI2008. *An Embedded Artificial Skin for Humanoid Robots*, 2008.
25. J.S. Heo, J.H. Chung and J.J. Lee. Tactile sensor arrays using fiber bragg grating sensors. *Sensors and Actuators A: Physical*, 126(2):312–327, 2006.
26. Robotics and Automation. *Conformable and Scalable Tactile Sensor Skin for Curved Surfaces*, 2006.
27. C. Lucarotti. Synthetic and bio-artificial tactile sensing: A review. *Sensors*, 13(2):1435–1466, 2013.
28. T.H. Speeter. A tactile sensing system for robotic manipulation. *The International Journal of Robotics Research*, 9(6):25–36, 1990.
29. N. Wettels. Biomimetic tactile sensor array. *Advanced Robotics*, 22(8):829–1849, 2008.
30. J. Han and A. Uchiyama. The measurement and analysis on the effect of olfactory stimulus on the human body. *IEEJ Transactions*, 122:1914–1915, 2002.
31. K. Nagamine, A. Nozawa, and H. Ide. Evaluation of emotions by nasal skin temperature on auditory stimulus and olfactory stimulus. *IEEJ Transactions EIS*, 124(9):1914–1915, 2004.
32. L.F. Bos, D.M.W. Nordgren and R.B. van Baaren. On making the right choice: The deliberation-without-attention effect. *Science*, 311(5763):1005–1007, 2006.
33. S. Ayabe. Nurtured odor preference. *The Japanese Journal of Ergonomics*, 35(2):68–70, 2004.
34. M. Funada, M. Shibukawa and S. Ninomija. Objective measurement of event related potentials' changes. *The Japanese Journal of Ergonomics*, 38:538–539, 2002.
35. Japan Mechanic Symposium. *Remarks on Emotion Recognition Using Brain-body Actuated System- Emotion Classification Using Support Vector Machines*, 2002.
36. SICE-ICASE International Joint Conference. *Proposal for the Extraction Method of Personal Comfort and Preference by the EEG Maps*, 2006.
37. H. Tagaito. An evaluation of user interface with emotions. *UNISYS Technology Review*, 2000.
38. M. Ohme, R.and Matuskin and T. Szczurko. Neurophysiology uncovers secrets of tv commercials. *DER MARKT*, 49(3-4):133–142, 2010.
39. G. Vecchiato, L. Astolfi, F.D.V. Fallani, D. Cincotti, Mattia, S. Salinari, R. Soranzo and F. Babiloni. Changes in brain activity during the observation of tv commercials by using eeg, gsr and hr measurements. *Brain Topogr*, 23(2):165–179, 2010.
40. S. Tasaki, T. Igarashi, N. Murayama and H. Koga. Relationship between biological signals and subjective estimation while humans listen to sounds. *IEEJ Transactions*, 122:1632–1638, 2002.

41. S. Shiba. *Factor Analysis Method.* University of Tokyo Press, 1976.
42. International Cognitive Technology Conference. *The Intelligent Room Project*, 1997.
43. Oxygen Project, website accessed 2020, 30 August, http://oxygen.lcs.mit.edu/Overview.html
44. "New IT from Japan 'Toward the Realization of Network Robots'," (in Japanese) Ministry of Internal Affairs and Communications Research Group on Network Robot Technology, 2003.
45. IEEE/RSJ IROS Workshop on Network Robot Systems. *Ubiquitous Networking Robotics In Urban Settings*, 2006.
46. Jawbone, website accessed 2020, 30 August, https://www.jawbone.com/
47. Fitbit inc, website accessed 2020, 30 August, https://www.fitbit.com/global/us/home
48. F. Foerster, M. Smeja, and J. Fahrenberg. Detection of posture and motion by accelerometry: A validation in ambulatory monitoring. *Computers in Human Behavior*, 15:571–583, 1999.
49. K. Aminian, P. Robert, E.E. Buchser, B. Rutschmann, D. Hayoz, and M. Depairon. Physical activity monitoring based on accelerometry: Validation and comparison with video observation. *Medical & Biological Engineering & Computing*, 37(3):304–308, 1999.
50. L. Bao and S.S. Intille. Activity recognition from user-annotated acceleration data. *Pervasive 2004, Lecture Notes in Computer Science 3001*, 2004.
51. T. Gu, L. Wang, Z. Wu, X. Tao and J. Lu. A pattern mining approach to sensor-based human activity recognition. *IEEE Transactions on Knowledge and Data Engineering*, 23(9):1359–1372, 2011.
52. IEEE International Symposium on Robot and Human Interactive Communication (Ro-Man2012). *Activity Recognition for Children Using Self-Organizing Map*, 2012.
53. R.W. Taylor, L. Murdoch, P. Carter, D.F. Gerrard, S.M. Williams, and B.J. Taylor. Longitudinal study of physical activity and inactivity in preschoolers: The flame study. *Medicine and Science in Sports and Exercise*, 41(1):96–102, 2009.
54. A. Must and R.S. Strauss. Risks and consequences of childhood and adolescent obesity. *International Journal of Obesity*, 23(2):2–11, 1999.
55. AAAI2005. *Activity Recognition from Accelerometer Data*, 2005.
56. A. Sugimoto, Y. Hara, T.W. Findley, and K. Yoncmoto. A useful method for measuring daily physical activity by a three-direction monitor. *Scandinavian Journal of Rehabilitation Medicine*, 29(1):37–42, 1997.
57. International Conference on Pervasive Computing. *Activity Recognition from User-Annotated Acceleration Data*, 2004.
58. WEKA official home page, Machine Learning Group at University of Waikato, 2015, http://www.cs.waikato.ac.nz/ml/weka/
59. Institute of Statistical Mathematics. *Authorship Identification Using Random Forests(in Japanese)*, volume 55, 2007.
60. L. Breiman. Random forests. *Machine Learning*, 45(1):5–32, 2001.
61. S.D. Oliner, D.M. Byrne and D. E. Sichel. How fast are semiconductor prices falling? *Review of Income and Wealth*, 64(3), 2018.
62. D.A. Rafeldt, S.A. Rawashdeh. and T.L. Uhl. Wearable imu for shoulder injury prevention in overhead sports. *Sensors*, 16(11):1847, 2016.
63. A. Delno, F. Lavagetto, I. Bisio and Sciarrone A. Enabling iot for in-home rehabilitation: Accelerometer signals classification methods for activity and movement recognition. *IEEE Internet of Things Journal*, 4(1):135–146, 2017.
64. Y. Han. 2d-to-3d visual human motion converting system for home optical motion capture tool and 3-d smart tv. *IEEE Systems Journal*, 9(1):131–140, 2015.

65. 7th Workshop on Positioning, Navigation and Communication. *Indoor Pedestrian Navigation Using an INS/EKF Framework for Yaw Drift Reduction and a Foot-mounted IMU*, 2010.
66. IECON 2018 - 44th Annual Conference of the IEEE Industrial Electronics Society. *Real-time Foot Clearance and Environment Estimation Based on Foot-Mounted Wearable Sensors*, 2018.
67. D. Guiraud, Z. Li and M. Hayashibe. Inverse estimation of multiple muscle activations from joint moment with muscle synergy extraction. *IEEE Journal of Biomedical and Health Informatics*, 19(1):64–73, 2015.
68. D. Torricelli. M. Schmid. A. Munoz-Gonzalez. J. Gonzalez-Vargas, F. Grandas, C. Caramia and J.L. Pons. Imu-based classification of parkinson's disease from gait: A sensitivity analysis on sensor location and feature selection. *IEEE Journal of Biomedical and Health Informatics*, 22(6):1765–1774, 2016.
69. K.I. Wong, Y.C Han, and I. Murray. Gait phase detection for normal and abnormal gaits using imu. *IEEE Sensors Journal*, 19(9):3439–3448, 2019.
70. IEEE Advanced Motion Control. *Novel Walking Assist Device Based on Generic Human Motion Tracking Criteria*, 2016.
71. IEEE International Conference on Mechatronics (ICM). *Walking Assistance System for Walking Stability by Using Human Motion Information*, 2017.

4 Model-Based Control for High-Tech Mechatronic Systems

Tom Oomen and Maarten Steinbuch

CONTENTS

4.1 INTRODUCTION

4.1.1 MOTION SYSTEMS

Motion systems are mechanical systems with actuators with the primary function to position a load. The actuator can be either hydraulic, pneumatic, or electric. The freedom in trajectory planning and motion profile is often limited when compared with general robotic systems. In addition, motion systems are closely related to active vibration isolation systems; a main difference is that active vibration systems regulate a motion to zero. Examples of motion systems include linear and rotational servo drives and also state-of-the art planar 6 DOF (Degree-of-Freedom) motion platforms. Typically, such motion systems can be represented well by linear models, which are possibly of high order due to flexible dynamics in the case of high performance and high DOF systems. Inexpensive motion systems typically have friction in the guidance. Backlash is often prevented by application of direct drive actuators. Sensor systems include encoders, capacitive sensors, and laser interferometers, with accuracies into the sub-nanometer range. For further aspects on the mechatronic design of such systems, see Schmidt et al. [38] and Fleming and Leang [16].

4.1.2 INDUSTRIAL STATE OF THE ART

The industrial state-of-the-art control of motion systems is summarized as follows. By appropriate system design, most systems are either decoupled or can be decoupled using static input-output transformations. Hence, most motion systems and their motion software architecture use single-input single-output (SISO) control design method and solutions. The feedback controller is typically designed using frequency domain techniques, in particular via manual loop-shaping. A typical motion controller has a proportional-integral-derivative (PID) structure, with a low-pass filter at high frequencies and one or two notch filters to compensate flexible dynamics [43,62]. In addition to the feedback controller, a feedforward controller is often implemented with acceleration, velocity, and friction feedforward for the reference signal. The setpoint itself is designed using a setpoint generator with jerk limitation profiles [31].

One of the most accurate, fast, and expensive motion systems available today is a wafer stage, which is part of a wafer scanner used in lithographic processes. The extreme accuracy and speed requirements necessitate and justify a further development of the state of the art, which is outlined next.

4.1.3 DEVELOPMENTS IN LITHOGRAPHY

Wafer scanners (see Figure 4.1) are the state-of-the-art equipment for the mass production of integrated circuits (ICs). During the production process, a photoresist is exposed on a silicon disc, called a wafer. During exposure, the image of the desired IC pattern, which is contained on the reticle, is projected through a lens on the photoresist. Subsequent chemical reactions enable etching of these patterns, which is repeated for successive layers. Typically, more than 20 layers are required for each

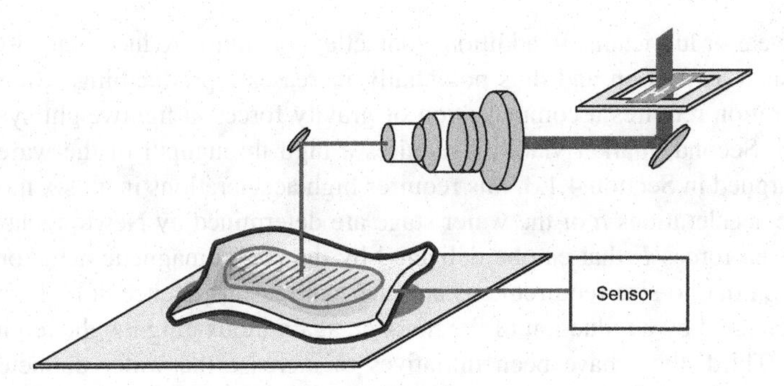

Figure 4.1 IC production. The lens system that exposes a wafer is being positioned using a sensor that measures the edge of the system.

wafer. Each wafer contains more than 200 ICs that are sequentially exposed. During this entire process, the wafer must extremely accurately track a predefined reference trajectory in six motion DOFs, with future requirements of tracking accuracy below 1 nm, velocity in the order of 1 $\frac{m}{s}$, and acceleration in the order of 10^2 $\frac{m}{s^2}$. These extreme speed requirements are imposed by throughput, which directly determines the market position of the machine. This highly accurate and fast motion task is performed by the wafer stage, which is one of the most accurate and expensive motion systems commercially available.

In the last decades, increasing demands with respect to computing power and memory storage have led to an ongoing dimension reduction of transistors. The minimum feature size associated with these transistors is called the critical dimension (CD) and is determined by the wavelength of light; see Martinez and Edgar [34], Hutcheson Hutcheson [30]. In Hutcheson [30], CDs of 50 nm have been achieved using deep ultraviolet (DUV) light with a wavelength of 193 nm through many enhancements of the production process. However, a technology breakthrough is required to reduce the wavelength of light and consequently improve the achievable CD.

Extreme ultraviolet (EUV) is a key technology for next-generation lithography [65,70]. At present, experimental prototypes with a 13.5 nm wavelength are reported in Hutcheson [30] and Arnold [3], and production systems are already in use [1]. The introduction of EUV light sources in lithography has far-reaching consequences for all subsystems of the wafer scanner, including the wafer stage. EUV does not transmit through any known material, including air. Hence, lenses used in DUV have to be replaced by mirrors in EUV. Moreover, the entire exposure has to be performed in vacuum.

4.1.4 DEVELOPMENTS IN PRECISION MOTION SYSTEMS

Due to the developments in lithographic production processes, next-generation precision motion systems are expected to be lightweight for several reasons. First, vacuum operation requires these systems to operate contactless to avoid pollution due to mechanical wear or lubricants. In addition, contactless operation reduces parasitic nonlinearities such as friction and thus potentially increases reproducibility. Since contactless operation requires a compensation of gravity forces, a lightweight system is essential. Second, market viability requires a high throughput of the wafer scanner. As is argued in Section 4.1.3, this requires high accelerations in all six motion DOFs. The accelerations a of the wafer stage are determined by Newton's law $F = ma$. Here, the forces F that can be delivered by the electromagnetic actuators are bounded, e.g., due to size requirements and thermal reasons. Hence, a high acceleration a is enabled by a reduction of the mass m, again motivating a lightweight system design. Third, there have been initiatives to increase the wafer diameter from 300 to 450 mm to increase productivity. This requires increased dimensions of the wafer stage, which again underlines the importance of a lightweight system design.

As a result of a lightweight system design, next-generation motion systems predominantly exhibit flexible dynamical behavior at lower frequencies. This has important consequences for control design, as is investigated next.

4.1.5 TOWARDS NEXT-GENERATION MOTION CONTROL: THE NECESSITY OF A MODEL-BASED APPROACH

The increasing accuracy and performance demands lead to the manifestation of flexible dynamical behavior at lower frequencies. On the other hand, due to these increasing demands, the controller has to be effective at higher frequencies. Combining these developments leads to a situation where flexible dynamical behavior is present within the control bandwidth. This is in sharp contrast to traditional positioning systems, where the flexible dynamical behavior can be considered as high-frequency parasitic dynamics, as is, e.g., the case in van de Wal et al. [71, Sec. 2.1, Assumption 1-3].

The presence of flexible dynamical behavior within the control bandwidth has significant implications for motion control design in comparison to the traditional situation:

i) Next-generation motion systems are inherently multivariable, since the flexible dynamical behavior is generally not aligned with the motion DOFs.
ii) Next-generation motion systems are envisaged to be designed with many actuators and sensors to actively control flexible dynamical behavior, whereas traditionally, the number of inputs and outputs equals the number of motion DOFs.
iii) A dynamical relation exists between measured and performance variables, since the sensors generally measure at the edge of the wafer stage system, while the performance is required on the spot of exposure on the wafer itself. In contrast, the flexible dynamical behavior is often neglected in traditional motion systems, leading to an assumed static geometric relation between measured and performance variables; see Figure 4.1 for a graphical illustration.

These implications of lightweight motion systems on the control design motivate a model-based control design, since:

i) A model-based design provides a systematic control design procedure for multivariable systems.
ii) A model is essential to investigate and achieve the limits of performance. Specifically, fundamental performance limitations are well-established for nominal models [58], and robust control provides a transparent trade-off between performance and robustness [60].
iii) A model-based design procedure enables the estimation of unmeasured performance variables from the measured variables through the use of a model.

As pointed out in van de Wal et al. [71], a model-based control design is far from standard in state-of-the-art industrial motion control, since the majority of such systems are controlled by manually-tuned SISO PID controllers in conjunction with rigid-body decoupling based on static input-output transformations. One of the main reasons is the fact that a manually tuned design achieves reasonable performance while only requiring easily available frequency response function (FRF) data instead of an accurate parametric model. The presence of flexible dynamical behavior in next-generation motion control necessitates and justifies the additional modeling effort required to deal with the situations sketched here.

4.1.6 CONTRIBUTION: FROM MANUAL TUNING TO MODEL-BASED SYNTHESIS

The main contribution of this chapter is an overview of a systematic control design procedure for motion systems that has proven its use in industrial motion control practise. A step-by-step procedure is presented that gradually extends SISO loop-shaping to the multi-input multi-output (MIMO) situation. This step-by-step procedure consists of (i) interaction analysis, (ii) decoupling, (iii) independent SISO design, (iv) sequential SISO design, and finally, (v) norm-based MIMO design. In the norm-based MIMO design, a model-based control is pursued that addresses the future motion control challenges from Section 4.1.5. As such, the present chapter provides a unified overview of ongoing research on motion feedback control [25,41,46,50,63]. The design of the feedforward controller is an important aspect but beyond the scope of this chapter; see Section 4.5 and Oomen [43] for results in this direction as well as Oomen [42] for learning control.

4.2 MOTION SYSTEMS

The dynamical behavior of motion systems is typically linear and dominated by the mechanics. Indeed, the actuator and sensor dynamics are typically relevant in higher frequency regions and are therefore ignored. These mechanics are typically described in the Laplace domain as [19,37,52]

$$G_m = \underbrace{\sum_{i=1}^{n_{RB}} \frac{c_i b_i^T}{s^2}}_{\text{rigid-body modes}} + \underbrace{\sum_{i=N_{rb}+1}^{n_s} \frac{c_i b_i^T}{s^2 + 2\zeta_i \omega_i s + \omega_i^2}}_{\text{flexible modes}}, \tag{4.1}$$

where n_{RB} is the number of rigid-body modes, the vectors $c_i \in \mathbb{R}^{n_y}$, $b_i \in \mathbb{R}^{n_u}$ are associated with the mode shapes, and $\zeta_i, \omega_i \in \mathbb{R}_+$. Here, $n_s \in \mathbb{N}$ may be very large and even infinite [29]. Note that the rigid-body modes are not suspended in Equation 4.1. In the case of suspended rigid-body modes, e.g., in the case of flexures [16,46], Equation 4.1 can directly be extended. As an example, Figure 4.2 shows a magnitude FRF and the underlying modes of a system.

Traditionally, motion systems are designed to be stiff, such that the flexible behavior is well above the intended closed-loop bandwidth, implying a rigid-body behavior in the control bandwidth. Also, the number of actuators n_u and sensors n_y is chosen to equal n_{RB}, and these are positioned such that the matrix $\sum_{i=1}^{n_{RB}} c_i b_i^T$ is invertible. In this case, matrices T_u and T_y can be selected such that

$$G = T_y G_m T_u = \frac{1}{s^2} I_{n_{RB}} + G_{\text{flex}} \tag{4.2}$$

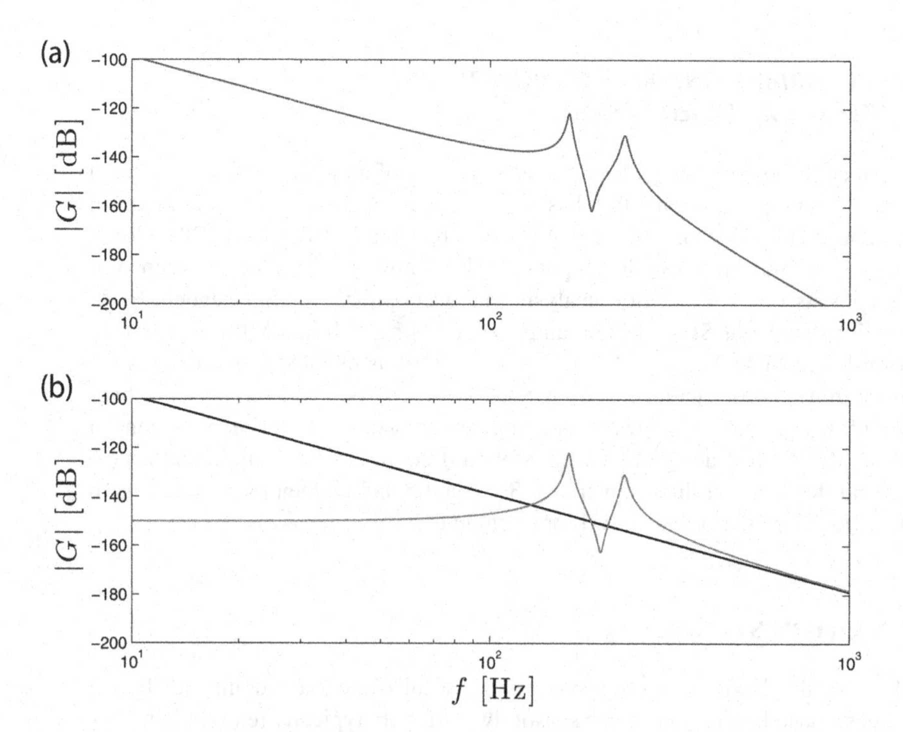

Figure 4.2 Separating rigid body and flexible dynamics. (a) Original system. (b) Separation in rigid-body dynamics (black) and flexible dynamics (grey).

where T_y is typically selected such that the transformed output y equals the performance variable z, as is defined in Section 4.4.1. Importantly, the selection of these matrices T_u and T_y can be done directly on the basis of FRF data; see Section 4.3.3.2 and, e.g., Stoev et al. [66]. As a result of this, often decentralized controllers can be designed, i.e., diagonal PID controllers with typically a few notch filters, as is outlined in the forthcoming sections.

As is argued in Section 4.1.5, future motion systems are envisaged to have more actuators and sensors to improve their performance. In Section 4.4.3, the modal representation (Equation 4.1) is employed to actively damp and stiffen certain flexible modes.

4.3 FEEDBACK CONTROL DESIGN

4.3.1 SYSTEM IDENTIFICATION – OBTAINING AN FRF

Motion control design is typically done in the frequency domain, since it allows the direct evaluation of performance and robustness in addition to intuitive tuning. Frequency domain tuning requires a model of the system. In the case where the motion system has already been realized, system identification is an inexpensive, fast, and accurate approach to model motion systems. In particular, the first step in motion systems typically involves an FRF identification using noise signals, single sine, swept sine, or multi-sine excitation [51]. It is important to note that most motion systems are unstable in open-loop, since these have rigid body dynamics leading to poles at zero in Equation 4.1.

Indeed, due to open-loop instability, safety requirements, or nonlinear behavior, it is often required to implement a feedback controller during the identification experiments. Consider the single DOF feedback configuration depicted in Figure 4.3. Here, K can be implemented during the identification experiment and is later redesigned to enhance system performance. For instance, K can be a low-bandwidth PD controller $K(s) = Ds + P$. Typically, the controller zero (ratio P/D) is taken to be sufficiently low. If the sign of the system at low frequencies is known, a simple procedure can be used to increase the gain while keeping the ratio P/D constant, or first D is increased, then P, while giving the motion system a modest setpoint. In particular, a so-called jogging motion enables overcoming the friction while tuning and identifying the system. The resulting low-bandwidth controllers are typically sufficient for system identification purposes. Once these experimental controllers have been designed, an identification experiment can be performed.

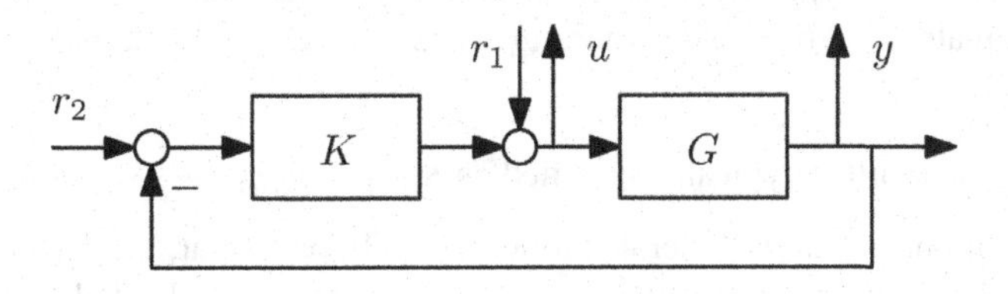

Figure 4.3 Standard single DOF feedback configuration.

Next, assume that an external excitation is applied for system identification. For the purpose of illustration, assume that $r_2 = 0$, whereas r_1 is the excitation signal. In the case that the measurement indeed is performed under closed-loop operating conditions, i.e, $K \neq 0$, then care must be taken when attempting to identify G in Figure 4.3. In particular, in the situation where the measured outputs are contaminated by noise and disturbances, and in addition, the external excitation r_1 is taken as a noise signal, as is common practise in motion control, then a direct identification of the plant, i.e., from u to y, may lead to biased estimates. In particular, poor signal-to-noise ratios will lead to an identified inverse controller. In particular, in the case of noise excitation, it is recommended to first identify the process-sensitivity ($GS : r_1 \mapsto y$) (see again Figure 4.3) and the sensitivity ($S : r_1 \mapsto u$). Subsequently, divide the two FRFs to obtain the FRF of G. This will reduce the bias in the estimate of the plant. The same procedure can be followed for MIMO systems, provided that appropriate matrix transformations are done for every frequency measurement point. Recently, important progress has been made to enhance the FRF by appropriate input design [50, Appendix A] and by non-parametric pre-processing; see Voorhoeve et al. [69] for an application on a motion system.

In the forthcoming sections 4.3.2 and 4.3.3, the nonparametric FRF model suffices for tuning. In Section 4.4, a parametric model is required. This parametric model is estimated based on the obtained FRF data.

4.3.2 LOOP-SHAPING – THE SISO CASE

The feedback controller can be directly designed based on the obtained FRF. First, manual loop-shaping for SISO systems is investigated. As an example, consider the measured FRF of a wafer stage that is presented in Section 4.4.2; see Figure 4.6.

The idea in loop-shaping is to select K that directly affects the loop-gain $L = GK$. Indeed, L can be directly manipulated by the choice of K using Bode and Nyquist diagrams [4, section 11.4], which is essentially due to the Laplace transform that replaces the time domain convolution by a simple multiplication. In turn, L directly connects to the closed-loop transfer functions in certain frequency ranges, including $S = \frac{1}{1+GK}$ and $T = \frac{GK}{1+GK}$ see also Figure 4.3. Loop-shaping typically consists of the following steps, which are typically re-tuned in an iterative fashion:

1. Pick a cross-over frequency f_{bw} (Hz), which is the frequency where $|L(2\pi f_{bw})| = 1$. Typically, $|S| < 1$ below f_{bw}, while $|T| < 1$ beyond f_{bw}. Furthermore, f_{bw} is typically selected in the region where the rigid-body modes dominate, i.e., $f_{bw} < \frac{\omega_i}{2\pi}$, where $\omega_i, i = n_{rb} + 1, \ldots, n_s$ is defined in Equation 4.1.
2. Implement a lead filter, typically specified as $K_{lead} = p_{lead} \frac{\frac{1}{2\pi\frac{1}{3}f_{bw}}+1}{\frac{1}{2\pi 3 f_{bw}}+1}$, such that sufficient phase lead is generated around f_{bw}. In particular, the zero is placed at $\frac{1}{3}f_{bw}$, while the pole is placed at $3f_{bw}$. Next, p_{lead} is adjusted such that $|GK_{lead}(2\pi f_{bw})| = 1$. This lead filter generates a phase margin at f_{bw}.

Indeed, the phase corresponding to the rigid-body dynamics in Equation 4.1 is typically −180 degrees, which leads to a phase margin of 0 degrees.

3. Check stability using Nyquist plot of GK and include possible notch filters in the case where the flexible modes endanger robust stability.
4. Include integral action through $K = K_{int} K_{lead}$, with $K_{int} = \frac{s+2\pi\frac{f_{bw}}{5}}{s}$. Since $GK \gg 0$ at low frequencies, $S \approx \frac{1}{GK}$, and the integrator pole at $s = 0$ inproves low-frequency disturbance attenuation while not affecting phase margin due to the zero at $\frac{1}{5} f_{bw}$.

The loop-shaping procedure is highly systematic and fast: it only requires a non-parametric FRF model, which is fast, inexpensive, and accurate, and the procedure is very systematic and intuitive to apply. However, if the system is multivariable, care should be taken to deal with interaction.

4.3.3 LOOP-SHAPING – THE MIMO CASE

Loop-shaping cannot be directly applied to MIMO systems due to interaction. Essentially, the interdependence between the decentralized PID controllers complicates the tuning of the overall multivariable controller, which at least has to be stabilizing. In this section, the manual loop-shaping procedure is gradually extended to the multivariable case. These results are essentially all based on the generalized Nyquist criterion.

In MIMO systems, closed-loop stability is determined by the closed-loop characteristic polynomial $\det(I + L(s))$. By graphically analyzing $\det(I + L(j\omega))$ similarly to the scalar Nyquist plot, closed-loop stability can be analyzed see Skogestad and Postlethwaite [59, Theorem 4.9] for details. This plot can be generated directly using the identified MIMO FRF of Section 4.3.1. However, in the case where the closed-loop system is unstable, it is not immediately obvious which element of the MIMO controller K to re-tune.

Note that a related test can be obtained in terms of the loci of $L(j\omega)$, i.e., the eigenvalues of L as a function of frequency. If both K and G are open-loop stable, then a sufficient condition for closed-loop stability is where all loci do not encircle the point $(-1,0)$. However, this does not resolve the design issue: the shaping of these eigenvalue loci is not straightforward if the plant has interaction. Furthermore, typical margins that are used in SISO systems, such as phase margins, are less useful. In particular, the phase margin based on characteristic loci implies a phase change at the same time in all loops simultaneously.

In the special case where the open loop transfer function matrix is diagonal, $L(s) = \text{diag}\{l_i(s)\}$, i.e. the open loop is decoupled, then

$$\det(I + L(s)) = \prod_{i=1}^{n} \det(1 + l_i(s)) \tag{4.3}$$

As a result, the characteristic loci of the open-loop transfer function matrix directly coincide with the Nyquist plots of the scalar loop gains $l_i(j\omega)$. The MIMO feedback control design complexity then reduces to that of a number of SISO feedback control design. Many classical MIMO control design methods aim at decoupling the open-loop function at some location in the feedback loop, e.g., at the plant input or plant output.

Since SISO loop-shaping is a systematic and very fast design procedure for motion systems, it is a preferred procedure even for MIMO systems. In practise, it is often attempted to decouple the system such that Equation 4.3 holds. If this is not possible, it may be tempting to directly go to a norm-based optimal design; however, this requires a parametric model. Such parameteric models, as is outlined in Section 4.4, are expensive and user-intensive to obtain. Therefore, several ideas from multivariable and robust control are exploited to extend loop-shaping towards MIMO systems. This leads to a design procedure for MIMO motion systems, consisting of the following steps:

1. Interaction analysis
2. Decoupling and possible independent SISO design
3. Multi-loop feedback control design with robustness for interaction
4. Sequential feedback control design
5. Norm-based control design

All except for the last step can be performed with a non-parametric model of the plant, i.e., an identified FRF. The norm-based control design requires a parametric model of the plant. Nonetheless, the last step may be essential to address the complex motion control problems envisaged in future systems; see Section 4.1.5.

4.3.3.1 Interaction Analysis

The first step in multivariable motion feedback design is interaction analysis. Especially two-way interaction is essential, since one-way interaction does not affect closed-loop stability.

Several approaches have been developed for interaction analysis. A well-known, easy to compute, and useful approach is the relative gain array. The frequency-dependent relative gain array (RGA) [11,60] is given by

$$\mathrm{RGA}(G(f)) = G(f) \times (G(f)^{-1})^T \tag{4.4}$$

where $\times$ denotes element-wise multiplication. Note that the RGA can be directly computed using FRF data. In addition, the RGA is independent of the feedback controller and invariant under scaling. The rows and columns of the RGA sum to 1 for all frequencies f (Hz). If $(RGA)(f) = I$, $\forall \omega$, then perfect decoupling is achieved.

4.3.3.2 Decoupling and Independent SISO Design

Static decoupling may be considered if the interaction is too severe to allow multi-loop SISO design. For motion systems, such transformations (see also Equation 4.2), can be obtained using kinematic models. Herein, combinations of the actuators are defined so that actuator variables act in independent directions at the center of gravity. Similarly, combinations of the sensors are defined so that each translation and rotation of the center of gravity can be measured independently. Such decouplings can be further refined using FRF data [66].

In certain situations, it may be desirable to decouple the plant at other frequencies [8] or to use a dynamic decoupling. In any situation, the effect of the decoupling transformations should be analysed using the interaction measures derived earlier. If the system is sufficiently decoupled using static transformations, then multi-loop SISO controllers can be designed.

4.3.3.3 Multi-loop Feedback Control Design with Robustness for Interaction

In cases where the interaction after decoupling is too large to successfully design multi-loop SISO controllers, robustness for interaction may be enforced. This approach employs concepts from robust control theory [72]. The objective is to design SISO controllers that are robust for interaction terms.

To this end, note that [23]

$$\det(I+GK) = \det(I+E_T T_d)\det(I+G_d K) \tag{4.5}$$

with $T_d = G_d K(I+G_d K)^{-1}$. Let $G_d(s)$ be the diagonal terms of G. Then, the non-diagonal terms of the plant $G_n d(s) = G(s) - G_d(s)$ can be considered in $E_T(s)$. As a result, MIMO closed-loop stability can be decomposed into stability of N non-interacting loops associated with $\det(I+G_d(s)K(s))$ and stability associated with $\det(I+E_T(s)T_d(s))$.

In $\det(I+E_T(s)T_d(s))$, T_d is the complementary sensitivity function of the N decoupled loops. If $G(s)$ is stable and $T_d(s)$ is stable, then the small gain theorem [59] implies that $\det(I+E_T T_d)$ does not encircle the origin if

$$\rho(E_T(j\omega)T_d(j\omega)) < 1, \forall\omega \tag{4.6}$$

where ρ is the spectral radius. Next, Equation 4.6 holds if

$$\mu_{T_d}(E_T(j\omega)T_d(j\omega)) < 1, \forall\omega \tag{4.7}$$

which in turn implies that

$$\overline{\sigma}(T_d(j\omega)) < \mu_{T_d}(E_T(j\omega))^{-1}, \forall\omega \tag{4.8}$$

where μ_{T_d} is the structured singular value [72] with respect to the diagonal (decoupled) structure of T_d. Since $\overline{\sigma}(T_d(j\omega)) = \max_i |T_{d,ii}(j\omega)|$, Condition 4.8 can be used as an additional bound on $T_{d,ii}$ when loop-shaping the multi-loop SISO controllers. In other words, once the structured singular value in Equation 4.8 is computed, the SISO approach in Section 4.3.2 can directly be followed, where an additional bound is enforced on all SISO complementary sensitivity functions $T_{d,ii}$ to provide robustness for interaction terms.

Clearly, the test in Equation 4.8 provides robustness for any choice of stabilizing controller in the other loops. Intuitively, the implication is that this may be conservative. In the next section, the particular choice of controller in a previous loop is explicitly taken into account, leading to a procedure for designing decentralized controllers for interaction.

4.3.3.4 Sequential Loop Closing

In the situation where the approach in Section 4.3.3.3, which provides robustness for interaction, leads to conservative designs, then the interaction terms and specific controller designs may be explicitly taken into account in subsequent design steps. This procedure is called sequential loop closing (SLC). Here, earlier designed controllers are explicitly taken into account in the next loops.

The idea is to start with an FRF of the open-loop system, which is assumed to be open-loop stable. Then, the first loop is closed, and a new FRF is computed of the resulting system, which includes effects of the designed controller. Then, one more loop is closed, and the resulting FRF is computed again. In particular, each SISO controller k_i, from $K = \text{diag}\{k_i\}, i = \{1,...,n\}$, is designed using the property [35]

$$\det(I+GK) = \prod_{i=1}^{n} \det(1+g^i k_i) \tag{4.9}$$

where for each ith design step, the equivalent plant g^i is defined as the lower fractional transformation:

$$g^i = \mathscr{F}_l(G, -K^i) \tag{4.10}$$

where $K^i = \text{diag}\{k_j\}$, $j = \{1,...,n\}$, $j \neq i$.

It can be directly proved that the closed-loop system is stable if every k_i is designed to be stabilizing. However, the robustness margins, which are typically enforced for each k_i, may change due to the closing of subsequent loops. As a consequence, the overall MIMO design may have very poor robustness margins [17].

To alleviate these drawbacks, the ordering of the design steps may be changed, since this may significantly influence the achieved performance. There is no general approach to determine the best sequence for design. This may lead to many design iterations, especially for large MIMO systems. Note that the loops should always be opened in the same ordering as they are closed; otherwise, closed-loop stability cannot be guaranteed.

The main advantage of the sequential loop closing approach is that it allows SISO loop-shaping while explicitly addressing interaction and only requiring a nonparametric plant model. If the sequential loop-closing approach still does not lead to satisfactory performance, then a full multivariable control design approach may be pursued, as is investigated next. Such an approach may become essential for envisaged future systems as described in Section 4.1.5, which exhibit a drastically increased complexity.

4.4 MODEL-BASED CONTROL DESIGN

A full model-based control design is a suitable approach in the case where the manual tuning approaches that are presented in the previous section do not lead to satisfactory results. In this section, a general framework for model-based motion control is outlined. The presented framework relies on parametric models of the system and optionally, a detailed description of the involved uncertainty. Due to the need for parametric models and uncertainty models, the proposed model-based control framework requires a significantly larger investment of time and effort in obtaining such models. Relevant situations where the additional investment may be justified and may be necessitated include (see also Section 4.1.5):

- Multivariable systems with large interaction and/or uncertainty
- Overactuation and oversensing, i.e., the use of additional actuators and sensors to actively control flexible dynamics
- Inferential control, i.e., the situation where not all performance variables are measured variables for feedback control

4.4.1 STANDARD PLANT APPROACH

A standard plant approach is pursued to deal with the relevant situations mentioned in Section 4.4. The considered standard plant is depicted in Figure 4.4. Here, P denotes the standard plant, and K is the to-be-designed controller. In addition, the vector w contains exogenous input variables, including disturbances, sensor noise, and set-points, whereas the vector z contains the regulated variables that are defined such that they are ideally zero, e.g., servo errors. Furthermore, the vector y contains the measured variables that are available to the feedback controller, whereas the vector u contains the manipulated variables.

The standard plant is general in the sense that it enables the formulation of general control problems, including an arbitrary number of regulated variables u and measured variables y for overactuation and oversensing, and a distinction between measured variables y and performance variables z, as is required in the inferential control situation.

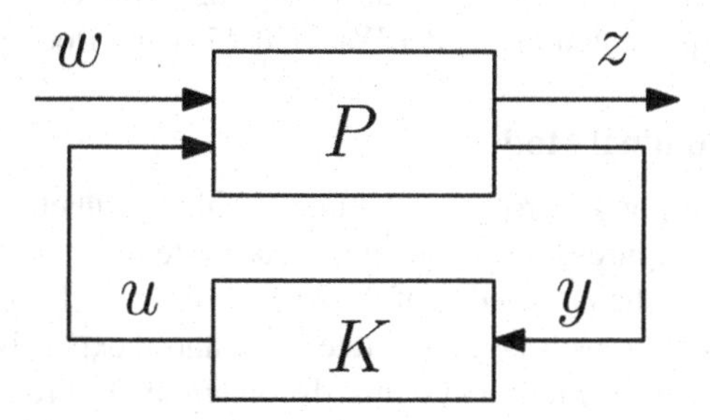

Figure 4.4 Standard plant configuration.

Once the motion control problem is posed in the standard plant framework, a large amount of literature [60,72], and software toolboxes [5] are available to synthesize the optimal controller C. Posing the control problem properly into the standard plant involves at least three aspects, which are investigated next: 1. weighting filter selection, 2. obtaining a parametric model of the motion system, and 3. uncertainty modeling.

4.4.1.1 Weighting Filter Selection for Performance Specification

To properly pose the control problem, weighting filters have to be selected. A criterion of the form

$$\mathcal{J}(P,K) = \|\mathcal{F}_l(P,K)\| \tag{4.11}$$

is adopted, where the goal is to compute

$$K^{\text{opt}} = \min_K \mathcal{J}(P,K) \tag{4.12}$$

Here, a specific norm $\|.\|$ has to be selected, e.g., $\mathcal{H}_2$ or $\mathcal{H}_\infty$. For motion control problems, the $\mathcal{H}_\infty$ norm has proven to be particularly useful, since it enables the specification of performance weights through loop-shaping ideas see (Sections 4.3.2 and 4.3.3), and it enables the design of robust controllers by explicitly addressing model uncertainty. The $\mathcal{H}_\infty$ norm for a stable system H is given by

$$\|H\|_\infty = \sup_\omega \bar{\sigma}(H(j\omega)) \tag{4.13}$$

The next step in specifying the performance goals is to select the internal structure of the standard plant. For instance, in the case where $y = z$, a common choice is

$$\mathcal{F}_l(P,K) = W \begin{bmatrix} G \\ I \end{bmatrix} (I + KG)^{-1} \begin{bmatrix} K & I \end{bmatrix} V \tag{4.14}$$

In Equation 4.14, G denotes the open-loop system, and W and V are weighting filters. These weighting filters enable the specification of various control goals and have to be user-specified. When using the $\mathcal{H}_\infty$ norm, these can be specified by means of desired loop-shapes of closed-loop transfer function matrices [60] or open-loop gains [67]. Suitable weighting filters for typical motion control problems that are in line with the guidelines in Sections 4.3.2 and 4.3.3 have been developed in, e.g., Schönhoff and Nordmann [56] and Steinbuch and Norg [62], whereas extensions to multivariable systems are presented in van der Wal et al. [71] and Boeren et al. [8].

4.4.1.2 Obtaining a Nominal Model

Besides the specification of the control goal through weighting functions, the optimization in Equation 4.12 requires knowledge of the true system denoted by G_o. In a model-based control design, the knowledge of the true system is reflected by means of a model $\hat{G}$. System identification is an accurate, fast, and inexpensive approach to obtain the required model for motion systems. To obtain an accurate model, the model should be tailored towards the control goal [21,57]. These results are further developed in Callafon and Van den Hof [12] and Oomen and Steinbuch [47] towards the identification of low-order models for control design. The main ingredient for these algorithms is the control criterion (Equation 4.11) and closed-loop FRFs, as is explained in Section 4.3.1.

4.4.1.3 Uncertainty Modeling

The model $\hat{G}$ obtained through the procedure in the previous section is not an exact description of the true system G_o, since i) motion systems generally contain many resonance modes [29], of which a limited number are included in the model; ii) parasitic nonlinearities are present, e.g., nonlinear damping [61]; and iii) identification experiments are based on finite time disturbed observations of the true system.

Robust control design [60,72] explicitly addresses these model errors by considering a *model set* through extending the nominal model $\hat{G}$ with a description of its uncertainty. This model set has to be chosen judiciously, since the resulting robust controller is typically designed to optimize the worst-case performance over all candidate models in this set. This has led to the development of model uncertainty structures for robust control, where the traditional uncertainty structures [Table 9.1 in 72], have been further developed towards coprime factor perturbations [32,36], and dual-Youla structures [2,12,14]; see Oomen and Bosgra [45] for recent developments.

4.4.2 CASE STUDY 1: MULTIVARIABLE ROBUST CONTROL

To illustrate the concepts of Section 4.4.1, a model-based controller is designed for a traditional motion control problem, i.e., non-inferential ($y = z$) and non-overactuated/non-oversensed ($\dim y = \dim u$, which in turn equals the number of motion DOF). In particular, the x and y translational DOF of the wafer stage in Figure 4.5 are considered. This wafer stage is the main high-precision motion system in a lithographic machine [3].

The following sequence of steps is applied:

1. An FRF is identified following the procedure in Section 4.3.1.
2. Weighting filters W and V are specified as outlined in Section 4.4.1.1. In view of the procedure in Section 4.3.2, a cross-over frequency of 90 Hz is set as target for both the x and y translational DOF.
3. A nominal model $\hat{G}$ is identified using the procedure outlined in Section 4.4.1.2. The resulting model is depicted in Figure 4.6. The model is control-relevant and of low order. Indeed, the model is accurate around the desired cross-over region, where the rigid body modes are accurately modeled. In addition, the first two resonance phenomena around 200 Hz are captured into the model, since these are known to endanger stability. In addition, the model is of order 8: four states correspond to two rigid body modes, whereas the other four states correspond to the inherently multivariable flexible modes around 200 Hz.
4. The uncertainty of the nominal model is taken into account; see Section 4.4.1.3. Here, the emphasis is on selecting the uncertainty such that

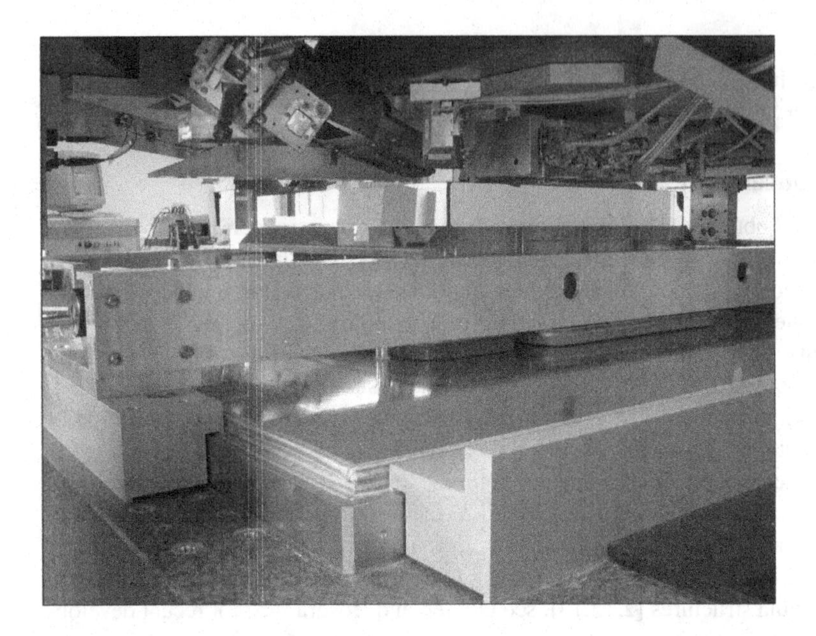

Figure 4.5 Experimental wafer stage system.

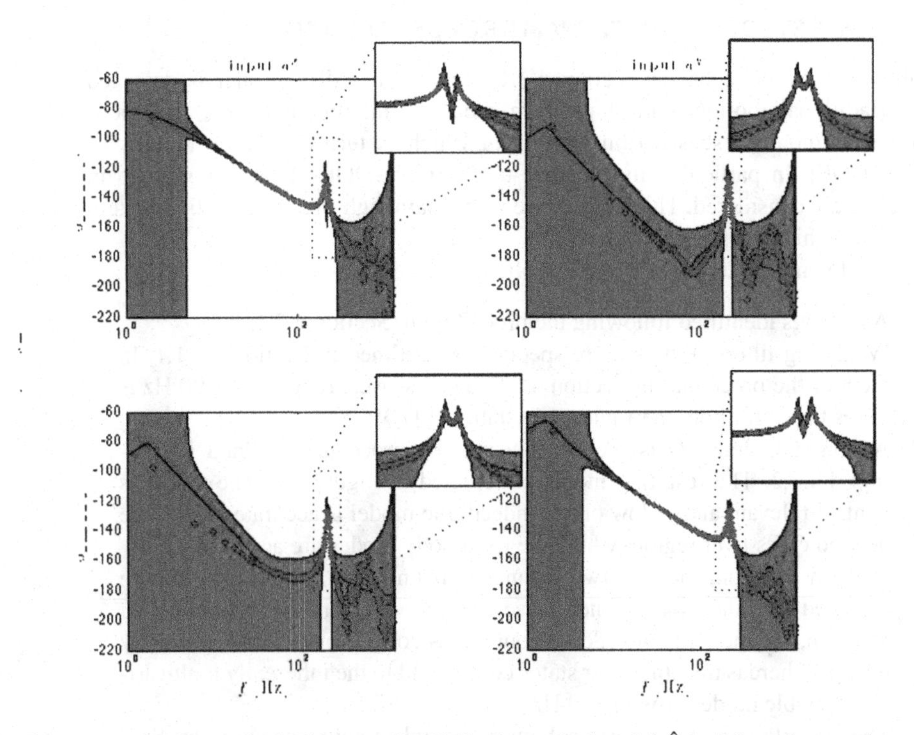

Figure 4.6 Bode magnitude diagram of FRF (dots), nominal model $\hat{G}$ (solid), and uncertain model set for robust control (filled area).

the true system behavior G_o is encompassed but that the uncertainty is selected as parsimoniously as possible. The resulting model set is depicted in Figure 4.6. Clearly, the model set is extremely tight in the cross-over region, which is well known to be important for subsequent control design [6].

5. Next, a robust controller K is synthesized using commercially available optimization algorithms [5]. Note that the resulting controller K is multivariable and obtained in one shot, which is in sharp contrast to the approach in Section 4.3.3.4. As a result, optimality is guaranteed for the multivariable system.
6. The resulting controller is implemented on the experimental system; see Figure 4.7 for the resulting power spectra. When compared with a standard manually tuned PID controller, the optimal robust controller automatically generates notch filters and leads to a factor of four in variance reduction. Further details of the approach and experimental results are presented in Oomen et al. [50].

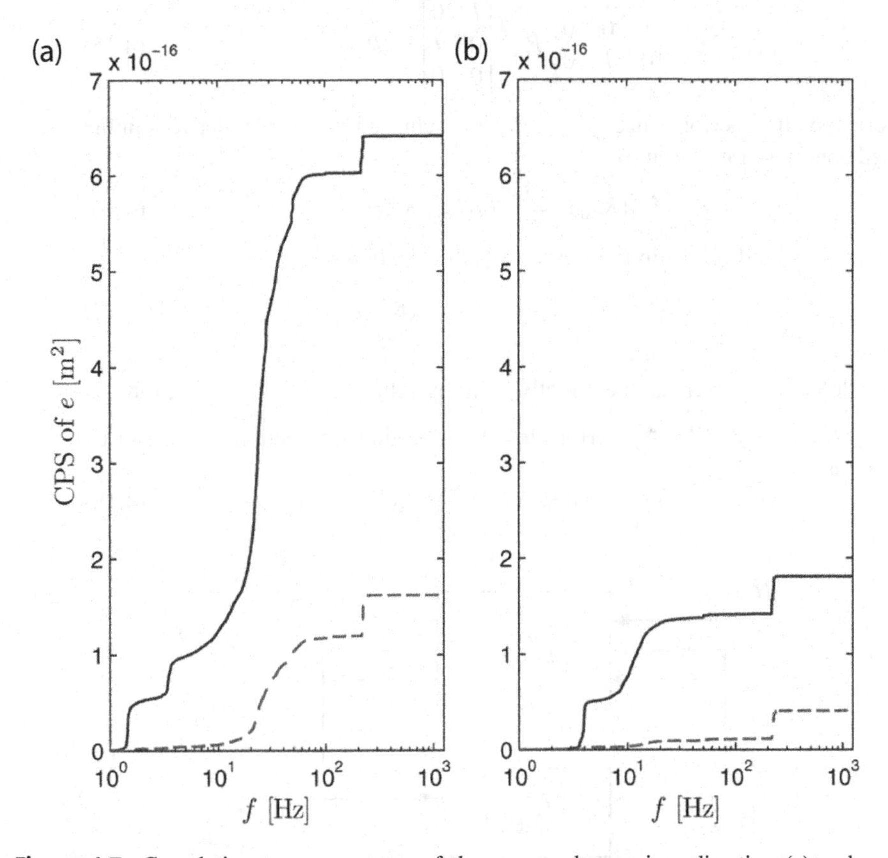

Figure 4.7 Cumulative power spectrum of the measured error in x-direction (a) and y-direction (b): manually tuned controller for reference (solid blue) and optimal robust controller (dashed red).

4.4.3 CASE STUDY 2: OVERACTUATION AND OVERSENSING

Besides the fact that the approach in Section 4.4.1 can deal with the traditional motion control design problem as shown in Section 4.4.2 [71], it can also systematically address control problems that cannot be solved immediately by manual design. An important case is the situation where additional actuators and sensors are used to enhance control performance. Indeed, when the sensors and actuators are equal to the number of motion DOF, the achievable performance is limited; see Boeren et al. [8] for a systematic $\mathscr{H}_\infty$-optimal approach to achieve optimal performance for this situation and Hoogendijk et al. [28] for a manual loop-shaping-based approach. The main rationale in overactuation and oversensing is that additional actuators and sensors, i.e., strictly more than motion DOFs, can significantly enhance performance.

The basic idea is to extend the setup of Figure 4.4 towards Figure 4.8, where additional inputs u_{ext} and outputs y_{ext} are introduced. In particular, consider the situation in Figure 4.8, then it can be immediately verified that

$$\begin{bmatrix} I & 0 & 0 \\ 0 & I & 0 \end{bmatrix} P_{\text{ext}} \begin{bmatrix} I & 0 \\ 0 & I \\ 0 & 0 \end{bmatrix} = P \tag{4.15}$$

with corresponding performance $\mathscr{J}_{\text{ext}}$, which is achieved by a controller K_{ext} in the extended configuration, given by

$$\mathscr{J}_{\text{ext}}(K_{\text{ext}}) := \|\mathscr{F}_\ell(P_{\text{ext}}, K_{\text{ext}})\| \tag{4.16}$$

The rationale behind the setup in Figure 4.8 is that the bound

$$\min_{K_{\text{ext}}} \mathscr{J}_{\text{ext}}(K_{\text{ext}}) \;\leq\; \mathscr{J}(K^\star) \tag{4.17}$$

holds by definition, which can be directly seen; by setting $C_{\text{ext}} = \begin{bmatrix} K^\star & 0 \\ 0 & 0 \end{bmatrix}$, it holds that $\mathscr{J}_{\text{ext}}(K_{\text{ext}}) = \mathscr{J}(K^\star)$. By performing the procedure of Section 4.4, it can be achieved that

$$\min_{K_{\text{ext}}} \mathscr{J}_{\text{ext}}(K_{\text{ext}}) \;\ll\; \mathscr{J}(K^\star) \tag{4.18}$$

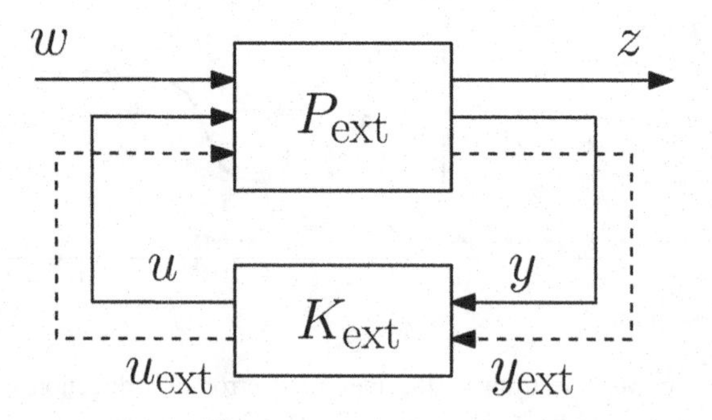

Figure 4.8 Extended standard plant configuration with additional control inputs and outputs.

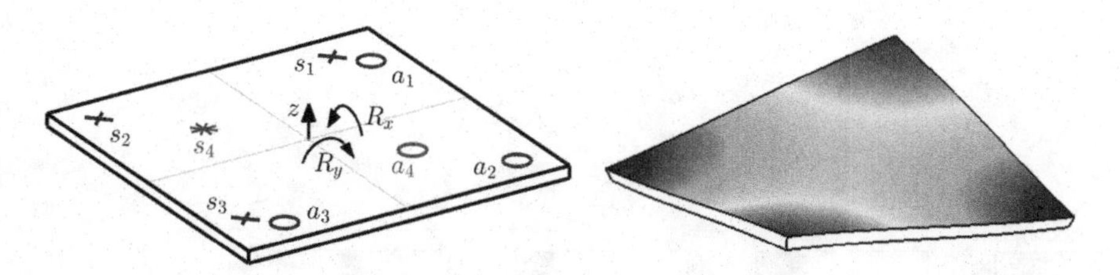

Figure 4.9 Inputs and outputs that can be used to counteract undesired torsion of the wafer stage.

if appropriate design and theoretical considerations are followed as suggested in van Herpen et al. [25].

A wafer stage case study is presented to show potential performance improvement of overactuation and oversensing. Herein, control design in the vertical direction is considered, i.e., the translational direction z and rotations R_x and R_y as depicted in Figure 4.9a. Also, four actuators are placed as indicated in Figure 4.9a. Specifically, a_1, a_2, and a_3 are placed below the corners of the stage, whereas a_4 is positioned at the middle of the line between the center and a_2. Moreover, corner sensors s_1, s_2, and s_3 are used as indicated in Figure 4.9a. Finally, a piezo sensor s_4, which measures strain of the wafer stage, is available at the middle of the line between the center and s_2. Since four actuators and sensors are available to control the three rigid-body DOFs, there is freedom left to actively control flexible dynamical behavior of the wafer stage.

A similar sequence of steps as in Section 4.4.2 is followed, yet specifically tailored to the overactuated and oversensed case. In particular, the system is rigid-body decoupled except for an additional input-output pair that is available for control of relevant flexible dynamical behavior. The first flexible mode of the system is a torsion bending of the stage, depicted in Figure 4.9b. This mode can effectively be controlled using the fourth input-output direction of P_{ext}, which is indicated by P_{flex} in Figure 4.10. On the one hand, the fourth output of P_{ext} is the piezo sensor s_4, which measures strain. As a consequence, no rigid-body displacements are observed at this output, while the structural deformations associated with the flexible dynamical behavior of the system are measured indeed. On the other hand, the fourth input of P_{ext} does not excite rigid-body behavior of the system.

The main idea is that the fourth loop can be closed to actively compensate the torsion mode in Figure 4.9b. To illustrate this, the equivalent plant as is defined in Equation 4.10 is depicted in Figure 4.11. The main result is an increase of the torsion mode frequency from 143 to 193 Hz as well as an increased damping of this loop. This enables enhanced performance for the remaining equivalent plant, since this torsion mode was performance limiting due to non-collocated dynamical behavior. In particular, active control of the torsion mode implies that a higher cross-over frequency can be achieved in the original motion DOF.

To reveal the potential performance enhancement with additional actuators, a controller for the conventional situation in Figure 4.4 to the extended control configuration in Figure 4.8, see also Figure 4.11. The resulting controllers are implemented on

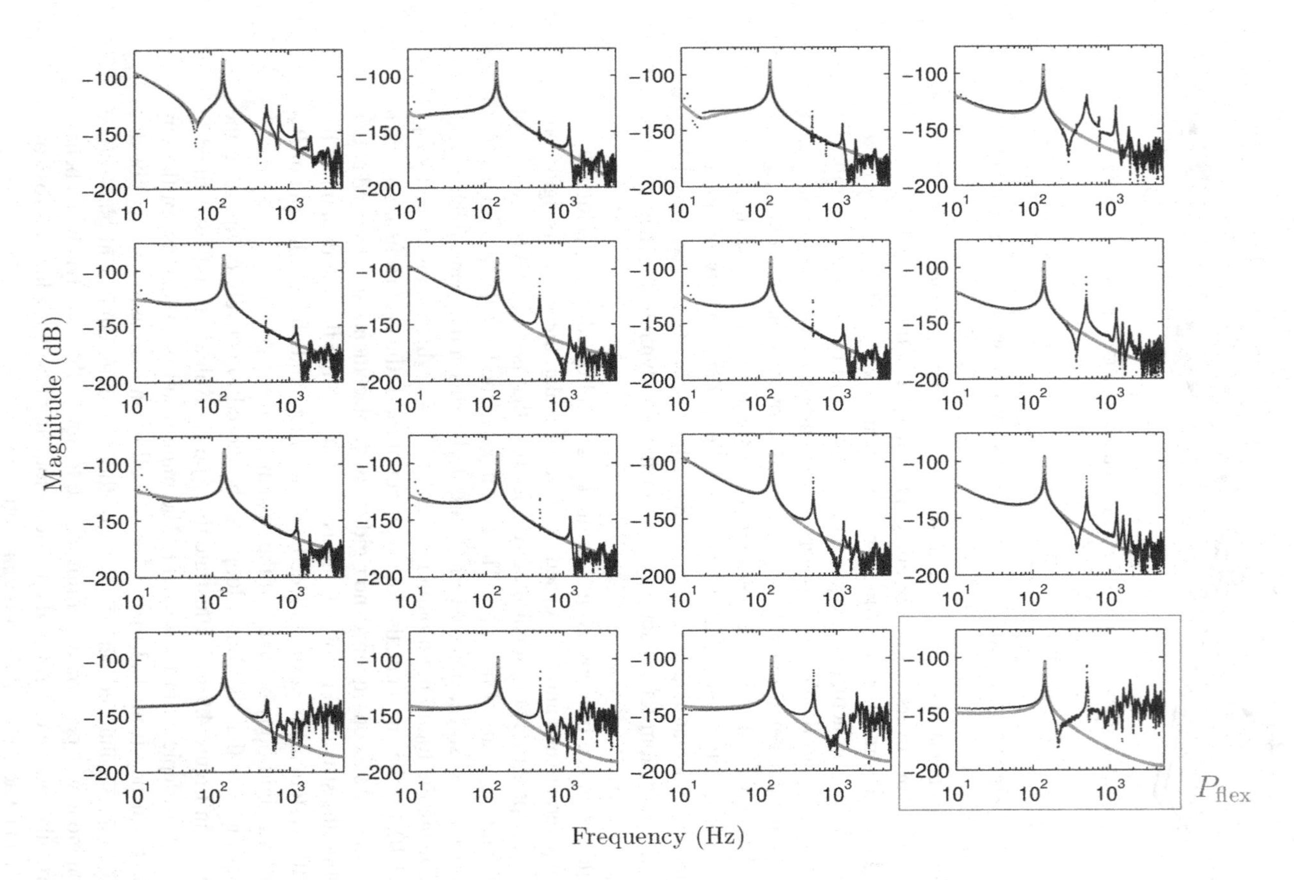

Figure 4.10 Identified FRF $G_{o,\text{ext}}$ (dots) and control-relevant eighth-order parametric model $\hat{G}_{\text{ext}}$ (solid).

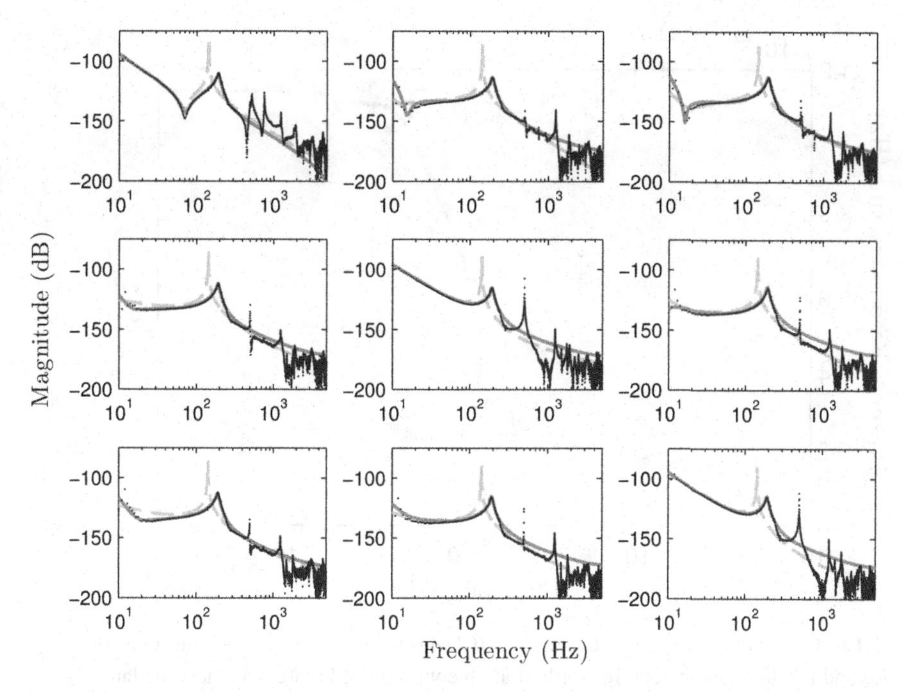

Figure 4.11 Equivalent plant $G_{o,\mathrm{eq}}$ (dotted) and parametric fit $\hat{P}_{\mathrm{eq}}$ (solid) under active control of the torsion loop. The initial model of motion DOFs $[z, R_x, R_y]$ Figure 4.10 is shown for comparison (dashed).

the wafer stage system, and stand-still errors are measured; see Figure 4.12. Clearly, the performance has been significantly increased by active control of the torsion mode. Further details of this approach are reported in van Herpen et al. [26]; see also Ronde et al. [55] for a related feedforward control approach.

4.4.4 CASE STUDY 3: INFERENTIAL CONTROL

As is argued in Section 4.1.5, increasing performance requirements lead to a situation where the measured variables y are not a good representation of the performance variables z. In motion systems, mechanical deformations may be present between the location where performance is desired and where the measurement takes place. This is shown in Figure 4.1. In this section, a third case study is presented to show potentially poor controller designs when this is not properly addressed. In addition, by means of a model-based approach, the performance variable z can be (possibly implicitly) inferred from the measured variables y.

The high-precision prototype motion system in Figure 4.13 has been developed for evaluating control strategies in next-generation motion systems that exhibit dominant flexible behavior. Four out of six motion DOFs of the movable steel beam have been fixed by means of leaf springs. Thus, the system can move in the x and φ direction indicated in Figure 4.14. The inputs consist of three current-driven voice-coil

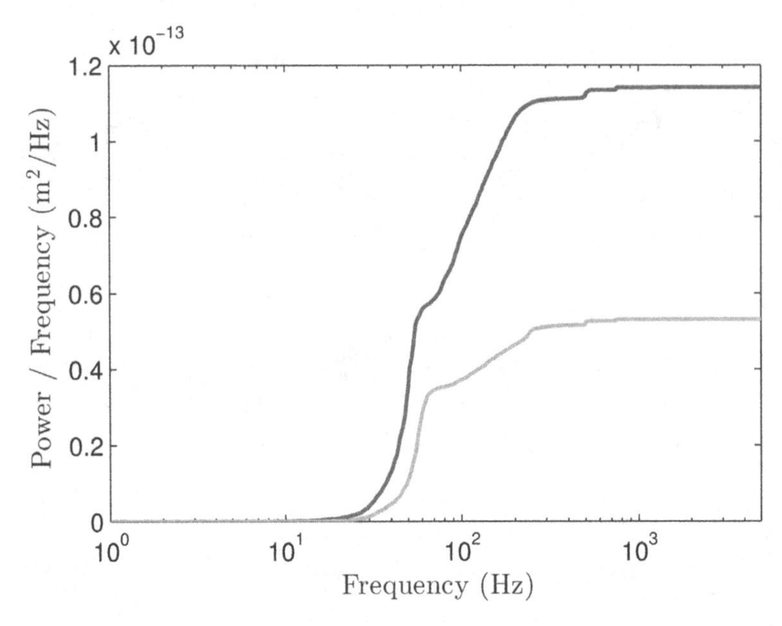

Figure 4.12 Cumulative power spectral density (CPSD) for the conventional controller of Figure 4.4 and for the extended control configuration controller of Figure 4.8 where the latter leads to a substantial reduction.

Figure 4.13 Photograph of the experimental flexible beam setup with three actuators and three sensors.

actuators, whereas the outputs are three contactless fiberoptic sensors with a resolution of approximately 1 μm. The following selection is made to show possible hazards in the inferential control situation and a solution. The goal is to control the translation of the center of the beam, i.e., at sensor location s_2; hence

$$z_p = s_2 \tag{4.19}$$

where s_2 is unavailable for the feedback controller and z_p will be further explained later. Regarding the measured variable, the center of the beam is determined by averaging the outer sensors s_1 and s_3, i.e.,

$$y_p = \begin{bmatrix} \frac{1}{2} & \frac{1}{2} \end{bmatrix} \begin{bmatrix} s_1 \\ s_3 \end{bmatrix} \tag{4.20}$$

which in fact corresponds to a sensor transformation based on static geometric relations, as is indicated in Figure 4.14. Consequently, a discrepancy between the measured variable y_p and performance variable z_p may exist due to internal deformations of the beam. Only the outer actuators are used for control, i.e.,

$$a_1 = a_3 = u_p, \quad a_2 = 0 \tag{4.21}$$

Comparing Equations 4.20 and 4.21 reveals that u_p and y_p are collocated. The resulting system is given by

$$\begin{bmatrix} z_p \\ y_p \end{bmatrix} = \begin{bmatrix} P_z \\ P_y \end{bmatrix} u_p = P u_p \tag{4.22}$$

where z_p denotes the point of interest of the system, y_p is the measured variable available for feedback control, and u_p is the manipulated plant input. Note that in the standard plant formulation of Figure 4.4, z_p will be part of z, in addition to, e.g., the input signal to avoid excessive control inputs [46]. The resulting control problem closely resembles the example in Figure 4.1 and standard geometric decoupling techniques [71].

Next, the procedure in Section 4.4.2 is followed again; i.e., it is for the moment assumed that $y_p = z_p$. The resulting controller is implemented on the beam system, where the measured response y_p is shown in Figure 4.15(b). The response is y_p, is as expected from simulations, and sufficiently fast to reach the setpoint. However, when

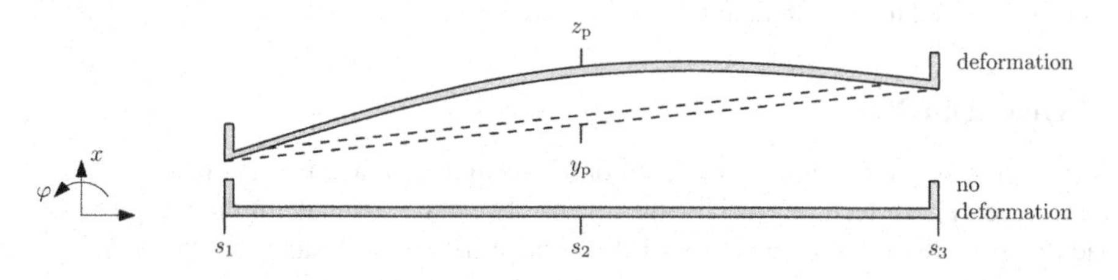

Figure 4.14 Schematic top view illustration of flexible beam setup.

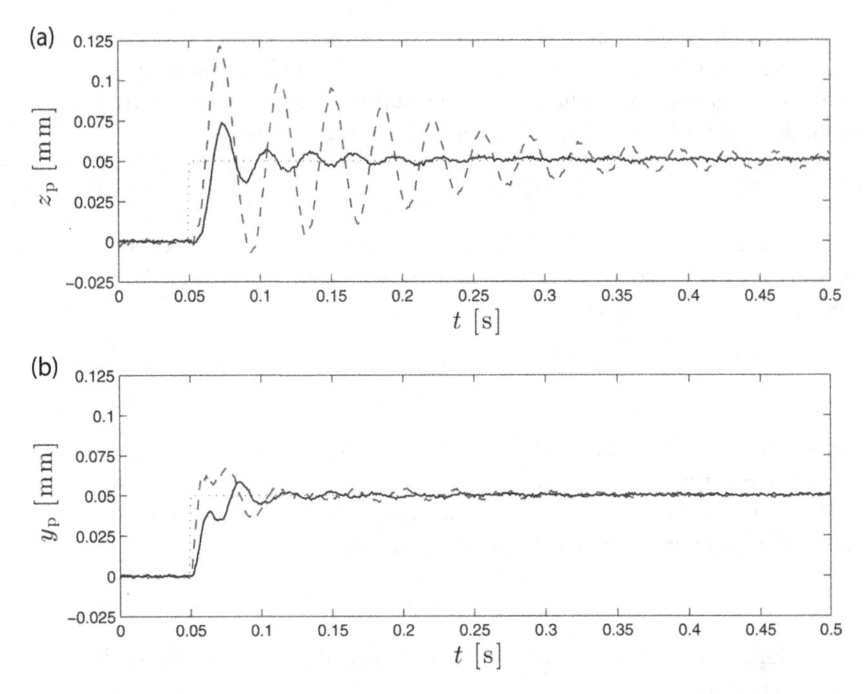

Figure 4.15 Experimental step responses: inferential motion controller (solid), traditional robust control design (dashed). Top: Performance variable z_p. Bottom: Measured variable y_p.

inspecting the measurement z_p, which is not available to the feedback controller but measured for performance validation, it is clear that the system exhibits large internal deformations. Hence, the true performance of the system is poor if one neglects the fact that a distinction exists between z_p and y_p. In addition, this poor performance cannot be directly observed from the feedback loop signals.

In Oomen et al. [46], an inferential motion control design framework is presented. This framework exploits measurements y_p and a model of P_z to *infer* the performance variables; i.e., it does not require a real-time measurement of the performance variables z_p. This framework is applied to design an inferential motion controller for the beam setup in Figures 4.13 and 4.14. The results are also depicted in Figure 4.15. Clearly, the performance in terms of z_p significantly increases. For further details, the reader is referred to Oomen et al. [46].

4.5 CONCLUSION

In this chapter, a step-by-step model-based control design approach is presented and applied to several high-tech industrial case studies. The main rationale of the design approach is to increase the complexity of the control design procedure and the associated modeling approach only if justified and necessitated by the performance and accuracy requirements. The step-by-step procedure consists of the following steps: (i) interaction analysis, (ii) decoupling, (iii) independent SISO design, (iv) sequential SISO design, and (v) norm-based MIMO design. The required model ranges from inexpensive FRF data in Steps (i)–(iv) towards a parametric model in Step (v).

Three recent case studies are presented, where the performance and accuracy requirements lead to a complex control design problem that necessitates a model-based optimal design. These include (i) a robust multivariable control design for a traditional yet high-performance wafer stage control problem, (ii) a case study on a prototype wafer stage to exploit overactuation and oversensing, and (iii) a case study on a prototype motion system to address an inferential control problem where the performance variables are not available for real-time feedback control.

Ongoing research mainly focuses on modeling. Regarding Steps (i)–(v), nonparametric identification is being enhanced to improve the FRF quality. This includes the use of multi-sine excitation signals (see Oomen et al. [50, appendix A] for initial results) and the use of preprocessing using LRM/LPM (see Voorhoeve et al. [69], van der Maas et al. [33], and Evers et al. [15] for initial results that are presently being applied to multivariable FRF estimation).

Regarding the norm-based MIMO control design step (v), the required modeling effort is a major obstruction for industrial application. Several aspects are important. First, regarding modeling for robust control, present research focuses on new approaches to identify "good" model sets for robust control. For mechatronic systems, important aspects include the identification of a control-relevant nominal model [57], the structure of model uncertainty (see, e.g., Sippe et al. [14] and Lanzon and Papageorgiou [32] for general results and Oomen and Bosgra [45] for recent developments), and the size of model uncertainty; approaches that are particularly suitable for mechatronic systems include, e.g., Geerardyn et al. [20] for local parametric models, Oomen et al. [49] for a data-driven approach, and Oomen and Bosgra [44] for a validation-based approach.

Second, the identification of accurate models of complex high-tech systems is numerically challenging. To enable the accurate and fast identification of nominal models, new approaches that enable a numerically reliable implementation are being developed [24,27]; see also Voorhoeve et al. [68] for a recent benchmark comparison.

Third, motion systems lead to position-dependent behavior due to the fact that the system inherently makes movements. At present, methods are being developed to quantify the position dependence through control-relevance (see Oomen [39, section 5.7] for early results) as well as explicitly modeling the position dependence by exploiting the physical model structure (Equation 4.1) in an LPV framework; see also Steinbuch et al. [64] and Wassink et al. [22] for general LPV modeling and motion control. Furthermore, identification of more general non-linear behavior is investigated in, e.g., Rijlaarsdam et al. [54].

Fourth, disturbances have a crucial role in disturbance attenuation; see Boerlage et al. [9] for initial research in this direction. Ongoing research focuses on control-relevant disturbance modeling in the framework of Oomen and Bosgra [45].

Fifth, sampled-data identification and control is being investigated. All the results presented in this chapter have tacitly been developed in the discrete time domain.

This neglects the important issue of intersample behavior; see Chan and Francis [13] for general results and Oomen et al. [48] for a motion control framework. Such sampled-data frameworks necessitate the modeling for sampled-data control, essentially requiring continuous time models [18]. In Oomen [40], a multirate framework is presented that addresses this to best extent for periodic sampling. Note that periodic sampling will always be restricted due to the Nyquist–Shannon sampling theorem yet at a higher frequency bound. This bound is for motion systems typically fairly high. If this is not sufficient, the approach can be further enhanced by measuring irregularly sampled data [18], which is straightforward to perform in certain motion systems.

Finally, the present chapter focused on feedback control design. For high-performance motion control, feedforward is equally important. For traditional feedforward tuning, the reader is referred to Steinbuch et al. [63, section 27.3] and Oomen [43]. Ongoing research includes iterative learning control and automated feedforward control tuning [42], both based on a tailored iterative learning control approach [10] and based on instrumental variable system identification [7]. Related approaches are presented in, e.g., Rijlaarsdam et al. [53], where a frequency domain approach is presented for nonlinear systems.

4.6 ACKNOWLEDGMENT

The authors would like to thank our very many industrial and academic collaborators, as well as former students, who have contributed to the overall framework and experimental results presented in this chapter. This work is part of the research programme VIDI with project number 15698, which is (partly) financed by the Netherlands Organisation for Scientific Research (NWO).

REFERENCES

1. S. Adee. EUV's underdog light source will have its day. *IEEE Spectr.*, 47(11):13–14, 2010.
2. Brian D.O. Anderson. From Youla-Kucera to identification, adaptive and nonlinear control. *Automatica*, 34(12):1485–1506, 1998.
3. Bill Arnold. Shrinking possibilities. *IEEE Spectr.*, 46(4):26–28, 50–56, 2009.
4. Karl J. ström and Richard M. Murray. *Feedback Systems: An Introduction for Scientists and Engineers.* Princeton University Press, Princeton, NJ, 2008. Available from: `http://www.cds.caltech.edu/~murray/amwiki/index.php/Main_Page`.
5. Gary Balas, Richard Chiang, Andy Packard, and Michael Safonov. *Robust Control Toolbox 3: User's Guide.* The Mathworks, Natick, MA, USA, 2007.
6. Gary J. Balas and John C. Doyle. Control of lightly damped, flexible modes in the controller crossover region. *J. Guid. Contr. Dyn.* 17(2):370–377, 1994.
7. Frank Boeren, Tom Oomen, and Maarten Steinbuch. Iterative motion feedforward tuning: A data-driven approach based on instrumental variable identification. *Contr. Eng. Prac.*, 37:11–19, 2015.

8. Frank Boeren, Robbert van Herpen, Tom Oomen, Marc van de Wal, and Maarten Steinbuch. Non-diagonal $\mathscr{H}_\infty$ weighting function design: Exploiting spatial-temporal deformations for precision motion control. *Contr. Eng. Prac.*, 35:35–42, 2015.
9. Matthijs Boerlage, Bram de Jager, and Maarten Steinbuch. Control relevant blind identification of disturbances with application to a multivariable active vibration isolation platform. *IEEE Trans. Contr. Syst. Techn.*, 18(2):393–404, 2010.
10. Joost Bolder, Tom Oomen, Sjirk Koekebakker, and Maarten Steinbuch. Using iterative learning control with basis functions to compensate medium deformation in a wide-format inkjet printer. *Mechatronics*, 24(8):944–953, 2014.
11. E. Bristol. On a new measure of interaction for multivariable process control. *IEEE Trans. Automat. Contr.*, 11(1):133–134, 1966.
12. R.A. de Callafon and P.M.J. Van den Hof. Suboptimal feedback control by a scheme of iterative identification and control design. *Math. Mod. Syst.*, 3(1):77–101, 1997.
13. Tongwen Chen and Bruce Francis. *Optimal Sampled-Data Control Systems*. Springer, London, UK, 1995.
14. Sippe G. Douma and Paul M.J. Van den Hof. Relations between uncertainty structures in identification for robust control. *Automatica*, 41(3):439–457, 2005.
15. Enzo Evers, Bram de Jager, and Tom Oomen. Improved local rational method by incorporating system knowledge: With application to mechanical and thermal dynamical systems. In *18th IFAC Symp. Sys. Id.*, pages 808–813, Stockholm, Sweden, 2018.
16. Andrew J. Fleming and Kam K. Leang. *Design, Modeling and Control of Nanopositioning Systems*. Switzerland, Springer, 2014.
17. J. Freudenberg and D. Looze. *Frequency Domain Properties of Scalar and Multivariable Feedback Systems, Lecture Notes in Control and Information Sciences*. Springer–Verlag, Berlin, Germany, 1988.
18. Hugues Garnier and Liu Wang, editors. *Identification of Continuous-Time Models from Sampled Data*. Advances in Industrial Control. Springer-Verlag, London, UK, 2008.
19. Wodek K. Gawronski. *Advanced Structural Dynamics and Active Control of Structures*. Springer, New York, NY, USA, 2004.
20. Egon Geerardyn, Tom Oomen, and Johan Schoukens. Enhancing $\mathscr{H}_\infty$ norm estimation using local LPM/LRM modeling: Applied to an AVIS. In *IFAC 19th Triennial World Congress*, pages 10856–10861, Cape Town, South Africa, 2014.
21. M. Gevers. Towards a joint design of identification and control ? In H.L. Trentelman and J.C. Willems, editors, *Essays on Control : Perspectives in the Theory and its Applications*, chapter 5, pages 111–151. Birkhäuser, Boston, MA, USA, 1993.
22. Matthijs Groot Wassink, Marc van de Wal, Carsten Scherer, and Okko Bosgra. LPV control for a wafer stage: Beyond the theoretical solution. *Contr. Eng. Prac.*, 13:231–245, 2003.
23. P. Grosdidier and M. Morari. Interaction measures for systems under decentralized control. *Automatica*, 22:309–319, 1986.
24. Robbert van Herpen, Okko Bosgra, and Tom Oomen. Bi-orthonormal polynomial basis function framework with applications in system identification. *IEEE Trans. Automat. Contr.*, 61(11):3285–3300, 2016.
25. Robbert van Herpen, Tom Oomen, Edward Kikken, Marc van de Wal, Wouter Aangenent, and Maarten Steinbuch. Exploiting additional actuators and sensors for nano-positioning robust motion control. *Mechatronics*, 24(6):619–631, 2014.
26. Robbert van Herpen, Tom Oomen, Edward Kikken, Marc van de Wal, Wouter Aangenent, and Maarten Steinbuch. Exploiting additional actuators and sensors for nano-positioning

robust motion control. In *Proc. 2014 Americ. Contr. Conf.*, pages 984–990, Portland, OR, USA, 2014.

27. Robbert van Herpen, Tom Oomen, and Maarten Steinbuch. Optimally conditioned instrumental variable approach for frequency-domain system identification. *Automatica*, 50(9):2281–2293, 2014.
28. R. Hoogendijk, M.F. Heertjes, M.J.G. van de Molengraft, and M. Steinbuch. Directional notch filters for motion control of flexible structures. *Mechatronics*, 24(6):632–639, 2014.
29. Peter C. Hughes. Space structure vibration modes: How many exist? Which ones are important? *IEEE Contr. Syst. Mag.*, 7(1):22–28, 1987.
30. G. Dan Hutcheson. The first nanochips. *Scient. Americ.*, 48–55, April 2004.
31. Paul Lambrechts, Matthijs Boerlage, and Maarten Steinbuch. Trajectory planning and feedforward design for electromechanical motion systems. *Contr. Eng. Prac.*, 13:145–157, 2005.
32. Alexander Lanzon and George Papageorgiou. Distance measures for uncertain linear systems: A general theory. *IEEE Trans. Automat. Contr.*, 54(7):1532–1547, 2009.
33. R. van der Maas, A. van der Maas, R. Voorhoeve, and T. Oomen. Accurate FRF identification of LPV systems: nD-LPM with application to a medical X-ray system. *IEEE Trans. Contr. Syst. Techn.*, 25(4):1724–1735, 2017.
34. Victor M. Martinez and Thomas F. Edgar. Control of lithography in semiconductor manufacturing. *IEEE Contr. Syst. Mag.*, 26(6):46–55, 2006.
35. D. Mayne. Sequential design of linear multivariable systems. *Proceedings of the IEE*, 126:568–572, 1979.
36. D.C. McFarlane and K. Glover. *Robust Controller Design Using Normalized Coprime Factor Plant Descriptions*, volume 138 of *LNCIS*. Springer-Verlag, Berlin, Germany, 1990.
37. S.O.Reza Moheimani, Dunant Halim, and Andrew J. Fleming. *Spatial Control of Vibration: Theory and Experiments*. World Scientific, Singapore, 2003.
38. Rob Munnig Schmidt, Georg Schitter, and Jan van Eijk. *The Design of High Performance Mechatronics*. Delft University Press, Delft, The Netherlands, 2011.
39. Tom Oomen. *System Identification for Robust and Inferential Control with Applications to ILC and Precision Motion Systems*. PhD thesis, Eindhoven University of Technology, Eindhoven, The Netherlands, 2010.
40. Tom Oomen. Controlling aliased dynamics in motion systems? An identification for sampled-data control approach. *Int. J. Contr.*, 87(7):1406–1422, 2014.
41. Tom Oomen. Advanced motion control for precision mechatronics: Control, identification, and learning of complex systems. *IEEE Trans. Ind. Appl.*, 7(2):127–140, 2018.
42. Tom Oomen. Learning in machines. *Mikroniek*, 6:5–11, 2018.
43. Tom Oomen. Control for precision mechatronics. In *Encyclopedia of Systems and Control*. 2nd edition, To Appear.
44. Tom Oomen and Okko Bosgra. Well-posed model uncertainty estimation by design of validation experiments. In *15th IFAC Symp. Sys. Id.*, pages 1199–1204, Saint-Malo, France, 2009.
45. Tom Oomen and Okko Bosgra. System identification for achieving robust performance. *Automatica*, 48(9):1975–1987, 2012.
46. Tom Oomen, Erik Grassens, and Ferdinand Hendriks. Inferential motion control: An identification and robust control framework for unmeasured performance variables. *IEEE Trans. Contr. Syst. Techn.*, 23(4):1602–1610, 2015.

47. Tom Oomen and Maarten Steinbuch. Identification for robust control of complex systems: Algorithm and motion application. In Marco Lovera, editor, *Control-Oriented Modelling and Identification: Theory and Applications*. IET, 2015.
48. Tom Oomen, Marc van de Wal, and Okko Bosgra. Design framework for high-performance optimal sampled-data control with application to a wafer stage. *Int. J. Contr.*, 80(6):919–934, 2007.
49. Tom Oomen, Rick van der Maas, Cristian R. Rojas, and Hkan Hjalmarsson. Iterative data-driven $\mathscr{H}_\infty$ norm estimation of multivariable systems with application to robust active vibration isolation. *IEEE Trans. Contr. Syst. Techn.*, 22(6):2247–2260, 2014.
50. Tom Oomen, Robbert van Herpen, Sander Quist, Marc van de Wal, Okko Bosgra, and Maarten Steinbuch. Connecting system identification and robust control for next-generation motion control of a wafer stage. *IEEE Trans. Contr. Syst. Techn.*, 22(1):102–118, 2014.
51. Rik Pintelon and Johan Schoukens. *System Identification: A Frequency Domain Approach*. IEEE Press, New York, NY, USA, second edition, 2012.
52. André Preumont. *Vibration Control of Active Structures: An Introduction*, volume 96 of *Solid Mechanics and Its Applications*. Kluwer Academic Publishers, New York, NY, USA, second edition, 2004.
53. David Rijlaarsdam, Pieter Nuij, Johan Schoukens, and Maarten Steinbuch. Frequency domain based nonlinear feed forward control design for friction compensation. *Mech. Syst. Sign. Proc.*, 27:551–562, 2012.
54. David Rijlaarsdam, Tom Oomen, Pieter Nuij, Johan Schoukens, and Maarten Steinbuch. Uniquely connecting frequency domain representations for given order polynomial Wiener-Hammerstein systems. *Automatica*, 48(9):2381–2384, 2012.
55. M.J.C. Ronde, M.G.E. Schneiders, E.J.G.J. Kikken, M.J.G. van de Molengraft, and M. Steinbuch. Model-based spatial feedforward for over-actuated motion systems. *Mechatronics*, 24(4):307–317, 2014.
56. Ulrich Schönhoff and Rainer Nordmann. A $\mathscr{H}_\infty$-weighting scheme for PID-like motion control. In *Proc. 2002 Conf. Contr. Appl.*, pages 192–197, Glasgow, Scotland, 2002.
57. Ruud J.P. Schrama. Accurate identification for control: The necessity of an iterative scheme. *IEEE Trans. Automat. Contr.*, 37(7):991–994, 1992.
58. María M. Seron, Julio H. Braslavsky, and Graham C. Goodwin. *Fundamental Limitations in Filtering and Control.* Springer-Verlag, London, UK, 1997.
59. S. Skogestad and I. Postlethwaite. *Multivariable Feedback Control, Analysis and Design.* Wiley, 2005.
60. Sigurd Skogestad and Ian Postlethwaite. *Multivariable Feedback Control: Analysis and Design.* John Wiley & Sons, West Sussex, UK, second edition, 2005.
61. Roy S. Smith. Closed-loop identification of flexible structures: An experimental example. *J. Guid. Contr. Dyn.*, 21(3):435–440, 1998.
62. M. Steinbuch and M.L. Norg. Advanced motion control: An industrial perspective. *Eur. J. Contr.*, 4(4):278–293, 1998.
63. Maarten Steinbuch, Roel J.E. Merry, Matthijs L.G. Boerlage, Michael J.C. Ronde, and Marinus J.G. van de Molengraft. Advanced motion control design. In W.S. Levine, editor, *The Control Handbook, Control System Applications*, pages 27–1/27–25. CRC Press, second edition, 2010.
64. Maarten Steinbuch, René van de Molengraft, and Aart-Jan van der Voort. Experimental modelling and LPV control of a motion system. In *Proc. 2003 Americ. Contr. Conf.*, pages 1374–1379, Denver, CO, USA, 2003.
65. Gary Stix. Getting more from Moore's. *Scient. Americ.*, 20–24, April 2001.

66. Julian Stoev, Julien Ertveldt, Tom Oomen, and Johan Schoukens. Tensor methods for MIMO decoupling and control design using frequency response functions. *Mechatronics*, 45:71–81, 2017.
67. Glenn Vinnicombe. *Uncertainty and Feedback: $\mathcal{H}_\infty$ Loop-Shaping and the ν-Gap Metric.* Imperial College Press, London, UK, 2001.
68. Robbert Voorhoeve, Tom Oomen, Robbert van Herpen, and Maarten Steinbuch. On numerically reliable frequency-domain system identification: new connections and a comparison of methods. In *IFAC 19th Triennial World Congress*, pages 10018–10023, Cape Town, South Africa, 2014.
69. Robbert Voorhoeve, Annemiek van der Maas, and Tom Oomen. Non-parametric identification of multivariable systems: A local rational modeling approach with application to a vibration isolation benchmark. *Mech. Syst. Sign. Proc.*, 105:129–152, 2018.
70. David Voss. Chips go nano. *Tech. Rev.*, 102(2):55–57, 1999.
71. Marc van de Wal, Gregor van Baars, Frank Sperling, and Okko Bosgra. Multivariable $\mathcal{H}_\infty/\mu$ feedback control design for high-precision wafer stage motion. *Contr. Eng. Prac.*, 10(7):739–755, 2002.
72. Kemin Zhou, John C. Doyle, and Keith Glover. *Robust and Optimal Control.* Prentice Hall, Upper Saddle River, NJ, USA, 1996.

5 Control and Manipulation

Bruno Siciliano and Luigi Villani

CONTENTS

This chapter deals with fundamental control problems in robotic manipulation. The motion control problem is surveyed as the basis for more advanced algorithms designed for controlling lightweight robots with non-negligible joint elasticity as well as complex robotics systems with a large number of degrees of freedom. The problem of controlling the physical interaction of robots with the environment or a human is addressed as well.

5.1 INTRODUCTION

The early years of robotics—prior to the 1980s—were largely focused on manipulator arms and simple factory automation tasks: materials handling, welding, painting. Cost of computation, lack of good sensors, and lack of fundamental understanding of robot control were the primary barriers to progress.

Robotics today is a much richer field with far-ranging applications. Also, the definition of what constitutes a robot has broadened dramatically. A key role in this evolution was played by the synergy of robotics and control. On the one hand, technological advancements in sensors, computation, and actuators have motivated the development of new control algorithms. On the other hand, advancements in control have enabled solutions to challenging problems in robotics.

The purpose of this chapter is to focus on fundamental control problems in robotics with special emphasis on robotic manipulation. The motion control problem is surveyed as the basis for more advanced algorithms designed for controlling lightweight robots with non-negligible joint elasticity as well as complex robotics systems with a large number of degrees of freedom (DOF), able to execute multiple tasks at the same time. Moreover, the problem of controlling the physical interaction of robots with the environment or a human is addressed. Future directions are sketched at the end of the chapter, where a list of recommended reading is provided.

5.2 MOTION CONTROL

According to the Robot Institute of America, a robot manipulator is fundamentally a positioning device designed to move material, parts, tools, or specialized devices through variable programmed motions for the performance of a variety of tasks. Thus, the motion control is a fundamental issue for the execution of manipulation tasks.

The problem of motion control is to determine the time behavior of the control inputs to achieve a desired motion. The control inputs are usually the motor currents but can be translated into torques or velocities for the purpose of control design. The desired motion is typically given by a reference trajectory generated by suitable motion planning algorithms. This problem is quite complex, since a manipulator is an articulated system, and, as such, the motion of one link influences the motion of the others. Moreover, the presence of non-negligible mechanical flexibility in the joints (and/or the links) of multi-DOF robots poses challenging control problems due to the possible excitation of vibrational phenomena.

In this section, we provide an overview of the main motion control techniques for rigid robot manipulators. Some hints about the control of robots with flexibility concentrated in the robot joints, which is a common situation in lightweight manipulators designed to work close to humans, are provided. Finally, a control approach based on coordinates suitable for specifying robotic tasks, i.e., the so-called task-space control, is presented.

5.2.1 JOINT SPACE CONTROL

The simplest control strategy for a n-DOF manipulator is to treat each robot joint as a single-input/single-output (SISO) system and design the controllers independently for each joint. Proportional, integral, derivative (PID) control is the most common method employed in this case. The objective of each joint controller is to ensure that the motor angle $q_m(t)$ follows a desired value $q_{md}(t)$. Coupling effects between joints due to varying configurations during motion are treated as disturbance inputs. This approach, known as *independent joint control*, works well for highly geared manipulators moving at relatively low speeds, since the large gear reduction and low speed tend to reduce the coupling effects.

Suitable linear decentralized feedforward compensation terms can be used to improve the tracking capability of desired trajectories with high values of speed and acceleration. To further improve performance, the interaction between the joints can be compensated by a *nonlinear centralized feedforward* action. This solution still consists of n independent joint control servos; each joint controller elaborates references and measurements that refer to the single joint. The disturbance torque due to the interaction between the various joints is compensated by a centralized feedforward action that depends on the manipulator dynamic model computed using the joint desired positions, velocities, and accelerations.

Advanced control methods for robots generally aim to take into account the dynamic coupling between joints by using complete or partial knowledge of the dynamic model of the manipulator in the *feedback control* action.

By neglecting friction, elasticity in the joints or links, and other effects, the dynamic model of a n-link robot manipulator can be written in the so-called Euler–Lagrange form as

$$B(q)\ddot{q}+C(q,\dot{q})\dot{q}+g(q)=\tau \tag{5.1}$$

where q is the $n\times 1$ joint vector, $B(q)$ is $n\times n$ inertia matrix, $C(q,\dot{q})\dot{q}$ is the $n\times 1$ vector of Coriolis and centrifugal torques, $g(q)$ is the $n\times 1$ vector of the gravitational torques, and τ is the $n\times 1$ vector of control torques.

The Euler–Lagrange equations fulfil a number of properties that can be suitably exploited for control design, namely:

1. Matrix $B(q)$ is symmetric and positive definite.
2. Matrix $C(q,\dot{q})$ can be chosen so that matrix $\dot{B}(q)-2C(q,\dot{q})$ is skew symmetric.
3. Equation 5.1 can be cast in a linear form with respect to a suitable $p\times 1$ vector π of dynamic parameters as $Y(q,\dot{q},\ddot{q})\pi=\tau$

An intuitive method of control, which ensures *tracking* of a desired time-varying joint trajectory $q_d(t)$, is the so-called *inverse dynamics*, or *feedback linearization* method, which computes the control torque τ as

$$\tau = B(q)r+C(q,\dot{q})\dot{q}+g(q) \tag{5.2}$$

$$r = \ddot{q}_d+K_d(\dot{q}_d-\dot{q})+K_p(q_d-q) \tag{5.3}$$

In view of Property 1, the closed loop equation in terms of the tracking error $e(t)=q_d(t)-q(t)$ satisfies the linear equation

$$\ddot{e}+K_d\dot{e}+K_pe=0$$

and therefore, the tracking error converges exponentially to zero for positive definite matrix gains K_d and K_p.

An interesting feature of the robot's dynamic model related to Property 2 is the so-called *passivity* property. Namely, the power supplied to the system $\dot{q}^T\tau$ is either stored as mechanical energy $E(t)$ or dissipated, which means that

$$\dot{q}^T\tau\geq\dot{E}(t)$$

This property can be exploited to design control laws with higher robustness in the presence of uncertainties in the dynamic model.

A first alternative to the inverse dynamics control law is the so-called *PD+ control*

$$\tau = B(q)\ddot{q}_d + C(q,\dot{q})\dot{q}_d + g(q) + K_d(\dot{q}_d - \dot{q}) + K_p(q_d - q) \tag{5.4}$$

containing a linear proportional derivative control action, which leads to a nonlinear closed loop equation. In this case, the global asymptotic convergence of the tracking error $e(t)$ can be proven by using the second Lyapunov stability method and exploits Properties 1 and 2. The advantage of the aforementioned control law is that in the case of *regulation* to a constant desired joint position q_d, Equation 5.4 reduces to

$$\tau = g(q) - K_d\dot{q} + K_p(q_d - q) \tag{5.5}$$

with positive definite matrices K_P and K_D, which is known as *PD control with gravity compensation.*

Another passivity-based motion tracking control law is the so-called *Slotine and Lie control,* defined by the equation

$$\tau = B(q)\ddot{q}_r + C(q,\dot{q})\dot{q}_r + g(q) + K_D\sigma = Y(q,\dot{q},\dot{q}_r\ddot{q}_r)\pi + K_D\sigma \tag{5.6}$$

where

$$\dot{q}_r = \dot{q}_d + \Lambda_e e \quad \sigma = \dot{q}_r - \dot{q},$$

with diagonal positive definite matrices K_D and Λ_e.

If the vector of the robot's dynamic parameters π is uncertain, it can be replaced by an estimate $\hat{\pi}$.

The tracking capabilities of the control law (Equation 5.6) can be improved by adopting a *robust control* algorithm, where the estimate $\hat{\pi}$ is chosen as $\hat{\pi} = \pi_0 + \delta\pi$, with

$$Y\delta\pi = \begin{cases} -\rho\dfrac{Y^T\sigma}{\|Y^T\sigma\|}, & \text{if } \|Y^T\sigma\| > \varepsilon \\[2ex] -\rho\dfrac{Y^T\sigma}{\varepsilon}, & \text{if } \|Y^T\sigma\| \leq \varepsilon \end{cases}$$

where $\rho > 0$ is a bound on the parameter's uncertainty. In this case, uniform ultimate boundedness of the tracking error $e(t)$ can be proven, where the size of the ultimate boundedness set depends on $\varepsilon > 0$.

Alternatively an *adaptive control* law can be adopted, where the estimate $\hat{\pi}$ can be computed by using the update law

$$\dot{\hat{\pi}} = K_\pi Y^T(q,\dot{q},\dot{q}_r\ddot{q}_r)\sigma$$

with K_π positive definite matrix gain. Global convergence of the tracking errors to zero and boundedness of the parameter estimates can be proven using a Lyapunov analysis. Moreover, if the desired trajectory satisfies a suitable condition of *persistency of excitation,* the estimated parameters converge to the true parameters.

5.2.2 CONTROL OF ROBOTS WITH ELASTIC JOINTS

The assumption that robot manipulators are rigid multi-body mechanical systems simplifies the control design but in some cases, may lead to performance degradation and even unstable behavior, if mechanical flexibility cannot be neglected. In dynamic modeling, flexibility can be assumed to be concentrated at the robot joints or distributed along the robot links. In this chapter, we consider only the case of manipulators with elastic joints, which is a common situation of many lightweight robots designed to work close to humans.

A robot with elastic joints can be modeled in the Euler–Lagrange form as

$$B(q)\ddot{q}+C(q,\dot{q})\dot{q}+g(q)+K(q-\theta) = 0 \tag{5.7}$$

$$M\ddot{\theta}+K(\theta-q)=\tau \tag{5.8}$$

where q is the $n \times 1$ vector of the link angular positions and θ is the $n \times 1$ vector of the motor angular positions. The quantity $\tau_J = K(\theta - q)$ is the vector of the elastic torques due to joint deformations, with K positive definite, diagonal joint stiffness matrix. Equation 5.7 is the so-called link equation describing the rigid multi-link dynamics, analogously to Equation 5.1. Equation 5.8 is the motor equation, where M is the positive definite, diagonal matrix of motor inertias (reflected through the gear ratio) and τ is the vector of the motor torques.

In the absence of gravity, the basic robotic task of moving between two arbitrary equilibrium configurations can be realized by using a simple decentralized PD control law using only feedback from the *motor variables*

$$\tau = -K_d\dot{\theta}+K_p(\theta_d-\theta) \tag{5.9}$$

with diagonal positive definite matrices K_P and K_D. In this case the equilibrium state is $\dot{q} = \dot{\theta} = 0$, and $q = q_d$, $\theta = \theta_d = q_d$, i.e., with no joint deflection at steady state. Global asymptotic stability of the equilibrium can be proven as a consequence of the passivity of the system.

In the presence of gravity, the equilibrium position of the motor associated to a desired link position q_d is $\theta_d = q_d + K^{-1}g(q_d)$. In this case, global asymptotic stability is obtained by adding an extra gravity-dependent term τ_g to the PD control law (Equation 5.9). This term must match the gravity load $g(q_d)$ at steady state, and the following choices are possible, with progressively better transient performance, in the hypothesis that the joint elastic torque dominates the gravity torque:

- *Constant* compensation $\tau_g = g(q_d)$, ensuring global regulation for large enough K_p
- *On-line* compensation $\tau_g = g(\tilde{\theta}), \tilde{\theta} = \theta - K^{-1}g(q_d)$, which allows approximate compensation of the gravity effects on the links during the transient
- *Quasi-static* compensation $\tau_g = \bar{g}(\theta)$, where $\bar{g}(\theta)$ is the numerical solution of the equation $g(q)+K(q-\theta) = 0$, which allows the lower bound on K_p to be removed.

It is important to notice that due to the presence of the joint elasticity, the state of the system is represented by four variables for each joint: the motor position and velocity, and the link position and velocity, i.e., quantities that are measured before and after the joint deformation. The use of the full state $(q, \dot{q}, \theta, \dot{\theta})$ in the feedback control laws allows, of course, faster and damped transient performance to be achieved. On the other hand, the use of controllers with a reduced set of measurements is particularly attractive. For this purpose, it can be shown that the only convenient partial feedback solution is that considered earlier, i.e., *feedback of the motor variables.* The other solutions, i.e. feedback of the link variables alone or the combination of motor positions and link velocities, and vice versa, must be avoided. The reason is related to the fact that the link variables and the control actuation are not physically *co-located.*

If joint *torque sensors* able to measure the elastic torque τ_J are available, the following torque feedback law can be used:

$$\tau = MM_\theta^{-1}u + (I - MM_\theta^{-1})\tau_J \tag{5.10}$$

where M_θ is a diagonal positive definite matrix and u is a new control input. Using Equation 5.10, the motor equation (5.8) becomes

$$M_\theta \ddot{\theta} + K(\theta - q) = u$$

where the matrix M_θ represents an apparent motor inertia that can be arbitrarily set. It can be easily understood that the smaller M_θ with relatively large K, the smaller the deviation from the rigid joint dynamics. In turn, torque feedback allows to neglect joint elasticity, and the control input u can be set as for a rigid robot with $\theta = q$.

5.2.3 TASK SPACE CONTROL

While the motion control problem for a robot manipulator is naturally formulated in terms of the n joint variables, i.e., in the joint space defined by the $n \times 1$ vector q, the robot's *task* is conveniently specified in terms of a vector of m coordinates x, which typically define the location of the manipulator's end effector or sometimes, of other points of interest of the robot's body. Typically, one has $n \geq m$, so that the joints can provide at least the DOF required for the end-effector task. If $n > m$ strictly, the manipulator is kinematically *redundant*. The mappings

$$x = k(q) \tag{5.11}$$

$$\dot{x} = J(q)\dot{q} \tag{5.12}$$

where Equation 5.12 can be obtained by differentiating Equation 5.11 with respect to time, are known as direct kinematics and differential kinematics, respectively. The matrix J is the $m \times n$ *task Jacobian*.

In the presence of redundant DOF, the same task can be executed with different joint motions, giving the possibility of better exploiting the workspace of the manipulator and ultimately resulting in a more versatile robotic arm.

In order to accomplish a given task, two kinds of control approaches can be considered, namely, joint space control and a task space control.

The *joint space control* approach is actually articulated in two subproblems. First, the manipulator inverse kinematics is solved to transform the desire task x_d into the corresponding joint motion q_d. Then, a joint space control scheme is designed that allows the actual motion q to track q_d.

In the *task space control* approach, the control torques are computed on the basis of the comparison between the desired task vector x_d and actual task vector x, and some kind of kinematic inversion is embedded in the feedback control loop. This implies that the algorithmic complexity of task space control schemes is usually higher than that of joint space control schemes, especially in the case of redundant robots.

Task space control approaches are usually preferred when the manipulator's task, usually defined in terms of end effector coordinates, may be subject to online modifications to accomodate unexpected events or to respond to sensor inputs. In particular, they are essential when physical interaction of the manipulator with the environment is of concern.

Task space control schemes can be split into two main categories: acceleration-based control and force-based control.

Acceleration-based control schemes start from the the second-order kinematics equation

$$\ddot{x} = J(q)\ddot{q} + \dot{J}(q)\dot{q}$$

which can be solved in terms of the joint space acceleration $\ddot{q}$ as

$$\ddot{q} = J^{\dagger}(\ddot{x} - \dot{J}(q)\dot{q}) + N\eta \tag{5.13}$$

where

$$J^{\dagger}(q) = W^{-1}J^{T}(JW^{-1}J^{T})^{-1}$$

is any $n \times m$ right generalized inverse of J with symmetric and positive definite weight matrix W and

$$N = I - J^{\dagger}J$$

is a $n \times n$ matrix, which projects the arbitrary vector η to the null space of J. The last term of Equation 5.13 provides the null-space accelerations corresponding to the joint motions that produce a change in the configuration of the manipulator without affecting its task accelerations.

In view of Equation 5.13, a simple task space inverse dynamics control can be defined by using control law (Equation 5.2) with vector r chosen as

$$r = \ddot{q}_{\text{task}} + \ddot{q}_{\text{null}} \tag{5.14}$$

with

$$\ddot{q}_{\text{task}} = J^{\dagger}(a - \dot{J}(q)\dot{q}), \quad \ddot{q}_{\text{null}} = N\eta \tag{5.15}$$

where

$$a = \ddot{x}_d + K_d(\dot{x}_d - \dot{x}) + K_p(x_d - x) \tag{5.16}$$

and η can be set to assign a null space task, i.e., a task set to satisfy an additional goal, besides the main task, projected in the null space of the main task through matrix N.

In the case $n = m$, the closed loop dynamics can be written in terms of the task space tracking error $\Delta x(t) = x_d(t) - x(t)$ as

$$\Delta\ddot{x} + K_d\Delta\dot{x} + K_p\Delta x = 0 \tag{5.17}$$

and the task space error converges exponentially to zero for positive definite matrix gains K_d and K_p.

In the case $n > m$, the aforementioned equation describes only the task space dynamics, while the stability of the closed loop dynamics depends also on the null space dynamics

$$N(\ddot{q} - \eta) = 0$$

The stability analysis for the null space dynamics is not easy, and different choices of the vector η can be made, depending on how redundancy is used. A practical choice just aimed at stabilizing the internal motions is $\eta = -k_\phi\dot{q}$, with $k_\phi > 0$.

Torque-based control schemes define the control torque as the sum

$$\tau = \tau_{\text{task}} + \tau_{\text{null}} \tag{5.18}$$

with τ_{task} and τ_{null} usually set as

$$\tau_{\text{task}} = J^T h_c + C(q,\dot{q})\dot{q} + g(q), \quad \tau_{\text{null}} = P^T\gamma \tag{5.19}$$

where $P = I - J^{\#}J$ is a null space projector matrix and $J^{\#}$ the dynamically consistent generalized inverse of J, obtained by setting $W = B(q)$. Vector h_c is a control generalized force acting on the task variables, while $P^T\gamma$ is a torque in charge of controlling the null-space motion.

The role of h_c and γ can be better understood by plugging Equation 5.18 into Equation 5.1 and multiplying both sides of the resulting equation once by $J^{\#T}$ and once by P^T. The resulting equations are, respectively, the task space dynamics

$$\Lambda_x(q)J\ddot{q} = h_c$$

with $\Lambda_x = (JB^{-1}J^T)^{-1}$, and the null space dynamics

$$P^T(B(q)\ddot{q} - \gamma) = 0$$

Therefore, by choosing

$$h_c = \Lambda_x(q)(a - \dot{J}(q)\dot{q})$$

with a as in Equation 5.16, the exponentially stable closed loop dynamics (Equation 5.17) is recovered. It is worth observing that the matrix $\Lambda_x(q)$ represents the inertia of the robot reflected in the task space.

It is not difficult to show that the torque-based control law is equal to the acceleration-based control law written using the inertia-weighted generalized inverse. Anyway, each of the two formulations has pros and cons, which are still debated in the robotics literature.

In the case of regulation to a constant desired task vector x_d, a simpler torque-based control law can be adopted by choosing τ_{task} in Equation 5.18 as

$$\tau_{\text{task}} = J^T K_p(x_d - x) - K_d\dot{q} + g(q) \tag{5.20}$$

which is a proportional derivative (PD) control with gravity compensation in task space.

A drawback common to both the control approaches is that the stability of the *null space dynamics* cannot be determined easily. One of the reasons is that differently from the task space dynamics, the null space dynamics is not expressed by a minimum number of equations, which should be $r = n - m$. This problem can be overcome by considering a $n \times r$ matrix $Z(q)$, such that $JZ = O$, and introducing a $r \times 1$ velocity vector ν, such that

$$\dot{q}_n = N\dot{q} = Z\nu \tag{5.21}$$

A convenient choice of ν is given by left inertia-weighted generalized inverse $\nu = Z^{\#}\dot{q} = (Z^T MZ)^{-1} Z^T M\dot{q}$. By this choice, the extended Jacobian matrix $J_E(q)$ defined as

$$\begin{pmatrix} \dot{x} \\ \nu \end{pmatrix} = J_E(q)\dot{q} = \begin{pmatrix} J(q) \\ Z^{\#}(q) \end{pmatrix} \dot{q} \tag{5.22}$$

is non-singular for full rank matrix J, and the inverse is $J_E^{-1}(q) = \left[J^{\#}(q)\ Z(q)\right]$.

Because of the unique relationship between the task space and the null space variables with the joint space variables given by Equation 5.22, it is possible to rewrite the dynamics equation of the robot in the new variables as

$$\Lambda_E(q)\begin{pmatrix} \ddot{x} \\ \dot{\nu} \end{pmatrix} + \mu_E(q,\dot{q})\begin{pmatrix} \dot{x} \\ \nu \end{pmatrix} + g_E(q) = J_E^{-T}\tau \tag{5.23}$$

where the quantities Λ_E, μ_E, and g_E have the same physical meaning of the corresponding quantities of the dynamic model in the joint space (Equation 5.1) and fulfil the noticeable properties of the Euler–Lagrange equations.

Therefore, control schemes similar to the joint space algorithms presented in Section 5.2.1 can be adopted. This approach is known as *extended task space control*. It can be shown that thanks to the particular choice of the weighted generalised inverses $J^{\#}$ and $Z^{\#}$, the inertial matrix Λ_E is block diagonal, i.e.,

$$\Lambda_E(q) = J_E^{-T} M J_E^{-1} = \begin{pmatrix} \Lambda_x & 0 \\ 0 & \Lambda_\nu \end{pmatrix} \tag{5.24}$$

with $\Lambda_x(q) = (JM^{-1}J^T)^{-1}$ and $\Lambda_v(q) = Z^T MZ$; thus, the task space dynamics and the null space dynamics are inertially decoupled. Moreover, the matrices $\dot{\Lambda}_x - 2\mu_x$ and $\dot{\Lambda}_v - 2\mu_v$ are skew symmetric, with

$$\mu_E = \begin{bmatrix} \mu_x & \mu_{xv} \\ -\mu_{xv}^T & \mu_v \end{bmatrix}$$

5.2.4 MULTI-PRIORITY CONTROL

The common feature of the task space control approaches considered here is that they allow the execution of a main task together with an additional goal projected in the null space of the main task. This *null space projection* guarantees that the additional goal has a lower priority with respect to the main task; i.e., the main task is fully executed while the additional goal is fulfilled only to the extent that it does not conflict with the main task.

In conventional manipulation structures, these additional goals or *secondary tasks* are typically used to improve the value of performance criteria during motion. The most frequently considered performance objective for trajectory tracking tasks is singularity avoidance, i.e., joint configurations where the Jacobian matrix is rank deficient. In these configurations, it is impossible to generate task velocities or accelerations in certain directions. Moreover, since task space control algorithms involve the inversion of the Jacobian matrix, very high joint velocities are produced in the vicinity of these configurations, which therefore, must be avoided. Another useful secondary task is that of keeping the robot away from joint limits, or from undesired regions of the workspace, due to the presence of obstacles.

For highly redundant systems, like humanoids or multi-legged robots, multiple tasks could be arranged in priority in order to try to fulfil most of them, hopefully all of them, simultaneously. For example, in the case of humanoid robots, we may want to give higher priority to the task of respecting joint limit constraints over the task of controlling the robot's center of gravity for balance and in the end, the task of controlling the hands' motion.

The priority order among the tasks can be guaranteed by using projection matrices, like those used in the previous subsection. In fact, all the task space control approaches presented earlier can be extended to handle a task hierarchy, although the generalization is not straightforward.

In the case of acceleration-based control, let x_k be the $m_k \times 1$ vector defining the k-th priority task, where $m_k < n$ and $\dot{x}_k = J_k(q)\dot{q}$, for $k = 1, \ldots, L$ are the corresponding differential kinematics equations. The task hierarchy is defined such that $k = 1$ is top priority, and $k_a < k_b$ implies that k_a is located higher in the priority order than k_b.

A multi-priority control can be designed by using control law (Equation 5.2) with vector r defined as

$$r = \sum_{k=1}^{L} \ddot{q}_k, \quad \ddot{q}_k = \bar{J}_k^{\dagger}(a_k - \dot{J}_k\dot{q} - b_k) \tag{5.25}$$

with $a_k = \ddot{x}_{k,d} + K_{k,d}(\dot{x}_{k,d} - \dot{x}_k) + K_{k,p}(x_{k,d} - x_k)$, $x_{k,d}$ being the desired value of x_k. The quantities b_k and $\bar{J}_k$ are chosen so that the tasks x_j, with $j \geq k$, do not generate accelerations in the space of all higher-priority tasks. In detail,

$$\begin{aligned} b_k &= J_k \sum_{i=1}^{k-1} \ddot{q}_i \\ \bar{J}_k &= J_k N_{k-1} \\ N_k &= \prod_{j=1}^{k} (I - \bar{J}_j^{\dagger} \bar{J}_j) \end{aligned}$$

with $b_1 = 0$ and $N_0 = I$.

It can be verified that any symmetric positive definite weight matrix W can be used for the computation of the generalized inverse $\bar{J}_k^{\dagger}$. Moreover, if the inertia-weighted generalized inverse is used, the acceleration-based multi-priority control becomes equivalent to a torque-based multi-priority control. Similarly, the extended task space formulation generalized to the case of multiple priority tasks results in an inertially decoupled dynamic system, as in Equation 5.24.

5.3 FORCE CONTROL

Control of the physical interaction between a robot manipulator and the environment is crucial for the successful execution of a number of practical tasks in both industrial and service robotics. Typical examples in industrial settings include polishing, deburring, machining, and assembly. A variety of examples can be found also in service robotics, as the case of a humanoid robot opening a door or pulling out a drawer or of a rehabilitative robot performing assisted training of patients with movement disorders.

During contact, the environment may set constraints on the geometric paths that can be followed by the robot's end effector (kinematic constraints), as in the case of cleaning a window or turning a door handle. In other situations, the interaction occurs with a dynamic environment, as in the case of collaboration with a human. In all cases, a pure motion control strategy is not recommended.

In the presence of interaction, the higher the environment stiffness and position control accuracy are, the more easily the contact forces may rise and reach unsafe values. This drawback can be overcome by introducing compliance, either in a passive or in an active fashion, to accommodate the robot motion in response to interaction forces.

Passive compliance may be due to the structural compliance of the links, joints, or end effector or to the compliance of the position servo. Soft robot arms with elastic joints or links are purposely designed for intrinsically safe interaction with humans. In contrast, active compliance is entrusted to the control system, denoted *interaction control* or *force control*. In same cases, the measurement of the contact force and moment is required, which is fed back to the controller and used to modify or even generate online the desired motion of the robot.

The passive solution is faster than active reaction commanded by a computer control algorithm. However, the use of passive compliance alone lacks flexibility and cannot guarantee that high contact forces will never occur. Hence, the most effective solution is to use active force control (with or without force feedback) in combination with some degree of passive compliance.

Often, a *force/torque sensor* is mounted at the robot wrist to measure the exchanged forces, but other possibilities exist; for example, force sensors can be placed on the fingertips of robotic hands; also, external forces and moments can be estimated via shaft torque measurements of joint torque sensors or from the currents of the joint actuators.

The force control strategies can be grouped into two categories: those performing direct force control and those performing indirect force control. The main difference between the two categories is that the former offer the possibility of controlling the contact force and moment to a desired value, thanks to the closure of a force feedback loop; the latter instead achieve interaction control via motion control without explicit closure of a force feedback loop.

5.3.1 INTERACTION OF THE END EFFECTOR WITH THE ENVIRONMENT

The case of interaction of the end effector of a robot manipulator with the environment is considered first, which is the most common situation in the applications.

The end effector pose can be represented by the position vector p_e and the rotation matrix R_e, corresponding to the position and orientation of a frame attached to the end effector with respect to a fixed base frame.

The end effector velocity is denoted by the 6×1 *twist* vector $v_e = \left(\dot{p}_e^T \, \omega_e^T\right)^T$, where $\dot{p}_e$ is the translational velocity and ω_e the angular velocity, and can be computed from the joint velocity vector $\dot{q}$ using the linear mapping

$$v_e = J_e(q)\dot{q}$$

The matrix J_e is the end effector Jacobian, also known as *geometric Jacobian.*

The force f_e and moment m_e applied by the end effector to the environment are the components of the *wrench* vector $h_e = \left(f_e^T \, m_e^T\right)^T$. The joint torques τ_e corresponding to h_e can be computed as

$$\tau_e = J_e(q)^T h_e$$

These torques contribute to the manipulator dynamics, which can be rewritten as

$$B(q)\ddot{q} + C(q,\dot{q})\dot{q} + g(q) = \tau - J_e(q)^T h_e \tag{5.26}$$

To control the interaction of the end-effector with the environment, it is useful to define the control torque τ in terms of generalized control forces acting on the end-effector and for redundant robots, control torques that do not contribute to the end-effector motion. Namely, the torque-based control scheme (Equation 5.18) can be adopted, with

$$\tau = J_e^T h_c + P_e^T \gamma \tag{5.27}$$

where P_e is the dynamically consistent null space projection matrix. Replacing Equation 5.27 in Equation 5.26 and multiplying both sides of the resulting equation by $J_e^{\#T}$ gives

$$\Lambda_e(q)\dot{v}_e + \mu_e(q,\dot{q})v_e + \eta_e(q) = h_c - h_e \tag{5.28}$$

known as dynamic model in the *operational space*, where $\Lambda_e(q) = (J_e M^{-1} J_e^T)^{-1}$ is the 6×6 equivalent inertia matrix seen by the end-effector, $\mu_e(q,\dot{q})v_e$ is the wrench including centrifugal and Coriolis effects, and $\eta_e(q)$ is the wrench of the gravitational effects.

Equation 5.28 can be seen as a representation of the Newton's Second Law of Motion where all the generalized forces acting on the joints of the robot are reported at the end-effector. In the case of redundant robots, this equation does not represent the overall robot dynamics but must be completed by the equations of the null space dynamics.

Moreover, the full specification of the system dynamics would require also the analytic description of the interaction force and moment h_e. This is a very demanding task from a modeling viewpoint.

The design of the interaction control and the performance analysis are usually carried out under simplifying assumptions. The following two cases are considered:

1. The robot and the environment are perfectly rigid, and purely kinematics constraints are imposed by the environment, known as *holonomic constraints*.
2. the robot—rigid or elastic—interacts with a passive environment.

It is obvious that these situations are only ideal. However, the robustness of the control should be able to cope with situations where some of the ideal assumptions are relaxed. In that case, the control laws may be adapted to deal with nonideal characteristics.

5.3.2 HOLONOMIC CONSTRAINTS

In this subsection, for simplicity, only the case of non-redundant robots, with square and non-singular geometric Jacobian matrix J_e, is considered.

It is assumed that the environment is rigid and frictionless and imposes kinematic constraints to the robot's end-effector motion. These holonomic constraints reduce the dimension of the space of the feasible end-effector velocities and of the contact forces and moments. In detail, in the presence of m independent constraints ($m < 6$), the end-effector velocity belongs to a subspace of dimension $6 - m$, while the end-effector wrench belongs to a subspace of dimension, and can be expressed in the form

$$v_e = S_v(q)\nu, \quad h_e = S_f(q)\lambda$$

where ν is a suitable $(6-m) \times 1$ vector and λ is a suitable $m \times 1$ vector. Moreover, the subspaces of forces and velocity are *reciprocal*, i.e.,

$$h_e^T v_e = 0, \quad S_f^T(q)S_v(q) = 0$$

The concept of reciprocity expresses the physical fact that in the hypothesis of rigid and frictionless contact, the wrench does not cause any work against the twist.

An interaction task can be assigned in terms of a desired end-effector twist v_D and wrench h_D that are computed as

$$v_D = S_v v_D, \quad h_D = S_f \lambda_D$$

by specifying vectors λ_D and v_D.

In many robotic tasks, it is possible to set an orthogonal reference frame, usually referred as a *task frame*, in which the matrices S_v and S_f are constant. The task frame can be chosen to be attached either to the end-effector or to the environment. Moreover, the interaction task is specified by assigning a desired force/torque or a desired linear/angular velocity along/about each of the frame axes.

In more complex and general situations, the matrices S_v and S_f can be found analytically, starting from the analytic expression of the holonomic constraints in the joint space

$$\phi(q) = 0 \tag{5.29}$$

where ϕ is an $m \times 1$ twice differentiable vector function, whose components are linearly independent at least locally in a neighborhood of the operating point. Constraints of the form in Equation 5.29 involving only the generalized coordinates of the system, are known as *holonomic constraints.*

Differentiation of Equation 5.29 yields

$$J_\phi(q)\dot{q} = 0 \tag{5.30}$$

where $J_\phi(q)$ is the $m \times 6$ *constraint Jacobian.* In the absence of friction, the generalized interaction forces are represented by a reaction wrench that tends to violate the constraints. This end-effector wrench produces reaction torques at the joints that can be computed using the principle of virtual work as

$$\tau_e = J_\phi^T(q)\lambda$$

where λ is an $m \times 1$ vector of *Lagrange multipliers*. The end-effector wrench corresponding to τ_e can be computed as

$$h_e = J_e^{-T}(q)\tau_e = S_f(q)\lambda \tag{5.31}$$

where

$$S_f = J_e^{-T}(q)J_\phi^{-T}(q) \tag{5.32}$$

Using Equations 5.31 and 5.32, Equation 5.30 can be rewritten in the form

$$J_\phi(q)J_e^{-1}(q)J_e(q)\dot{q} = S_f^T v_e = 0 \tag{5.33}$$

which, by virtue of Equation 5.31, is equivalent to the reciprocity condition $h_e^T v_e$. At this point, matrix S_v can be computed from the equality $S_f^T(q)S_v(q) = 0$.

5.3.3 HYBRID FORCE/MOTION CONTROL

The reciprocity of the velocity and force subspaces naturally leads to a control approach, known as *hybrid force/motion control*, aimed at controlling simultaneously both the contact force and the end-effector motion in two reciprocal subspaces.

In the presence of holonomic constraints, the external wrench can be written in the form $h_e = S_f \lambda$, and it is possible to compute the dynamics of the constrained system in terms of $6 - m$ second-order equations

$$\Lambda_v(q)\dot{v} = S_v^T \left[h_c - \beta_e(q,\dot{q})\right]$$

where $\Lambda_v = S_v^T \Lambda_e S_v$ and $\beta_e(q,\dot{q}) = \mu_e(q,\dot{q})v_e + \eta_e(q)$, assuming constant matrices S_v and S_f. Moreover, the vector λ can be computed as

$$\lambda = S_f^{\dagger}(q)\left[h_c - \beta_e(q,\dot{q})\right]$$

$S_f^{\dagger}$ being a suitable pseudoinverse of matrix S_f. The aforementioned equation reveals that the contact force is a constraint force, which instantaneously depends on the applied input wrench h_c.

An inverse dynamics inner control loop can be designed by choosing the control wrench h_c as

$$h_c = \Lambda_e(q) S_v \alpha_v + S_f f_\lambda + \beta_e(q,\dot{q})$$

where α_v and f_λ are properly designed control inputs, which leads to the equations

$$\dot{v} = \alpha_v, \quad \lambda = f_\lambda$$

showing a complete decoupling between motion control and force control.

Then, the desired force $\lambda_D(t)$ can be achieved by setting

$$f_\lambda = \lambda_D(t)$$

but this choice is very sensitive to disturbance forces, since it contains no force feedback. Alternative choices are

$$f_\lambda = \lambda_D(t) + K_{P\lambda}\left[\lambda_D(t) - \lambda(t)\right]$$

or

$$f_\lambda = \lambda_D(t) + K_{I\lambda}\int_0^t \left[\lambda_D(\tau) - \lambda(\tau)\right] \mathrm{d}\tau$$

where $K_{P\lambda}$ and $K_{I\lambda}$ are suitable positive-definite matrix gains. The proportional feedback is able to reduce the force error due to disturbance forces, while the integral action is able to compensate for constant bias disturbances.

Motion control is achieved by setting

$$\alpha_V = \dot{v}_D(t) + K_{Pv}\left[v_D(t) - v(t)\right] + K_{Iv}\int_0^t \left[v_D(\tau) - v(\tau)\right] \mathrm{d}\tau$$

where K_{Pv} and K_{Iv} are suitable matrix gains. It is straightforward to show that asymptotic tracking of $v_D(t)$ and $\dot{v}_D(t)$ is ensured with exponential convergence for any choice of positive definite matrices K_{Pv} and K_{Iv}.

Notice that the implementation of force feedback requires the computation of vector λ from the measurement of the end-effector wrench h_e as $S_f^{\dagger} v_e$. Analogously, vector v can be computed from v_e as $S_v^{\dagger} v_e$, $S_v^{\dagger}$ being a suitable pseudoinverse of matrix S_v.

The hypothesis of rigid contact can be removed, and this implies that along some directions, both motion and force are allowed, although they are not independent. Hybrid force/motion control schemes can be defined also in this case.

5.3.4 IMPEDANCE CONTROL

The interaction force of the ideal situation considered in the previous subsection, being a constraint reaction force, is not a state variable of the system. In that case, the stability of the controlled system can be analyzed by considering only the robot dynamics and the kinematic constrains.

In more general situations, the environment is a dynamic system itself, and the overall coupled robot–environment dynamics must be considered for stability analysis and control design.

A physical interaction can be modeled as an exchange of mechanical power at an *interaction port* in the form $p(t) = h^T v$, h being the generalized interaction force and v the generalized velocity at the port.

In this perspective, a mechanical *impedance* at an interaction port can be defined as a dynamic operator that determines an output force in response to an input velocity at the same port. Vice versa, a mechanical *admittance* is a dynamic operator that determines an output velocity in response to an input force.

For linear systems, admittance is the inverse of impedance, and both can be represented as transfer functions in Laplace domain. Moreover, impedance (admittance) is analogous to electrical impedance (admittance) if we replace force with voltage and velocity with current.

In modeling the physical interaction between the robot and the environment, if one system is modeled as an impedance, the other must be modeled as an admittance, and vice versa. The roles of impedance and admittance are interchangeable, with some exceptions.

It is important to notice that the value of the two port variables characterizing the interaction, i.e, the force h and the velocity v, depends on both robot and environment dynamics, as can be easily verified considering the electric circuit analogy. On the other hand, a given impedance (or admittance) dynamic behavior can be imposed on the robot by the control independently of the dynamics of the other interacting system.

The aim of *impedance control* is to control neither the force nor the velocity during the interaction, but the dynamic relationship between force and velocity, namely, the robot's impedance or admittance.

Therefore, a successful impedance control must satisfy the following goals:

- *Robust stability* regardless of the environment dynamics, or for a certain class of environments
- *Performance*, i.e., minimal deviation of the actual robot impedance (or admittance) from the desired one.

Concerning the first goal, from the passive control theory, it is known that if the dynamics of the environment is passive, the stability of the interaction can be guaranteed, provided that the dynamics imposed on the robot is passive as well. This property, which derives from the Nyquist stability criterion for feedback interconnected linear systems extended to nonlinear passive systems using a Lyapunov-like analysis, provides only a sufficient condition. Therefore, passivity is sought to ensure robustness, but sometimes it may be too conservative, and stability can be achieved also if the interacting systems are not passive.

With reference to the second goal, a number of different schemes have been proposed to reshape the robot's natural impedance or admittance.

The simplest approach is that of imposing a suitable static relationship between the deviation of the end-effector position and orientation from a desired pose and the force exerted on the environment by using the control law

$$h_c = K_P \Delta x_{de} - K_D v_e + \eta_e(q) \tag{5.34}$$

where K_P and K_D are suitable matrix gains, and Δx_{de} is a suitable error between a desired and the actual end-effector position and orientation. The position error component of Δx_{de} can be simply chosen as $p_D - p_e$. Concerning the orientation error component, different choices are possible, which are not all equivalent, but this issue is outside the scope of this chapter.

The control input (Equation 5.34) corresponds to a wrench (force and moment) applied to the end-effector, which includes a gravity compensation term $\eta_e(q)$, a viscous damping term $K_D v_e$, and an elastic wrench provided by a virtual spring with stiffness matrix K_P (or equivalently, compliance matrix K_P^{-1}) connecting the end-effector frame with a frame of desired position and orientation. This control law is known as *stiffness control* or *compliance control*, and when the viscous damping torque is chosen in the joint space as $-K_D \dot{q}$, it coincides with the task space control law (Equation 5.20).

Using passivity, it is possible to prove the stability in the case of interaction with a passive environment. Moreover, at steady-state, the robot's end-effector has the desired elastic behavior

$$K_P \Delta x_{de} = h_e$$

It is clear that if $h_e \neq 0$, then the end-effector deviates from the desired pose, which is usually denoted as *virtual pose*.

The closed loop system (Equations 5.35 with 5.34) can be written as

$$\Lambda_e(q)\dot{v}_e + \mu_e(q,\dot{q})v_e + K_D v_e - K_P \Delta x_{de} = h_e \tag{5.35}$$

corresponding to a 6-DOF nonlinear and configuration-dependent mechanical impedance of mass-spring-damper type, with inertia (mass) matrix $\Lambda_e(q)$, adjustable damping K_D, and stiffness K_P, producing the external wrench h_e.

One problem of the aforementioned impedance control law, which does not require feedback of the interaction force and moment h_e, is that is not possible to modify the natural inertia of the robot (seen from the end-effector) or, the friction torques which have not been considered here but which can be easily added to the term $\mu_e(q,\dot{q})v_e$. Hence, the performance of the control scheme is poor in the case that the natural robot inertia and friction are dominant with respect to the desired damping and stiffness. This happens, for example, when trying to impose a low impedance on a bulky industrial robot.

A partial solution to this problem is to use feedback of the interaction force and moment h_e in the following control law, known as *impedance control*:

$$h_c = \Lambda_e(q)\alpha + \mu_e(q,\dot{q})v_e + \eta_e(q) + h_e \tag{5.36}$$

where α is chosen as

$$\alpha = \dot{v}_d + K_M^{-1}(K_D\Delta v_{de} + K_P\Delta x_{de} - h_e)$$

The following expression can be found for the closed loop system:

$$K_M\Delta\dot{v}_{de} + K_D\Delta v_{de} + K_P\Delta x_{de} = h_e \tag{5.37}$$

representing the equation of a 6-DOF configuration independent mass-spring-damper system with adjustable inertia (mass) matrix K_M, damping K_D, and stiffness K_P. This ideal behavior can be obtained only in the hypothesis that friction forces, included into the term $\mu_e(q,\dot{q})v_e$, can be completely cancelled out using Equation 5.36, which is unlikely to happen.

An alternative control scheme able to cope with friction and unmodeled dynamics is known as *admittance control.* In this case, an inner motion control loop is used to ensure tracking of a reference motion trajectory, and thus, the robot becomes an impedance producing force in response to motion. The reference trajectory for the inner motion control loop is produced by an outer controller, which uses feedback of the interaction control force and moment h_e. This latter control system is, therefore, an admittance and can be designed so as to produce, in closed loop with the motion-controlled robot, the dynamic behavior described by Equation 5.37. The main difference, with respect to the impedance control scheme, is that friction rejection is not demanded to model-based cancellation but is robustly performed by the inner motion control loop.

One common feature of the impedance and admittance control schemes described before is that they allow the natural inertia of the robot reflected to the end-effector to be modified thanks to the feedback of the interaction force and moment h_e.

In the case of rigid robots, the closed loop stability is ensured provided that the impedance matrices in the impedance equation Equation 5.37 are positive semidefinite in case of contact with a generic passive environment. For robots with non-negligible joint elasticity, considering a simple linear 1-DOF case, it can be shown that passivity is guaranteed only if the desired mass of the impedance is higher than the mass of the robot at the link side. The reason is related to the fact that the interaction force, used in the feedback control law, and the control actuation are not physically *co-located*. The same result can be extended to the case of multi-variable liner systems, but it is not easy to prove for nonlinear systems. In this latter case, the passivity can be easily shown only if the robot's inertia is left unchanged. Therefore, to guarantee passivity in the nonlinear multi-variable case, usually the mass of the desired impedance is set equal to the natural inertia of the robot reflected to the end-effector. This conservative choice may often limit the performance.

5.3.5 MULTI-PRIORITY INTERACTION

A significant problem for robots with a high number of DOF is that of achieving a decoupled behavior among a hierarchy of tasks with different priorities, also in the presence of interaction.

An external force applied anywhere to the robot's body produces an external torque τ_{ext}, which contributes to the robot's dynamics as follows:

$$B(q)\ddot{q} + C(q,\dot{q})\dot{q} + g(q) = \tau - \tau_{ext} \tag{5.38}$$

By adopting the extended task space control formulation presented in Subsection 5.2.3, the inverse dynamics control law (Equation 5.2) with r chosen as

$$r = J_E^{-1}\left(\begin{pmatrix}\ddot{x}_c \\ \dot{v}_c\end{pmatrix} - \dot{J}_E\dot{q}\right) = J^{\#}(\ddot{x}_c - \dot{J}\dot{q}) + Z(\dot{v}_c - \dot{Z}^{\#}\dot{q}) \tag{5.39}$$

yields the closed-loop dynamics

$$\ddot{q} = J^{\#}(\ddot{x}_c - \dot{J}\dot{q}) + Z(\dot{v}_c - \dot{Z}^{\#}\dot{q}) - B^{-1}(q)\tau_{ext} \tag{5.40}$$

Multiplying both sides of Equation 5.40 by $Z^{\#}$, and considering that $\dot{v} = \dot{Z}^{\#}\dot{q} + Z^{\#}\ddot{q}$, the null space closed-loop equation is obtained as

$$\dot{v} = \dot{v}_c - Z^{\#}B^{-1}\tau_{ext} \tag{5.41}$$

On the other hand, multiplying both sides of Equation 5.40 by J gives the task space closed loop equation

$$\ddot{x} = \ddot{x}_c - JB^{-1}\tau_{ext} \tag{5.42}$$

Notice that the null space velocity vector v is, in general, non-integrable, and thus, a null space position error cannot be easily defined. However, a projected joint space error $Z^T\tilde{q}$ can be used to define the null space command acceleration

$$\dot{v}_c = \dot{v}_d + \Lambda_v^{-1}((\mu_v + B_v)\tilde{v} + Z^T k_q\tilde{q}) \tag{5.43}$$

with $k_q > 0$, B_v symmetric and positive definite matrix, and $\tilde{v} = v_d - v$. The configuration-dependent quantities $\Lambda_v = Z^T MZ$ and μ_v are, respectively, the inertia matrix and the Coriolis/centrifugal matrix in the null space defined in Subsection 5.2.3. The corresponding closed loop equation is

$$\Lambda_v\dot{\tilde{v}} + (\mu_v + B_v)\tilde{v} + Z^T k_q\tilde{q} = Z^T\tau_{ext} \tag{5.44}$$

where $Z^T \tau_{ext}$ is the projection of the external torque on the null space. Equation 5.44 can be interpreted as an impedance equation defined in the null space, with inertia Λ_v, damping B_v, and projected elastic torque $Z^T k_q \tilde{q}$.

On the other hand, the following choice for the task space command acceleration

$$\ddot{x}_c = \ddot{x}_d + \Lambda_x^{-1}((\mu_x + D)\dot{\tilde{x}} + K\tilde{x}) \tag{5.45}$$

produces the closed-loop equation

$$\Lambda_x \ddot{\tilde{x}} + (\mu_x + D)\dot{\tilde{x}} + K\tilde{x} = J^{\#T} \tau_{ext} \tag{5.46}$$

which also represents an impedance equation in task space with inertia Λ_x, damping D, and stiffness torque K.

The parameters of the two impedance equations (Equations 5.44 and 5.46) can be set depending on the expected behavior in response to forces and moments applied anywhere to the robot.

Consider, for example, the following scenario: a robot assistant composed by an anthropomorphic manipulator mounted on a mobile wheeled platform carries a glass with some water, and for some reason, a human applies a force to the robot's elbow. If the robot is compliant, and the compliance is not selective, the force may produce a change in the glass orientation, and the water can flow out. On the other hand, if the robot is controlled so that, in the task space defined as the glass (position and) orientation, the impedance is set high (i.e., low compliance), while in the null space, the impedance is set low (i.e., high compliance), then the robot's structure will comply mainly in the null space, keeping the (position and) orientation of the glass close to the desired one.

5.4 FUTURE DIRECTIONS AND RECOMMENDED READING

The motion control problem for robot manipulators was a challenging research topic from about the mid-1980s until the mid-1990s, when researchers started to exploit the structural properties of manipulator dynamics, such as feedback linearizability, passivity, linearity in the parameters, and others. Many of these methods are well assessed and can be found in standard robotics textbooks, such as Siciliano et al. [33]. An incomplete list of fundamental references where more detailed treatment of these topics can be found is: Arimoto and Miyazaki [4] for PD control with gravity compensation, Paden and Panja [27] for PD+ algorithm, Slotine and Li [35] and Ortega and Spong [24] for passivity-based adaptive control, and Abdallah et al. [1] and Spong [37] for robust control.

Research in control of robot manipulators with flexible transmissions started in the same years and is still continuing. The assumptions leading to the dynamic model for robots with elastic joints considered in this chapter were introduced in Spong [36]. The three versions of PD controller with gravity compensation were presented in Tomei [38], De Luca et al. [11] and Albu-Schäffer et al. [3]. To achieve robust control performance, special interest has been devoted to joint torque feedback

Albu-Schäffer and Hirzinger [2] and [25]. A detailed treatment of dynamic modeling and control issues for flexible robots can be found in De Luca and Book [10].

The presence of joint elasticity, at the beginning, was seen as a limiting factor for performance. Nowadays, joint elasticity is considered an explicit advantage for safe physical human–robot interaction and for locomotion. A relatively new and active area of research is the control of actuators with on-line controlled variable stiffness and damping, which exhibit properties that look appealing to build the next generation of robots and humanoids.

Another challenging research area is the control of robotic structures with a large number of joints, like hyper-redundant robots, cooperating manipulators, multifingered hands, vehicle-manipulator systems, or multiarm/multilegged robots, such as humanoid robots. The motion control problem in task space is particularly relevant for such robotic structures, because the high number of DOF at their disposal may be suitably exploited to satisfy a certain number of different tasks at the same time. This problem was initially addressed for simple manipulation structures [17] under the name of operational space control, using a torque-based approach, and then extended in more recent papers to more complex dynamics, like humanoids [18]. The multi-priority acceleration-based approach was considered in Sadeghian et al. [30], while an interesting comparison between the two methods was presented in Nakanishi et al. [21]. The decomposition of the robot dynamics in the task space and null space can be found in Oh et al. [23], and the extension to the multi-priority case is presented in Ott et al. [26].

The performance of a force-controlled robotic system or in general, of interaction control depends on the dynamics of the environment, which is very difficult to model and identify accurately. Hence, the standard performance indices used to evaluate a control system, i.e., stability, bandwidth, accuracy, and robustness, cannot be defined by considering the robotic system alone, as for the case of robot motion control, but must be always referred to the particular contact situation at hand. For this reason, although force control in industrial applications can be considered as a mature technology, standard design methodologies are not yet available.

Force control techniques are employed also in medical robotics, haptic systems, telerobotics, humanoid robotics, and micro and nano robotics. An interesting field of application is related to human-centered robotics, where control plays a key role in achieving adaptability, reaction capability, and safety. Robots and biomechatronic systems based on the novel variable impedance actuators, with physically adjustable compliance and damping, capable of reacting softly when touching the environment, necessitate the design of specific control laws [39]. The combined use of exteroceptive sensing (visual, depth, proximity, force, and tactile sensing) for reactive control in the presence of uncertainty represents another challenging research direction [28].

The two major paradigms of force control (impedance and hybrid force/motion control) presented in this section have a number of different implementations, deriving from the specific needs of the applications or exploiting more advanced control methods that have been developed in the last three decades. A description of the state of the art of the first decade is provided in Whitney [41], whereas later advancements are presented in Gorinevsky et al. [14] and Siciliano and Villani [34]. An extensive treatment of this topic with related bibliography can be found in Villani and De Schutter [40].

The original hybrid force/position control concept was introduced in Raibert and Craig [29], based on the concepts of natural and artificial constraints proposed in Mason [19]. The inclusion on constraints in the manipulator dynamics was considered in Yoshikawa [42] and McClamroch and Wang [20], while the task frame formalism was systematically developed in De Schutter and Van Brussel [13], Bruyninckx and De Schutter [5] and De Schutter et al. [12].

The original idea of a mechanical impedance model used for controlling the interaction between the manipulator and the environment was presented in Hogan [15], while the use of controlled stiffness or compliance was introduced in Salisbury [32]. The stability of impedance and admittance control based on passivity for rigid and elastic joint robots is discussed in Colgate and Hogan [9], Newman [22], Hogan and Buerger [16] and Ott et al. Ott-08. A less conservative stability concept for interaction control, with the name of complementary stability, was proposed in Buerger and Hogan [6]. For spatial rotations and translations, the specification of task-relevant impedance (and especially stiffness) is particularly challenging [7,8].

Finally, the interaction control problem of redundant robots in a multi-priority framework is considered in Sadeghian et al. [31] and Ott et al. [26].

REFERENCES

1. C. Abdallah, D. Dawson, P. Dorato, and M. Jamshidi. Survey of robust control for rigid robots. *IEEE Control Syst. Mag.*, 11(2):24–30, 1991.
2. A. Albu-Schäffer and G. Hirzinger. A globally stable state feedback controller for flexible joint robots. *Adv. Robot.*, 15(8):799–814, 2001.
3. A. Albu-Schäffer, C. Ott, and G. Hirzinger. A unified passivity based control framework for position, torque and impedance control of flexible joint robots. *Int. J. Robot. Res.*, 26:23–39, 2007.
4. S. Arimoto and F. Miyazaki. Stability and robustness of PID feedback control for robot manipulators of sensory capability. In M. Brady and R. Paul, editors, *Robotics Research: The First International Symposium*, pages 783–799, Cambridge, MA, MIT Press, 1984.
5. H. Bruyninckx and J. De Schutter. Specification of force-controlled actions in the "task frame formalism" – a synthesis. *IEEE Trans. Robot. Autom.*, 12(4):581–589, 1996.
6. S.P. Buerger and N. Hogan. Complementary stability and loop shaping for improved human-robot interaction. *IEEE Trans. Rob.*, 23(2):232–244, 2007.
7. F. Caccavale, P. Chiacchio, A. Marino, and L. Villani. Six-DOF impedance control of dual-arm cooperative manipulators. *IEEE/ASME Trans. Mechatronics*, 13(5):576–586, 2008.
8. F. Caccavale, C. Natale, B. Siciliano, and L. Villani. Six-DOF impedance control based on angle/axis representations. *IEEE Trans. Robot. Autom.*, 15:289–300, 1999.
9. E. Colgate and N. Hogan. The interaction of robots with passive environments: Application to force feedback control. In K.J. Waldron, editor, *Advanced Robotics: 1989*, pages 465–474. Springer, Berlin Heidelberg, D, 1989.
10. A. De Luca and W. Book. Robots with flexible elements. In B. Siciliano and O. Khatib, editors, *Springer Handbook of Robotics*, pages 287–319. Springer, Berlin, D, 2008.

11. A. De Luca, B. Siciliano, and L. Zollo. PD control with on-line gravity compensation for robots with elastic joints: Theory and experiments. *Automatica*, 41(10):1809–1819, 2005.
12. J. De Schutter, T. De Laet, J. Rutgeerts, W. Decré, R. Smits, E. Aerbeliën, K. Claes, and H. Bruyninckx. Constraint-based task specification and estimation for sensor-based robot systems in the presence of geometric uncertainty. *Int. J. Robot. Res.*, 26(5):433–455, 2007.
13. J. De Schutter and H. Van Brussel. Compliant robot motion I. A formalism for specifying compliant motion tasks. *Int. J. Robot. Res.*, 7(4):3–17, 1988.
14. D. Gorinevsky, A. Formalsky, and A. Schneider. *Force Control of Robotics Systems*. CRC Press, Boca Raton, FL, 1997.
15. N. Hogan. Impedance control: An approach to manipulation. Parts I–III. *ASME J. Dyn. Syst. Meas. Contr.*, 107:1–24, 1985.
16. N. Hogan and S.P. Buerger. Impedance and interaction control. In T.R. Kurfess, editor, *Robotics and Automation Handbook*, pages 19.1–19.24. CRC Press, New York, NY, 2004.
17. O. Khatib. A unified approach for motion and force control of robot manipulators: The operational space formulation. *IEEE J. Robot. Autom.*, 3:43–53, 1987.
18. O. Khatib, L. Sentis, J.H. Park, and J. Warren. Whole-body dynamic behavior and control of human-like robots. *Int. J. Hum. Robot.*, 1:29–43, 2004.
19. M.T. Mason. Compliance and force control for computer controlled manipulators. *IEEE Trans. Syst. Man Cybern.*, 11:418–432, 1981.
20. N.H. McClamroch and D. Wang. Feedback stabilization and tracking of constrained robots. *IEEE Trans. Autom. Contr.*, 33(5):419–426, 1988.
21. J. Nakanishi, R. Cory, M.J. Peters, and S. Schaal. Operational space control: A theoretical and empirical comparison. *Int. J. Robot. Res.*, 27:737–757, 2008.
22. W.S. Newman. Stability and performance limits of interaction controllers. *ASME J. Dyn. Syst. Meas. Control*, 114(4):563–570, 1992.
23. Y. Oh, W. Chung, and Y. Youm. Extended impedance control of redundant manipulators based on weighted decomposition of joint space. *J. Rob. Syst.*, 15(5):231–258, 1998.
24. R. Ortega and M.W. Spong. Adaptive motion control of rigid robots: A tutorial. *Auomatica*, 25:877–888, 1989.
25. C. Ott, A. Albu-Schäffer, A. Kugi, and G. Hirzinger. On the passivity based impedance control of flexible joint robots. *IEEE Trans. Rob.*, 24:416–429, 2008.
26. C. Ott, A. Dietrich, and A. Albu-Schäffer. Prioritized multi-task compliance control of redundant manipulators. *Automatica*, 53(0):416–423, 2015.
27. B. Paden and R. Panja. Globally asymptotically stable PD+ controller for robot manipulators. *Int. J. Control*, 47(6):1697–1712, 1988.
28. M. Prats, A.P. Del Pobil, and P.J. Sanz. *Robot Physical Interaction through the combination of Vision, Tactile and Force Feedback - Applications to Assistive Robotics*, volume 84 of *Springer Tracts in Advanced Robotics*. Springer, Berlin Heidelberg, D, 2013.
29. M.H. Raibert and J.J. Craig. Hybrid position/force control of manipulators. *ASME J. Dyn. Syst. Meas. Contr.*, 103:126–133, 1981.
30. H. Sadeghian, L. Villani, M. Keshmiri, and B. Siciliano. Dynamic multi-priority control in redundant robotic systems. *Robotica*, 31(7):1155–1167, 2013.
31. H. Sadeghian, L. Villani, M. Keshmiri, and B. Siciliano. Task-space control of robot manipulators with null-space compliance. *IEEE Trans. Robot. Autom.*, 30(2):493–506, 2014.
32. J.K. Salisbury. Active stiffness control of a manipulator in Cartesian coordinates. In *19th IEEE Conf. Decis. Contr.*, Albuquerque, NM, pages 95–100, 1980.

33. B. Siciliano, L. Sciavicco, L. Villani, and G. Oriolo. *Robotics: Modelling, Planning and Control.* Springer, London, 2009.
34. B. Siciliano and L. Villani. *Robot Force Control.* Kluwer Academic Publishers, New York, 1999.
35. J.-J.E. Slotine and W. Li. On the adaptive control of robot manipulators. *Int. J. Robot. Res.*, 6(3):49–59, 1987.
36. M.W. Spong. Modeling and control of elastic joint robots. *ASME J. Dyn. Syst. Meas. Control*, 109(4):310–319, 1987.
37. M.W. Spong. On the robust control of robot manipulators. *IEEE Trans. Autom. Contr.*, 37:1782–1786, 1992.
38. P. Tomei. A simple pd controller for robots with elastic joints. *IEEE Trans. Autom. Contr.*, 36(10):1208–1213, 1991.
39. B. Vanderborght, A. Schäffer, A. Bicchi, E. Burdet, D.G. Caldwell, R. Carloni, M. Catalano, O. Eiberger, W. Friedl, G. Ganesh, M. Garabini, M. Grebenstein, G. Grioli, S. Haddadin, H. Hoppner, A. Jafari, M. Laffranchi, D. Lefeber, F. Petit, S. Stramigioli, N. Tsagarakis, M. Van Damme, R. Van Ham, L.C. Visser, and S. Wolf. Variable impedance actuators: A review. *Robot. Auton. Syst.*, 61(12):1601–1614, 2013.
40. L. Villani and J. De Schutter. Force control. In B. Siciliano and O. Khatib, editors, *Springer Handbook of Robotics*, pages 161–185. Springer, Berlin, D, 2008.
41. D.E. Whitney. Force feedback control of manipulator fine motions. *ASME J. Dyn. Syst. Meas. Cont.*, 99:91–97, 1977.
42. T. Yoshikawa. Dynamic hybrid position/force control of robot manipulators – description of hand constraints and calculation of joint driving force. *IEEE J. Robot. Aut.*, 3:386–392, 1987.

6 Navigation, Environment Description, and Map Building

Henry Carrillo, Yasir Latif, and José A. Castellanos

CONTENTS

6.1 THE ROBOTIC NAVIGATION PROBLEM

Navigation, according to the standardized definition given in the IEEE standard 172-1983 [40], is the process of directing a vehicle so as to reach an intended destination. A *robot*, taking also its standardized definition [41], is an actuated mechanism programmable in two or more axes with some degree of autonomy. Also, a robot should be able to move at will in its workspace, aiming to perform an intended task. Ideally, a robot should be completely autonomous so as to perform the task it was designed for without human intervention. The problem of guiding a robot towards a particular location, most of the time in order to perform an intended task, is the one of robot navigation.

The capability of navigating an environment is a critical ability for a mobile robot [78,79], especially if the robot is performing an intended task autonomously. A common precondition to effectively complete any task is to accurately reach *a priori* the initial position established in the workspace where this task will be performed. For example, a mobile manipulator aiming at opening a door using visual servoing needs to position its manipulator within the range of the doorknobs. Guiding a mobile robot to a particular initial position is done by solving a particular instance of the robotic navigation problem.

The robotic navigation problem is a meta-problem that involves key problems for mobile robots. Endowing a robot with the capability of autonomously navigating an environment requires the solution of at least three basic tasks in mobile robotics, namely *localization*, *mapping*, and *motion control* [42,90]. Localization is the problem of determining the position of the robot within a given map. Mapping refers to the problem of integrating the robot's sensor information into a coherent representation. Motion control is the problem of how to steer the robot to a particular location, therefore giving the robot the ability of performing active behavior. We will discuss each of these problems in the following sections.

6.1.1 WHERE AM I GOING?

Before the robot can start moving, it has to decide where to go. Deciding a single or multiple goal locations involves solving by the robot the question *Where Am I Going?* under the restriction that the aforementioned locations should help in solving the high-level problem the robot is dealing with, of which the navigation is just a part.

Giving an answer to this question requires endowing the robot with the ability of decision making and high-level deliberation so as to be able to choose between different alternatives. In practice, solving this question is difficult, and usually, the goal location is given by a human in the loop or a hand-crafted mission planner [78], although there have been several attempts to automate the decision process for some specific task such as optimal coverage or exhaustive exploration.

Some examples of those attempts to automate the decision process are the geometry-based approaches presented by Gónzales-Baños and Latombe [29] and Tovar et al. [99], which use distinctive features in the environment and their geometrical relations to define the places to visit by the robot in order to fulfill a task of area coverage. Another example is the use of soft computing techniques, e.g., fuzzy logic, neural networks, and genetic algorithms, among others, to decide the places to visit. Kosmatopoulos, based on these soft computing techniques, proposed a framework named Cognitive-based Adaptive Optimization (CAO) [50], which has been used to select goal positions for a swarm of robots performing optimal surveillance of a given area [23]. A final example of automating the decision process of where to go during navigation can be given by the work of Carlone et al. [12] and Carrillo et al. [13], which use information theoretic measurements to select goal destinations to improve the task of mapping an environment.

6.1.2 HOW DO I GET THERE?

After selecting the goal destination, a robot needs to devise how to get there, i.e, solving the question *How Do I Get There?* This is the so-called robotic motion planning problem [16,46]. Solutions to this problem have been studied for several decades [56] in the robotic context. One practical solution is to compute a set of motion controls that should be executed by the robot; this involves generating a path from the current position of the robot to the goal destination and also following the path.

Generating the path is known as the robotic path planning problem and requires an operative model of the environment. This is needed to take into account obstacles during the path generation. Computing a model of the environment is done through a mapping process, which will be discussed in Section 6.1.3, but in the following, we will assume that an operative model of the environment is given.

Following a given path requires synthesizing a controller with a control law that fulfills the kinematic and dynamic restrictions of the robot [46,89]. Usually, the control law requires as input the current position of the robot; this can be obtained through the localization process, which we will discuss in Section 6.1.3, but in the followings we will assume that the current position of the robot is known at all times.

Before delving into the process of generating a path and following it, we shall define some preliminary terms. A complete specification of all the physical points belonging to a robot is named *configuration* [16] and denoted as q. The space of all possible configurations for the robot in question is named the *configuration space* [89] and denoted as $\mathscr{C}$. Every configuration q is a point in $\mathscr{C}$.

Let us consider, for illustration purposes, a robot modeled as a single physical point in two dimensions but having a circular shape of radius r. A configuration q of the robot can be represented as a point $[x, y, \theta] \in SE(2)$ [89]. A configuration q is *valid* if it does not collide with an obstacle in the environment. If we assume the robot to have a circular shape with radius r and describe it abstractly as a point, its configuration will be valid if it does not collide with the obstacles inflated by the robot's radius r [16] [90]. The configuration space of the robot will be defined by $\mathbb{R}^2 \times SO(2)$, the dimension of which is 3 [89].

6.1.2.1 Robotic Path Planning

Moving a robot from a configuration q_{start} to a configuration q_{goal} requires several steps, among them the computation of a list of sequential valid configurations from q_{start} to q_{goal}. This problem is known as robotic path planning.

Path planning for mobile robots has been studied for many years [56] with a plethora of methods available. In the following, we will focus on methods tailored for autonomously moving a robot between two configurations [61]. Broadly speaking, robotic path planning can be split into two high-level steps: (*i*) representing the configuration space and (*ii*) finding the incremental list of optimal valid configurations.

Representing the Configuration Space. The most common way of representing a configuration space is by using a discrete structure, although continuous structure also exists [24]. The idea is to store in the structure a graph that connects configurations in the free space, which will allow the robot to move without collisions. The use of discrete structures is often preferred, because it allows a seamless integration with the next step, which is a search. Examples of discrete structures for representing configuration spaces are the visibility graph [56,65], which is a graph comprising all the vertices of the obstacles as well as the start and desired goal position of the robot. The edge of this graph connects all the vertices that are visible between each other. Traversing the graph guarantees a collision-free path in the environment. A detailed account of the construction of a visibility graph can be found in Choset et al. [16].

A related approach to build a graph is the Generalized Voronoi Diagram (GVG) [59]. In this approach, the graph is built using the points that have equal distance to the two closest obstacles. A deep tutorial on building a graph using GVG can be found in Choset et al. [16]. An advantage of GVG over the visibility graph is that maximum clearance to obstacles is guaranteed which is deemed a good property in practical scenarios where the localization of the robot is uncertain, but both approaches have a major drawback in that they do not consider the kinematic constraints of the robot; i.e., both methods consider robots to be free of holonomic constraints.

Among the discrete structures available for representing configuration spaces, the *state lattice* proposal of Pivtoraiko and Kelly [84,85] allows kinematics constraints of the robot in the search space of the path to be integrated in a principled manner; an example of this approach can be found in Likhachev and Ferguson [61] or Likhachev [60].

The key idea behind state lattice is *not* to assume four or eight connected neighborhoods between cells, as not all robots can move like that, but to define the connectivity according to feasible motions according to the kinematic constraints of the robot. The discrete structure defined by a state lattice is therefore a hyperdimensional grid. State lattice representations first discretize the control space of the robot, as shown in Figure 6.1, and then connect the states maintaining the continuity of all state variables Pivtoraiko et al. [85]. A deep explanation of state lattice is available in [85].

Building the Sequence of Configurations. After setting up a representation of the configuration state, the next step is to look for a sequential list of feasible configurations that the robot should follow to get from q_{start} to q_{goal}. The methods to build the sequences of configurations are tightly coupled with the representation used [90]. For the case of state lattice representations, graph-based search approaches are the best fit [85].

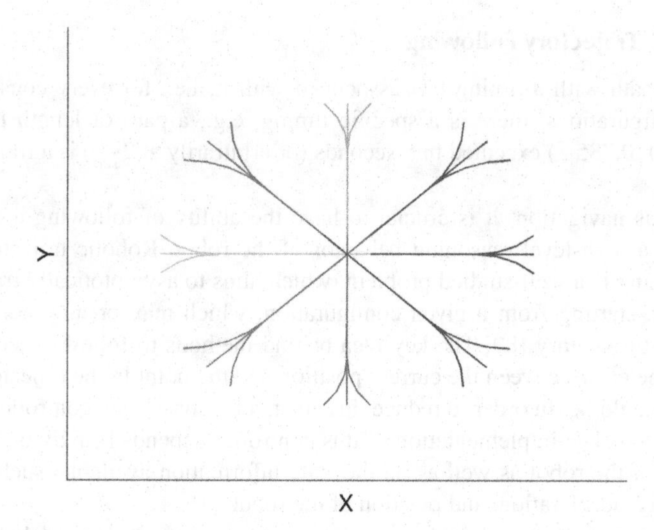

Figure 6.1 Samples of control actions for a unicycle robot to be used in a state lattice. Each line represents a sampled control in the heading direction. For each sample in the heading, three more in x and y are taken. The resulting joint samples define the feasible kinematic transitions between configurations.

Graph-based search techniques stem from the artificial intelligence community but have a pervasive use in robotics, especially in path planning [48,57]. The key idea of a graph search algorithm is to search the set of actions of size n that solve the problem of connecting two given states, e.g., q_{start} and q_{goal}, with the minimum cost. The cost is task dependent, and it is encoded in the discrete structure representation. Usually, the cost will depend on how likely is that the occupancy grid used for the configuration q_i is occupied and also on the robot's ability of performing the transition from one configuration to another; this is based on Likhachev and Ferguson [61].

Classical graph search algorithms such as Dijkstra [22], A* [36], or D*-lite [48] can be used to find the sequence of configurations, but they cannot guarantee to find a solution in a given time, which is a desirable property for autonomous navigation. Recent graph search algorithms such as ARA* [64] or AD* [62,63] have this property, which allows setting a maximum time for the search in the graph in which the algorithm will give a solution of the planning problem, although it may not be optimal.

In summary, the methodology of planning a path to navigate with a mobile robot can be simplified to two steps. First, the configuration space of the robot has to be built. For example, a state lattice representation can be used, which uses information from an occupancy grid representation of the world and kinematic constraints of the robot to set up the cost of transitioning between states. Second, an incremental sequence of feasible configurations between the start and end positions has to be built. For example, it is possible to use graph-based search techniques, like ARA*, which allow a maximum allotted time to be set up for the construction of a feasible sequence of configurations.

6.1.2.2 Robotic Trajectory Following

A trajectory is a path with a timing law associated with it; i.e., for every consecutive pair of configurations, there is a specific timing; e.g., a path of length two ($[0,0,0.785]^{\mathsf{T}};[1,1,0.785]^{\mathsf{T}}$) executed in 5 seconds (or arbitrarily at $2\frac{m}{sec}$) is a trajectory [19,89].

For autonomous navigation, it is crucial to have the ability of following a trajectory, given by a high-level reasoning behavior of the robot. Robotic trajectory following or tracking is a well-studied problem, which aims to asymptotically track a given trajectory, starting from a given configuration, which may or may not be part of the desired trajectory [89].The key idea behind methods to follow trajectories is to reduce the error between the current position and the point in the trajectory where the robot should be. In order to reduce this error, a feedback loop controller is usually implemented. The implementation of this controller depends heavily on the kinematic model of the robot as well as on the state information available, such as the current velocity, acceleration, and position of the robot.

With the purpose of illustrating this point, let us examine a robot modeled as an unicycle. It is well known that for any given configuration q_a, a model of the unicycle robot expressed in Cartesian coordinates cannot be stabilized in another given configuration q_b with a static smooth feedback controller [1,9,11], although if the model is expressed in different coordinates, e.g., polar coordinates, it is indeed possible [1,2]. Nevertheless, if we have a path planner that only outputs pairs of consecutive kinematic feasible configurations, such as the one described in Likhachev and Ferguson [61] and discussed in the previous subsection, this will enable the use of the unicycle model in Cartesian coordinates. Also, the stabilization problem does not appear if non-stationary motion is imposed, e.g., tracking a moving virtual cart [11,92].

A classical methodology for trajectory tracking is presented by Kanayama [45], which relies on following a moving virtual cart and closing the control loop with a PID controller [46]. There are more sophisticated approaches to trajectory following, such as Aicardi et al. [1] or Astolfi [2], which guarantee stability and controllability in broader conditions. Also, in Choset et al. [16] or Siegwart et al. [90], it is possible to find detailed accounts of different methods for trajectory tracking.

6.1.3 WHERE HAVE I BEEN? AND WHERE AM I?

Knowing what the world looks like is an important ability for a robot. This is known as mapping and involves the robot answering the question *Where Have I Been?* Obtaining a map of an environment allows a robot to have an operative model of the environment from which it can plan actions to fulfill intended tasks. Crucially, a map is needed to plan paths to be used in navigation.

The process of mapping involves the creation of a relation between the world and an internal representation maintained by the robot. We will discuss these internal representations in Section 6.2. The relation between the world and an internal representation can be one-to-one, as in the case of metrical mapping [27], or many-to-one, as in topological [67] or semantic mapping [86]. In any case, the process of mapping requires sensing the world and integrating this information into a coherent reference frame. Each chunk of information needs to be referenced accurately against the world; i.e., the position of the sensor with respect to the reference frame needs to be known. This requirement causes the processes of mapping and localization to be highly coupled, which is unfortunate, as the process of localization requires a map of the environment to work.

The process of localization aims at estimating the current position of the robot at any given time. This process involves the robot answering the question *Where Am I?* The localization of the robot is always given with respect to a reference frame that is attached to a map of the world. One major problem in performing localization is that a mobile robot cannot accurately measure its own position, a process that is known as dead-reckoning [46], because of the inherent noise in the proprioceptive sensors and the usually incomplete model of the kinematics of the robot. In order to overcome this problem, the position of the robot has to be inferred from external data [97], which is unfortunate, as data available to the robot stems from the map, which makes the processes of localization and mapping highly coupled.

In some scenarios, it is possible to have an external reference that via triangulation, can help to estimate the position of the robot accurately. Examples of this are global positioning systems (GPS) [18] or radio transmitters [35]. But in general, those external references are not available, sometimes because it is not possible to modify the environment to install them or just because the reference signal is not accessible, as happens to GPS in outer space or in indoor areas.

Given the highly coupled relation between mapping and localization, the vast majority of research in those topics is focused on solving the two problems in a joint manner, which is know in robotics as the problem of Simultaneous Localization and Mapping (SLAM). We will discuss this approach in Section 6.4.

6.2 ENVIRONMENT REPRESENTATIONS FOR ROBOTIC NAVIGATION

In order to move in the world, robots need a representation of the environment. This representation is needed, among other things, to solve problems such as path planning. In general, one may argue that an operative model of the environment is needed so as to fulfill intended tasks with a robot. A robot needs to know its surroundings in order to deliberate and select the best action needed at the moment. But, what is meant by an environment representation for a robot? A lucid definition borrowed from neuroscience [28] and adapted to robotics [58] is:

> A robot is said to represent an aspect of the environment when there is a functioning isomorphism between some aspect of the environment and a robot process that adapts the robot behavior to it.

Let us illustrate with a concrete example the direct relation between the robot's task and the environment emerging from this definition: if a robot needs to traverse an environment without colliding with obstacles, it needs to adapt its behavior according to the state (free or occupied) of its surroundings. Therefore, the robot needs to find a one-to-one correspondence between the state (free or occupied) of the surrounding regions of the environment and its internal representation.

In the following, we will concentrate on environment representations useful for robotic autonomous navigation. In this scenario, environment representations can be broadly partitioned into two groups:

- *Topological representations*—This type of representation encodes a list of recognizable locations of the environment by the robot. Moreover, the representation contains collision-free connections between the locations, making it possible to recover connectivity properties between the visited locations. Examples of this type of representation are given in the works of Choset [17] and Ranganathan [87].
- *Metrical representations*—This type of representation keeps a list of geometrical relationships between the objects in the environment. Usually the geometrical or metric information is stored with respect to a global reference frame, making it easy to reason about all objects in a common framework. The important idea when selecting which particular object in the environment to process is that it should help to describe in a compact manner the metrical information of the environment, so that a robot can traverse it safely. Commonly, the objects used to describe the environment are either distinguishable features, e.g., lines, corners, or cylinders, or fixed-shape geometry regions, e.g., squares.

In the remainder of this section, we will concentrate on the metrical representations detailing the two most common object implementations: feature-based maps and occupancy grid–based maps. We focus on them because it has been proven practically that they are more suitable for robotic autonomous navigation.

To conclude this part, it is worth pointing out that it is also possible to combine both representations into what is called a metric-topological representation as presented by Bosse [8] or Blanco [7]. The main advantage of this type of representation is that it breaks down the problem of updating a monolithic metrical representation in submaps, which has proven to be favorable for map building algorithms based on filtering techniques [15]. On the down side, this type of representation makes it cumbersome to plan a path globally across submaps.

6.2.1 FEATURE-BASED MAPS

Feature-based maps seek to model the environment by unique characteristics that can be extracted from the sensor data. The characteristics can be planes, lines, or points, among others. An advantage of this type of representation is that they do not occupy much storage space and can be transmitted easily between robots; this is because the features can be described very compactly. A drawback of feature-based maps is that they do not define clearly which space in the world is empty, occupied, or unknown,

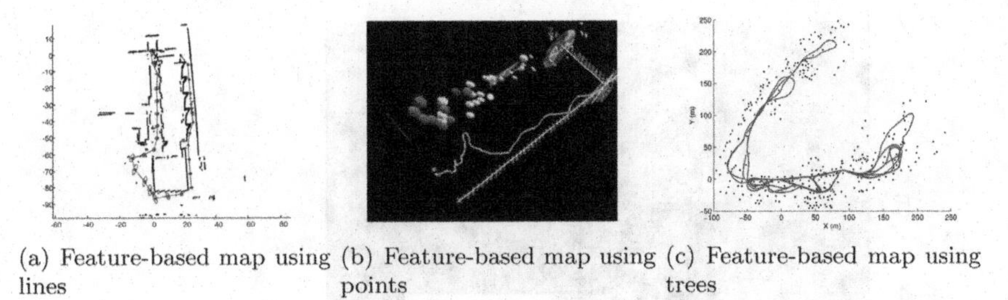

(a) Feature-based map using lines (b) Feature-based map using points (c) Feature-based map using trees

Figure 6.2 Each figure depicts an example of a feature-based map. (a) This map uses lines as features [15]. (b) This map uses points extracted from visual cues [20]. (c) This map uses tress as features. In the map, the tree's radii are visualized [34].

which is essential for robotic autonomous navigation. Also, the association between features over time is problematic, a problem known as data association. To ease the data association problem, a prior structure of the environment can be assumed, but often, this is not available.

An example of feature-based maps can be found in the work of Nguyen et al. [81], where lines, extracted from two-dimensional (2-D) light detection and ranging (LIDAR) sensors, are used as features. In Siegwart et al. [90] and Siciliano and Khatib [88], there are comprehensive tutorials about line extractions from 2-D LIDAR sensors for use in feature-based maps. Another common feature used for this type of representation is points; usually, these are extracted from visual cues or physical corners in the environment. An example of this type of map can be found in the work of Davison et al. [20], where visual cues extracted from a monocular camera are used as features. A final interesting example is given by Guivant and Nebot [34], which uses trees as features. The trees are extracted from 2-D LIDAR data and are identified using their radii. Figure 6.2 depicts examples of the feature-based maps mentioned.

6.2.2 OCCUPANCY GRID–BASED MAPS

Occupancy grid methods represent the world as a discrete grid m, which defines the map. Each cell c of the discrete map m has the same size, and there are L cells, which are denoted as $m_1, m_2, \cdots, m_L$. Each cell has a predefined geometry shape, usually square. Each cell also has an associated probability function $P(m_L)$, which describes the uncertainty of being occupied. Each cell being a random variable, the occupancy grid m is a random field [100]. In order to make inference tractable in the random field m, independence is assumed between the cells; hence,

$$P(m) = \prod_{i=1}^{L} P(m_i) \tag{6.1}$$

Now, to compute the map m, the problem narrows down to computing the probability of each individual cell. Moravec and Elfes [73] presented a probabilistic incremental approach based on a Bayesian formulation to update each cell of a grid

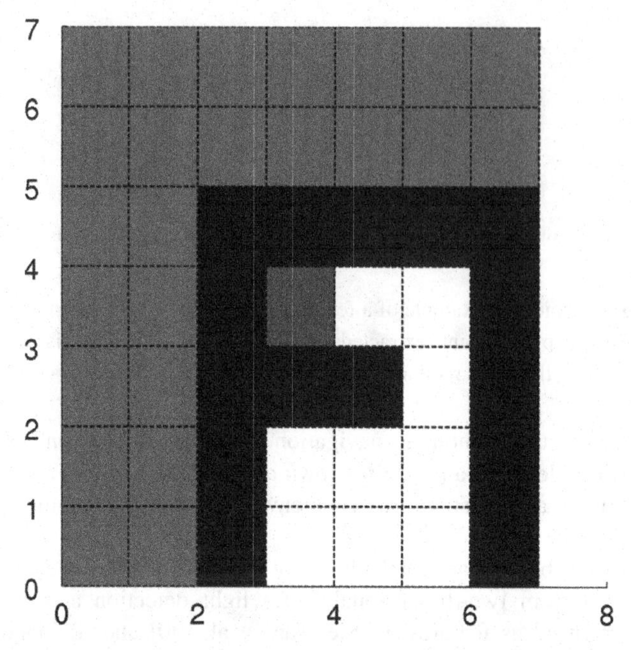

Figure 6.3 Pictorial representation of an occupancy grid. The probability of each occupancy cell is encoded in gray scale. Black cells denote high probability, i.e., they are occupied. White cells denotes low probability, i.e., they are free. Cells with probability of 0.5 are unknown, i.e., gray cells. Cells with a certain belief are darker or lighter gray; e.g., the cell at $(3,3)$ has a probability of 0.8, and hence, it is darker gray.

map m. Accessible derivations of the formulae can be found in Thrun et al. [97] or Stachniss [93]. In Figure 6.3, there is a pictorial representation of an occupancy grid representation.

Key properties about occupancy grid maps are:

- The computation of $P(m)$ is highly dependent on the type of sensor used, the quality of the measurement, and the localization of the sensor.
- It allows discrimination between free, occupied, and unknown space.
- It can occupy a lot of storage space if it is implemented naively, especially on 3-D environments. Clever computational implementation methods are available [37].

An example of 2-D occupancy grid–based maps can be found in the pioneering work of Moravec and Elfes [73,96]. In both cases, squared fixed-sized grids are used to tessellate the environment. An example of 3-D occupancy grid–based maps can be found in Hornung et al. [37]. Figure 6.4 shows examples of 2-D and 3-D occupancy grid–based maps.

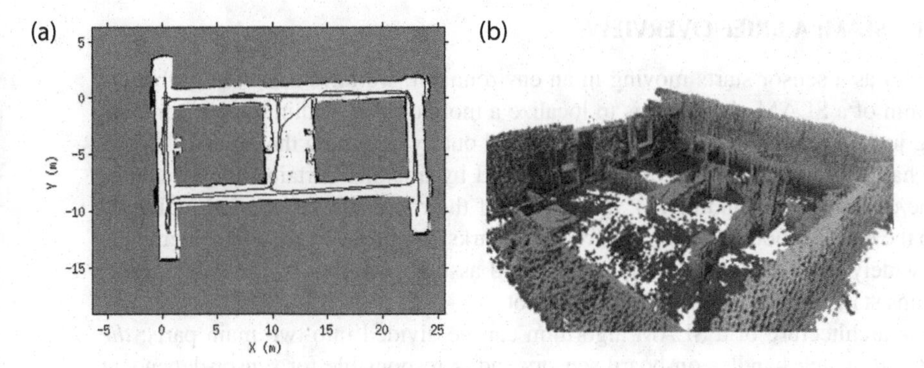

Figure 6.4 Example of occupancy grid–based maps. (a) Example of 2-D occupancy grid–based maps [73]. (b) Example of 3-D occupancy grid–based maps [37].

6.3 SIMULTANEOUS LOCALIZATION AND MAPPING – SLAM

In order to carry out various high-level tasks, a mobile robot needs the knowledge of its current location within the environment. This knowledge depends on the robot's ability to localize itself with respect to a representation of the environment (a "map"), if such a representation is available. Similarly, this environment representation can be constructed from on-board sensors if the robot location is known. Therefore, localization needs a map, and mapping needs localization. The mathematical framework for concurrently doing both of these tasks, i.e., localization and mapping, is termed "Simultaneous Localization and Mapping" (SLAM). Better localization leads to better mapping, which in turn, leads to better localization.

SLAM forms the back-bone of mobile robotics, as it is a prerequisite for higher-level tasks such as path planning and navigation. It takes the raw measurements from on-board sensors and creates a coherent representation of the environment that can then be used not only to localize the robot but to plan and navigate. SLAM has received a lot of attention in the robotics community, and many algorithms that address different aspects of the problem have been proposed over the years.

An important aspect of the SLAM problem, which differentiates it from sensor-based open loop odometry, is that of bounding the uncertainty in the robot position by detecting revisits to already explored locations, a phenomenon termed "loop closing". Closing loops correctly reduces the overall uncertainty in the estimate of the robot location and also improves the precision of the environment representation.

In this section, we start by presenting a brief overview of the different approaches that have been employed over the years to solve the SLAM problem. Next, we pose in a more mathematical manner the SLAM problem and discuss the three most common paradigms to solve the SLAM problem. We also look into the problem of loop closing and its verification, which is of paramount importance for a SLAM algorithm. Finally, we delve into the problem of pose graph reduction, which aims to keep a tractable representation of the SLAM structure over time.

6.3.1 SLAM: A BRIEF OVERVIEW

As soon as a sensor starts moving in an environment, we have a SLAM problem.[1] The aim of a SLAM algorithm is to localize a mobile robot within the environment using just the available sensors. This is carried out by assuming that the world does not change as the robot moves through it, and by tracking certain salient features of the environment (landmarks), the motion of the robot can be inferred indirectly from the relative motion of the observed landmarks. We have already encountered the first widely used assumption: the static world assumption, that is, the environment remains static during the motion of the robot.

The architecture of a SLAM algorithm can be divided into two main parts: *the front-end*, which handles on-board sensors and is responsible for sensor-dependent tasks such as data association and loop closure detection, and *the back-end*, which takes the information generated by the front-end and provides an updated map estimate using an estimation algorithm.

6.3.2 SLAM FRONT-END

The front-end is the sensor-dependent component of a SLAM algorithm and deals with various sensors to extract the information required to calculate an updated estimate of the map. It serves as a layer of abstraction between the raw sensor measurement and the estimation part of the SLAM algorithm.

Over the years, a vast number of sensors with different capabilities have been employed to give the robot a better understanding of its surroundings. The initial SLAM solution used sonar range finders that provided range and (very noisy) bearing measurements. Algorithms that extract features such as corners or lines representing flat walls [81] are a part of the front-end. The operation of the front-end is heavily sensor dependent, and the algorithms that are classified as front-end algorithms often deal with a specific sensing modality. Calculating quantities such laser odometry using a laser range finder, the incremental change in the camera position between two frames, position estimation using inertial measurement units (IMUs), fusion of multiple sensor measurements, etc., are all done at the front-end. With each new sensing capability, new front-end algorithms are developed to extract away the sensor and extract the needed measurements (such as relative motion constraints).

The first major task of the front-end is continuous data association; that is, given two consecutive sensor measurements, the front-end algorithm is tasked with finding correspondence between them. These correspondences can be point features representing landmarks in the case of monocular cameras or complete laser scans in the case of laser odometry. Correct data association is important for obtaining a consistent estimate of the map. The simple strategy of associating nearest neighbors can fail in the presence of clutter. Approaches that establish a consensus between correspondences, such as Neira and Tardós [80], have been proposed and tested successfully.

---

[1] Quotation attributed to Prof. John Leonard, MIT.

Another related task assigned to the front-end is that of loop closure detection; that is, based on the sensor readings and comparing them with all the previous measurements, the front-end has to decide when the robot has returned to an already explored region of the space. This task will be explored further in Section 6.3.8.1.

6.3.3 SLAM BACK-END

Once the front-end has generated the needed information (such as relative measurements, data association decisions, and loop closings), the back-end runs an estimation algorithm that calculates an updated estimate of the map based on this information. Estimation algorithms for SLAM can be broadly divided into two classes: a) filtering techniques that estimate the current position of the robot and all the landmarks given all the previous information. The focus of these algorithms is on calculating the current estimate of the robot position along with the observed landmarks, and b) smoothing methods that update a part or the whole of the trajectory and the map at each step.

6.3.3.1 Filtering Techniques

The first approaches to address the SLAM problem proposed to solve it as an extended Kalman filtering (EKF) estimation problem [74,75,91]. In this approach, the current robot position is updated based on the previous robot position and landmark estimates in light of the new observation of some of the landmarks, using a probabilistic framework. This approach makes certain assumptions about the problem to make it computationally tractable. The first assumption is that of Gaussianity, that is, the noise in the system has a Gaussian distribution. Second, the Markov assumption allows the use of the previous estimate of the robot pose as being a "good enough" estimate for the whole trajectory. Third, SLAM is a nonlinear problem, and the EKF approach solves a linearized version at each step of the problem. An even more restricting issue is that the computational complexity, which for a filter with a state of size n is $O(n^2)$, becomes practically unfeasible even in small environments.

Exploring ways of alleviating the problems arising from these assumptions, different approaches had been proposed that improve on the initial EKF-based solution. Particle filters [47,95] maintain multiple representation of the map in order to deal with errors arising due to linearization. Efficient implementations [33,71] also have a reduced cost of $O(nk)$ per step, where k is the number of particles maintained in the system. However, the performance of the system depends on the number of particles, and theoretically guaranteed performance is only achievable in the limiting case when the number of particles tends to infinity.

Another approach to minimize the effects of nonlinearities is the unscented Kalman filter (UKF) [68]. Rather than naively propagating the covariance matrix of the SLAM state, UKF proposes propagating a set of well-defined points, which leads to a better approximation of the covariance matrix. This keeps the filter consistent over a longer duration than the traditional EKF framework by reducing the effects of nonlinearities. A detailed introduction to filtering-based techniques can be found in Durrant-Whyte and Bailey [25] and Bailey et al. [4].

6.3.3.2 Graph-Based Approaches

While filtering-based approaches estimate the most recent robot position along with all the landmarks, smoothing-based approaches maintain and update the whole robot trajectory as well as the map at each step. Prominent among smoothing techniques is graph SLAM. The first application of the graph-based approach to the SLAM problem was proposed by Lu and Milios [66]. The basic idea behind this formulation is to pose the problem as a network of constraints, which is then solved to minimize the error in the network using a least squares formulation. As the robot moves in the environment, each robot position and observed landmark forms a "vertex" in the graph, and relative constraints between different vertices form edges. Vertices represent quantities that need to be estimated from the relative measurements (edges). The main goal of the graph-based SLAM formulation is to find a configuration of the vertices that leads to the minimum error in the graph.

A special case of the graph-based approach is the pose graph SLAM, in which the graph consists of just the robot positions and relative measurements between them. The input to the pose graph algorithm is a set of relative constraints between robot positions and if available, an initial estimate of these robot positions. The relative constraints between robot positions can come from two sources: the first are the sequential constraints that are derived from the robot's motion between two consecutive time instances, also referred to as "odometry", and the second set of constraints originates from closing loops. Loop closures are generated by algorithms that detect revisits based on sensor readings. These are non-sequential in nature, as they associate different positions in time. Without loop closing constraints, the uncertainty in the positions estimated from odometry grows unbounded. Finally, the output of the graph-SLAM algorithm is the estimated position of the robot positions as well as a measure of the confidence about the estimates (uncertainty).

The graph SLAM problem is inherently sparse due to the small number of relative observations made from each robot positions. By exploiting this sparsity, efficient solutions for solving the graph SLAM problem have been proposed [32,43,51].

6.3.4 SLAM: PROBLEM DEFINITION

SLAM algorithms permit robots to localize themselves within a map while concurrently constructing it. Typical implementations of SLAM algorithms [25,88,97] assume that the robot can move at will, i.e., it is a mobile robot, and that the robot can take measurements of the environment.

Underlying all SLAM algorithms is the estimation problem of recovering the robot position and the environmental representation of the world, given the movements commanded to the robot and the measurements taken from the environment. We could also aim to recover the history of robot positions, which has been shown to improve the estimation's result due to a better conditioning of the problem [21].

Moreover, it allows useful environmental representations for robotic navigation to be maintained as occupancy grid maps.

Assuming realistic conditions, the estimation problem at hand is nonlinear and stochastic due to the functions that govern the motion of mobile robots and the inherent noise in the acquisition of measurements from sensors. A nonlinear stochastic estimation problem can be solved using Bayesian inference [6,53]. Optimal nonlinear stochastic estimation consists in computing the conditional probability density function (`pdf`) of the state vector given all the previous information [6]. The conditional `pdf` is equivalent to an *infinite dimensional vector* [6], which renders realizations of optimal nonlinear stochastic filters non-computationally tractable [52]; therefore, approximations need to be considered.

Using the inherent probabilistic framework [83,94] to Bayesian inference, let us write the terminology of the SLAM problem, assuming the motion of the robot to be discrete and to be currently at time k :

- The history of positions is a random vector $\mathbf{X}_{0:k} = [x_0, x_1, \cdots, x_k]$, where each $x_i, \forall i \in [0, \cdots, k]$ is a random variable (RV).
- The map is a random vector $\mathbf{M} = [m_1, m_2, \cdots, m_n]$, and each component $m_i, \forall i \in [1, \cdots, n]$ is an RV.
- The history of control commands is a random vector $\mathbf{U}_{0:k} = [u_0, u_1, \cdots, u_k]$, where each $u_i, \forall i \in [0, \cdots, k]$ is an RV.
- The history of acquired measurements is a random vector $\mathbf{Z}_{0:k} = [z_0, z_1, \cdots, z_k]$, where each $z_i, \forall i \in [0, \cdots, k]$ is an RV.

Using Bayesian inference, the estimation problem of SLAM narrows down to finding out the posterior distribution of the joint density of RVs $\mathbf{X}_{0:k}$ and $\mathbf{M}$ conditioned on the history of control commands and acquired measurements from the environment [25,26,97], which is

$$P(\mathbf{X}_{0:k}, \mathbf{M} | \mathbf{U}_{0:k}, \mathbf{Z}_{0:k}) \tag{6.2}$$

Inferring the posterior of Equation 6.2 is a hard task. Some assumptions about the nature of the probabilities involved are needed in order to make the inference problem solvable [6,52]. A common assumption is to consider that the probability distribution related to the motion of the robot fulfills the Markov property [83,97], and hence [25,97]:

$$P(\mathbf{x}_k | \mathbf{x}_{k-1}, \mathbf{u}_k) \tag{6.3}$$

meaning that robot position at time k ($\mathbf{x}_k$), depends only on the preceding position ($\mathbf{x}_{k-1}$) and the control applied ($\mathbf{u}_k$). Also, it is commonly assumed that the measurements acquired are conditionally independent given the position of the robot and the map, i.e., [25,97]:

$$P(\mathbf{z}_k | \mathbf{x}_k, \mathbf{M}) \tag{6.4}$$

Applying Bayes' rule to Equation 6.2 and using the two aforementioned assumptions leads to a recursive equation, which allows us to infer $P(\mathbf{x}_k, \mathbf{M})$ as [25]:

$$P(\mathbf{x}_k, \mathbf{M} | \mathbf{U}_{0:k}, \mathbf{Z}_{0:k}, \mathbf{x}_0) = \frac{P(\mathbf{z}_k | \mathbf{x}_k, \mathbf{M}) P(\mathbf{x}_k, \mathbf{M} | \mathbf{U}_{0:k}, \mathbf{Z}_{0:k-1}, \mathbf{x}_0)}{P(\mathbf{z}_k | \mathbf{Z}_{0:k-1}, \mathbf{U}_{0:k})} \tag{6.5}$$

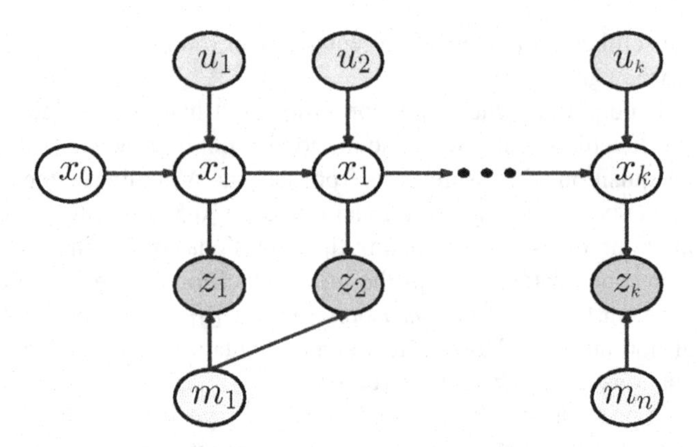

Figure 6.5 Pictorial representation of a possible dynamic Bayesian network (DBN) from a SLAM problem. The RVs of interest are $[x_0, x_1, \cdots, x_k] = \mathbf{X}_{0:k}$ and $[m_1, m_2, \cdots, m_n] = \mathbf{M}$.

with

$$P(\mathbf{x}_k, \mathbf{M}|\mathbf{U}_{0:k}, \mathbf{Z}_{0:k-1}, \mathbf{x}_0) = \sum_{i=0}^{k-1} P(\mathbf{x}_i|\mathbf{x}_{i-1}, \mathbf{u}_i) \times P(\mathbf{x}_{i-1}, \mathbf{M}|\mathbf{U}_{0:i-1}, \mathbf{Z}_{0:i-1}, \mathbf{x}_0) \quad (6.6)$$

Given the recursive nature of Equation 6.5 and the conditional independence assumption between the RVs, it is possible to pose the SLAM estimation problem as inference over a dynamic Bayesian network (DBN) [76]. In Figure 6.5, there is a pictorial representation of a possible DBN [49,77] of a SLAM problem. The gain in posing the problem as a DBN is that there are fast and robust methods to perform the inference over the desired RVs [49].

In the following, three of the main paradigms to perform the Bayesian inference over Equations 6.2 and 6.5 [88] will be described.

6.3.5 EKF-BASED SLAM ALGORITHMS

The SLAM algorithms based on the EKF usually infer Equation 6.5 and not the full posterior. This is because of the high computational complexity of estimating a big state vector with the EKF. The EKF [6], like the Kalman filter, follows the assumption that the underlying distributions in the estimation problem are Gaussian; therefore,

$$P(\mathbf{x}_k|\mathbf{x}_{k-1}, \mathbf{u}_k) \sim \mathcal{N}(f(\mathbf{x}_{k-1}, \mathbf{u}_k), \mathbf{Q}_k) \quad (6.7)$$

$$P(\mathbf{z}_k|\mathbf{x}_k, \mathbf{M}) \sim \mathcal{N}(h(\mathbf{x}_k, \mathbf{M}), \mathbf{R}_k) \quad (6.8)$$

$$P(\mathbf{x}_{k-1}, \mathbf{M}|\mathbf{U}_{0:k-1}, \mathbf{Z}_{0:k-1}, \mathbf{x}_0) \sim \mathcal{N}([\mathbf{x}_{k-1}, \mathbf{M}], \mathbf{P}_{k-1}) \quad (6.9)$$

where $f(\mathbf{x}_{k-1}, \mathbf{u}_k)$ is a nonlinear function that describe, the motion of the robot, $\mathbf{Q}_k$ is a covariance matrix that parameterizes the Gaussian distribution and models the noise in the motion, $h(\mathbf{x}_k, \mathbf{M})$ is a nonlinear function describing the observation model of the robots and $\mathbf{R}_k$ is the covariance matrix associated with the observation model that accounts for the noise in the observation process. The standard EKF can be applied using this motion and observation model, and the formulae to implement the SLAM algorithm can be found elsewhere [25,97].

The first algorithm of SLAM was based on the EKF [91] more than 25 years ago. It is by now considered a well-studied approach [25], from which it is known that it has a high computational complexity, which grows with the size of the map. Its implementation requires a feature-based map (Section 6.2.1), but more importantly, it is known that its use in large environments leads to an inconsistent estimation [4,14], which is not desirable for a SLAM algorithm.

6.3.6 RAO–BLACKWELLIZED PARTICLE FILTERS–BASED SLAM ALGORITHMS

The key idea of Rao–Blackwellized Particle Filters (RBPF)–based SLAM algorithms is to factorize the posterior distribution of the SLAM problem (Equation 6.2) into [25,33]

$$P(\mathbf{X}_{0:k},\mathbf{M}|\mathbf{U}_{0:k},\mathbf{Z}_{0:k}) = P(\mathbf{M}|\mathbf{X}_{0:k}\mathbf{U}_{0:k},\mathbf{Z}_{0:k})P(\mathbf{X}_{0:k}|\mathbf{U}_{0:k},\mathbf{Z}_{0:k}) \tag{6.10}$$

The advantage of this factorization is that $P(\mathbf{M}|\mathbf{X}_{0:k}\mathbf{U}_{0:k},\mathbf{Z}_{0:k})$ can be computed analytically if the history of positions and measurements is given [72]. Therefore, the remainder term $P(\mathbf{X}_{0:k}|\mathbf{U}_{0:k},\mathbf{Z}_{0:k})$ is now the only issue.

A posterior probability function can be approximated using a technique known as particle filters [30], which works by sampling from the distribution we seek to approximate and incrementally constructing its approximation by handpicking the samples according to their importance. The importance of a sample or weight is the ratio between the likelihood of the sample and the proposal distribution:

$$w_k^{(i)} = \frac{P(\mathbf{x}_{0:k}^{(i)}|\mathbf{Z}_{0:k},\mathbf{U}_{0:k})}{\pi(\mathbf{x}_{0:k}^{(i)}|\mathbf{Z}_{0:k},\mathbf{U}_{0:k})} \tag{6.11}$$

where $\pi(\mathbf{x}_{0:k}|\mathbf{Z}_{0:k},\mathbf{U}_{0:k})$ is the approximation being built. Particle filters techniques guarantee convergence with an infinite number of samples, but this leads to computational intractability; therefore, handpicking the samples according to their weights is important for a good approximation.

Particle filters–based SLAM algorithms are easily the most used solution to empower a mobile robot with SLAM capabilities. Among the most popular implementations are Grisetti et al. [33] and Montemerlo and Thrun [70]. Bailey [5] studied the latter approach to RBPF SLAM and reported that like the EKF-based SLAM algorithm, the particle filter solution will become inconsistent, although the consistency problem could be solved by using a high number of particles, but this leads to computational intractability.

6.3.7 GRAPH-BASED SLAM ALGORITHMS

Graphical models [77] in which nodes represent RVs of interest and the edges represent the conditional probability between the nodes allow computationally tractable methods of computing the joint distribution of RVs under consideration. Modeling the SLAM process as a DBN allows recent Bayesian inference techniques to be leveraged to estimate the posterior in Equation 6.2 [49]. The idea behind graphical models is to use the conditional independence assumption and the chain rule of probability to ease the underlying inference problem.

In Figure 6.5, there is a pictorial representation of a possible realization of a SLAM process. Using the chain rule and the conditional independence assumption, Equation 6.2 turns to be proportional to [21,26]:

$$P(\mathbf{X}_{0:k}, \mathbf{M} | \mathbf{U}_{0:k}, \mathbf{Z}_{0:k}) \propto P(\mathbf{x}_0) \prod_{i=1}^{k} P(\mathbf{x}_i | \mathbf{x}_{i-1}, \mathbf{u}_i) \prod_{i=1}^{k} P(\mathbf{z}_i | \mathbf{x}_i, \mathbf{m}_{o(i)}) \tag{6.12}$$

The relation in Equation 6.12 is the starting point of graph-based SLAM algorithms. A popular approach [31,44,88] is to pose the problem as a batch maximum likelihood estimation, in which the aim is to find the set of parameters $\{\mathbf{X}^{\star}_{0:k}, \mathbf{M}^{\star}\}$ that best explain the joint distribution in Equation 6.12, which is

$$\{\mathbf{X}^{\star}_{0:k}, \mathbf{M}^{\star}\} = \underset{\mathbf{X}_{0:k}, \mathbf{M}}{\arg\max}\, P(\mathbf{X}_{0:k}, \mathbf{M} | \mathbf{U}_{0:k}, \mathbf{Z}_{0:k}) \tag{6.13}$$

$$= \underset{\mathbf{X}_{0:k}, \mathbf{M}}{\arg\min} - \log\left(P(\mathbf{X}_{0:k}, \mathbf{M} | \mathbf{U}_{0:k}, \mathbf{Z}_{0:k})\right) \tag{6.14}$$

$$= \underset{\mathbf{X}_{0:k}, \mathbf{M}}{\arg\min} - \log\left(P(\mathbf{x}_0) \prod_{i=1}^{k} P(\mathbf{x}_i | \mathbf{x}_{i-1}, \mathbf{u}_i) \prod_{i=1}^{k} P(\mathbf{z}_i | \mathbf{x}_i, \mathbf{m}_{o(i)})\right) \tag{6.15}$$

This follows from the monotonicity of the logarithm function. If we assume, as in the EKF-based SLAM algorithms, that the motion model $P(\mathbf{x}_k | \mathbf{x}_{k-1}, \mathbf{u}_k)$ and the observation model $P(\mathbf{z}_k | \mathbf{x}_k, \mathbf{M}_{o(k)})$ are Gaussian distributed,

$$P(\mathbf{x}_k | \mathbf{x}_{k-1}, \mathbf{u}_k) \sim \mathcal{N}(f(\mathbf{x}_{k-1}, \mathbf{u}_k), \mathbf{Q}_k) \tag{6.16}$$

$$\propto \exp\left(-\frac{1}{2}\left((f(\mathbf{x}_{k-1}, \mathbf{u}_k) - \mathbf{x}_k)^{\mathsf{T}} \mathbf{Q}_k^{-1} (f(\mathbf{x}_{k-1}, \mathbf{u}_k) - \mathbf{x}_k)\right)\right) \tag{6.17}$$

$$\propto \exp\left(-\frac{1}{2} \|(f(\mathbf{x}_{k-1}, \mathbf{u}_k) - \mathbf{x}_k)\|^2_{\mathbf{Q}_k^{-1}}\right) \tag{6.18}$$

$$P(\mathbf{z}_k | \mathbf{x}_k, \mathbf{M}_{o(k)}) \sim \mathcal{N}(h(\mathbf{x}_k, \mathbf{M}_{o(k)}), \mathbf{R}_k) \tag{6.19}$$

$$\propto \exp\left(-\frac{1}{2}\left((h(\mathbf{x}_k, \mathbf{M}_{o(k)}) - \mathbf{z}_k)^{\mathsf{T}} \mathbf{R}_k^{-1} (h(\mathbf{x}_k, \mathbf{M}_{o(k)}) - \mathbf{z}_k)\right)\right) \tag{6.20}$$

$$\propto \exp\left(-\frac{1}{2} \|(h(\mathbf{x}_k, \mathbf{M}_{o(k)}) - \mathbf{z}_k)\|^2_{\mathbf{R}_k^{-1}}\right) \tag{6.21}$$

where the last step is just to conform to the standard notation in estimation literature [6]. If we replace Equations 6.18 and 6.21 in Equation 6.15 we obtain

$$\{\mathbf{X}^{\star}_{0:k}, \mathbf{M}^{\star}\} = \underset{\mathbf{X}_{0:k},\mathbf{M}}{\arg\min} \left\{ \sum_{i=1}^{k} -\frac{1}{2} \|(f(\mathbf{x}_{k-1}, \mathbf{u}_k) - \mathbf{x}_k)\|^2_{\mathbf{Q}_k^{-1}} + \sum_{i=1}^{k} -\frac{1}{2} \|(h(\mathbf{x}_k, \mathbf{M}_{o(k)}) - \mathbf{z}_k)\|^2_{\mathbf{R}_k^{-1}} \right\} \quad (6.22)$$

$$= \underset{\mathbf{X}_{0:k},\mathbf{M}}{\arg\min} F(\star) \quad (6.23)$$

which is a standard minimization problem that can be solved with a least squares approach. There are several methods that specifically handle this minimization problem for SLAM [44,51,98], which work mainly by solving the underlying nonlinear least squares problem iteratively. Regarding the consistency of the estimation, Huang [38,39] has shown that graph-based SLAM approaches based on least squares have greater consistency than EKF-based SLAM. The main reason is because of the frequent relinearization necessary to solve the nonlinear estimation problem found in the iterative procedures.

6.3.8 LOOP CLOSURE DETECTION AND VERIFICATION

6.3.8.1 Loop Closure Detection

Detecting loops is an important aspect of the SLAM problem, as it bounds the uncertainty accumulated over time due to sensor noise. The advantage of closing a loop can be seen in Figure 6.6, where incorporating the loop closure constraints leads to a consistent map estimate.

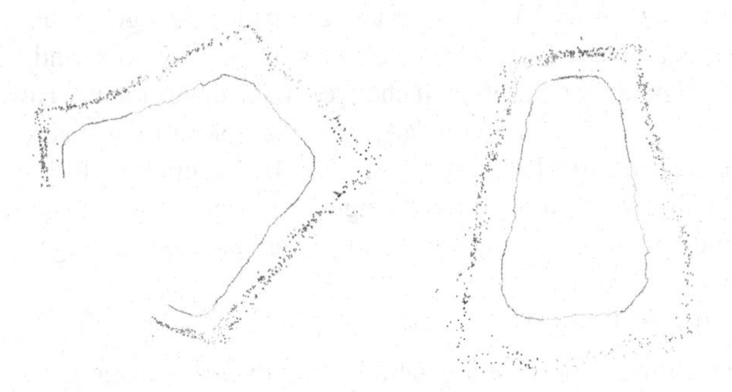

Figure 6.6 An example of updating the map estimate with a correctly detected loop. The map is constructed using a monocular camera and the loop detected at the end of the trajectory correctly aligns it with the start, leading to a consistent map estimate. (a) Before loop closure. (b) After loop closing. (Adapted from Williams, B., et al., *Robotics and Autonomous Systems (RAS)*, 2009.)

Figure 6.7 Example of correct loop closure detection (New College dataset) using the algorithm presented in Latif et al. [55].

The naive way of detecting loops is to look for previous sensor readings that match the current one; however, this approach becomes computationally expensive really quickly. Many approaches have been proposed that exploit properties of various types of sensors in order to detect when the loop has been closed; however, the underlying idea is the same: to efficiently compare the current observation with all the previous observations. An example of correct loops detected using a monocular sequence in the New College dataset can be seen in Figure 6.7.

The initial work in loop closure detection dealt with point features, and the algorithms focused on matching features in the current observation with the features already present in the map in order to detect revisits. Later, especially for the monocular camera case, the problem of loop closure detection has been posed as an image retrieval task under the assumption that the environment does not change significantly during the time that the robot makes a revisit. Loop closure algorithms extract features from images that have invariance properties, such as viewpoint and illumination invariance, which helps in dealing with changes in the environment. However, such descriptors can only provide limited invariance. Examples of these algorithms can be found in Williams et al. [101] and Latif et al. [55]. In recent years, loop closing algorithms that deal with severe illumination changes have been investigated. These algorithms aim to identify loop closures over seasonal and day-night changes [69].

6.3.8.2 Loop Closure Verification

Features extracted from images provide a generalization over the image space. This generalization, while providing certain desirable properties such as view and illumination invariance, can also lead to ambiguity. Moreover, self similarities in the environment can deceive the loop closing algorithm into falsely declaring two physically distinct places as a revisit. This is the phenomenon known as "perceptual aliasing", wherein the sensor signature coming from two different locations appears the same to the loop closing algorithm. An example of perceptual aliasing can be seen

Figure 6.8 Examples of false positive loop closure detection (KITTI dataset) using a Bag-of-words based algorithm.

in Figure 6.8. This example comes from an urban environment in which the robot moves along different streets in the city. Algorithms that work on matching low-level descriptors can easily find the same descriptors in almost all the images because of the similar structures present in them. This leads them to incorrectly label different places as being the same and hence, suggest a loop closure between them.

Perceptual aliasing has great implications for the SLAM back-end. The algorithm used by the back-end is a form of non-linear least squares optimization, in the case of the graph SLAM formulation, and thus even a single outlier can corrupt the map estimate unboundedly, an example of which can be seen in Figure 6.9. At the loop

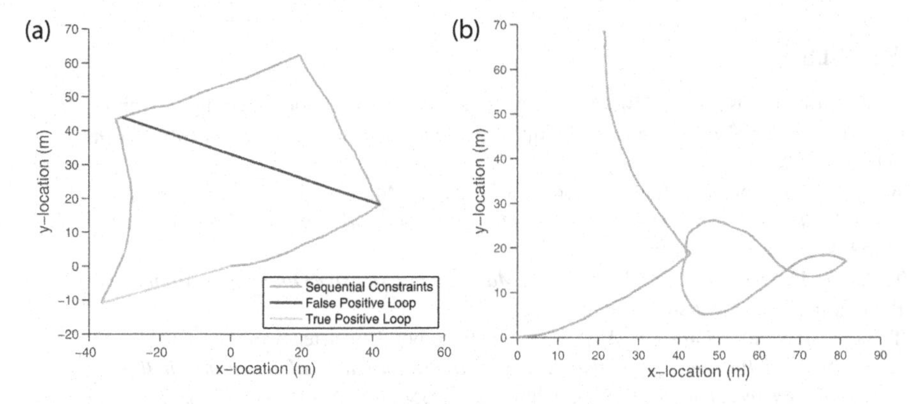

Figure 6.9 Toy example illustrating the effect of a single incorrect loop on the map estimate. **(a)** Pose graph with two loop closing hypotheses. **(b)** Optimized map using the incorrect loop closure.

closure detection step, heuristics such as temporal coherence (loop closures should happen in a sequence) and geometrical checks (similar features should occur in similar locations in the image) can also fail in highly self-similar environments. Global consistency needs to be established at the back-end, as it contains global information that can help in distinguishing between correct and incorrect loop closures. Examples of existing techniques for performing loop closure verification are Latif et al. [54] and Olson and Agarwal [82].

6.4 CONCLUSIONS AND FURTHER READING

In this chapter, we have given an overview of the robotic navigation problem. Solving this problem involves answering key issues of mobile robotics. We have also given an overview of available solutions to these key issues. In particular, we have discussed methods for endowing a robot with the ability of answering the questions: *Where Am I Going?*, *How Do I Get There?*, *Where Have I Been?*, and *Where Am I?*

Answering the first question, *Where Am I Going?*, implies carrying out a high-level process of deliberation by the robot. The aim of this process is to choose among alternative actions to fulfill an intended task. Murphy's texbook [78] gives an overview on the topic. Hertzberg and Chatilla [88] gives a comprehensive tutorial. The second question, *How Do I Get There?*, condenses an all-time problem of mobile robotics: robot motion. A classical textbook on the topic is by Latombe [56]. A more recent book is by Choset et al. [16]. The third and fourth questions, *Where Have I Been?* and *Where Am I?*, relate to the problems of mapping and localization using a robot. These two problems are interlaced, and in practice, both have to be solved together. A solution to these problems can be given by a SLAM algorithm. We have discussed what problems this type of algorithm tackles, its main assumptions, and several paradigms for its implementation. A classical text book on SLAM is by Thrun et al. [97], which covers and expands most of the topics presented in this chapter related to SLAM. A more recent survey on SLAM is by Cadena et al. [10].

REFERENCES

1. M. Aicardi, G. Casalino, A. Bicchi, and A. Balestrino. Closed Loop Steering of Unicycle like Vehicles Via Lyapunov Techniques. *Robotics and Autonomous Systems (RAS)*, 2(1):27–35, March 1995.
2. A. Astolfi. Exponential Stabilization of a Wheeled Mobile Robot Via Discontinuous Control. *Journal of Dynamic Systems, Measurement, and Control*, 121(1):121–126, March 1999. doi: 10.1115/1.2802429.
3. T. Bailey. *Mobile Robot Localisation and Mapping in Extensive Outdoor Environments*. PhD thesis, The University of Sydney, 2002.
4. T. Bailey, J. Nieto, J. Guivant, M. Stevens, and E. Nebot. Consistency of the EKF-SLAM Algorithm. In *Proceedings of the IEEE/RSJ International Conference on Intelligent Robots and Systems (IROS)*, Beijing, China, pages 3562–3568, October 2006.
5. T. Bailey, J. Nieto, and E. Nebot. Consistency of the FastSLAM Algorithm. In *Proceedings of the IEEE International Conference on Robotics and Automation (ICRA)*, Orlanado, FL, pages 424–429, May 2006.

6. Y. Bar-Shalom, T. Kirubarajan, and X.R. Li. *Estimation with Applications to Tracking and Navigation.* John Wiley & Sons, Inc., New York, NY, USA, 2002.
7. J.-L. Blanco, J.-A. Fernandez-Madrigal, and J. Gonzalez. Toward a Unified Bayesian Approach to Hybrid Metric–Topological SLAM. *IEEE Transactions on Robotics (TRO)*, 24(2):259–270, April 2008.
8. Michael Bosse, Paul Newman, John Leonard, and Seth Teller. Simultaneous Localization and Map Building in Large-Scale Cyclic Environments Using the Atlas Framework. *The International Journal of Robotics Research (IJRR)*, 23(12):1113–1139, 2004.
9. R.W. Brockett. Asymptotic Stability and Feedback Stabilization. In R.W. Brockett, R.S. Millman, and H.J. Sussmann, editors, *Differential Geometric Control Theory*, pages 181–191. Birkhauser, Boston, MA, USA, 1983.
10. C. Cadena, L. Carlone, H. Carrillo, Y. Latif, D. Scaramuzza, J. Neira, I. Reid, and J. Leonard. Past, Present, and Future of Simultaneous Localization and Mapping: Toward the Robust-Perception Age. *IEEE Transactions on Robotics (TRO)*, 32(6):1309–1332, 2016.
11. C. Canudas De Wit and O.J. Sordalen. Exponential Stabilization of Mobile Robots with Nonholonomic Constraints. *IEEE Transactions on Automatic Control*, 37(11):1791–1797, November 1992.
12. L. Carlone, Jingjing Du, M. Kaouk, B. Bona, and M. Indri. Active SLAM and Exploration with Particle Filters Using Kullback-Leibler Divergence. *Journal of Intelligent & Robotic Systems*, 75(2):291–311, October 2014.
13. H. Carrillo, P. Dames, K. Kumar, and J.A. Castellanos. Autonomous Robotic Exploration Using Occupancy Grid Maps and Graph SLAM Based on Shannon and Rényi Entropy. In *Proceedings of the IEEE International Conference on Robotics and Automation (ICRA)*, Seattle, WA, USA, May 2014.
14. J.A. Castellanos, J. Neira, and J.D. Tardós. Limits to the Consistency of EKF-based SLAM. In *5th IFAC Symposium on Intelligent Autonomous Vehicles (IAV)*, Lisbon, Portugal, 2004.
15. J.A. Castellanos, R. Martinez-Cantin, J.D. Tardós, and J. Neira. Robocentric Map Joining: Improving the Consistency of EKF-SLAM. *Robotics and Autonomous Systems (RAS)*, 55(1):21–29, 2007.
16. H. Choset, K.M. Lynch, S. Hutchinson, G.A. Kantor, W. Burgard., L.E. Kavraki, and S. Thrun. *Principles of Robot Motion: Theory, Algorithms, and Implementations.* MIT Press, Cambridge, MA, June 2005.
17. H. Choset and K. Nagatani. Topological Simultaneous Localization and Mapping (SLAM): Toward Exact Localization Without Explicit Localization. *IEEE Transactions on Robotics (TRO)*, 17(2):125–137, April 2001.
18. G. Cook. *Mobile Robots: Navigation, Control and Remote Sensing.* John Wiley and Sons, Inc., Hoboken, NJ, 2011.
19. P. Corke. *Robotics, Vision and Control: Fundamental Algorithms in MATLAB.* Springer Tracts in Advanced Robotics. Springer, Berlin, Germany, 2011.
20. A.J. Davison, I.D. Reid, N.D. Molton, and O. Stasse. MonoSLAM: Real-Time Single Camera SLAM. *IEEE Transactions on Pattern Analysis and Machine Intelligence (PAMI)*, 29(6):1052–1067, June 2007.
21. F. Dellaert and M. Kaess. Square Root SAM: Simultaneous Localization and Mapping via Square Root Information Smoothing. *The International Journal of Robotics Research (IJRR)*, 25(12):1181–1203, December 2006.
22. E.W. Dijkstra. A Note on Two Problems in Connexion with Graphs. *Numerische Mathematik*, 1(1):269–271, 1959.

23. L. Doitsidis, S. Weiss, A. Renzaglia, M.W. Achtelik, E. Kosmatopoulos, R. Siegwart, and D. Scaramuzza. Optimal Surveillance Coverage for Teams of Micro Aerial Vehicles in GPS-denied Environments Using Onboard Vision. *Autonomous Robots (AR)*, 33(1-2):173–188, August 2012.
24. G. Dudek and M. Jenkin. *Computational Principles of Mobile Robotics*. Cambridge University Press, New York, NY, USA, 2nd edition, 2010.
25. H. Durrant-Whyte and T. Bailey. Simultaneous Localization and Mapping: Part I. *IEEE Robotics and Automation Magazine*, 13(2):99–110, June 2006.
26. J.A. Fernández-Madrigal and J.L. Blanco. *Simultaneous Localization and Mapping for Mobile Robots: Introduction and Methods*. IGI Global, Hershey, PA, USA, 1st edition, 2012.
27. U. Frese, P. Larsson, and T. Duckett. A Multilevel Relaxation Algorithm for Simultaneous Localization and Mapping. *IEEE Transactions on Robotics (TRO)*, 21(2):196–207, April 2005.
28. C.R. Gallistel. *The Organization of Learning*. Bradford books. MIT Press, Cambridge, MA, 1993.
29. H.H. Gonzalez-Banos and J.C. Latombe. Navigation Strategies for Exploring Indoor Environments. *The International Journal of Robotics Research (IJRR)*, 21(10–11):829–848, October 2002.
30. N.J. Gordon, D.J. Salmond, and A.F.M. Smith. Novel Approach to Nonlinear/non-Gaussian Bayesian State Estimation. *IEE Proceedings F: Radar and Signal Processing*, 140(2):107–113, April 1993.
31. G. Grisetti, R. Kuemmerle, C. Stachniss, and W. Burgard. A Tutorial on Graph-based SLAM. *IEEE Intelligent Transportation Systems Magazine*, 2(4):31–43, January 2010.
32. G. Grisetti, R. Kümmerle, C. Stachniss, U. Frese, and C. Hertzberg. Hierarchical Optimization on Manifolds for Online 2D and 3D Mapping. In *Proceedings of the IEEE International Conference on Robotics and Automation (ICRA)*, Anchorage, AK, pages 273 –278, May 2010.
33. G. Grisetti, C. Stachniss, and W. Burgard. Improved Techniques for Grid Mapping With Rao-Blackwellized Particle Filters. *IEEE Transactions on Robotics (TRO)*, 23(1):34–46, February 2007.
34. J.E. Guivant, F.R. Masson, and E.M. Nebot. Simultaneous Localization and Map Building Using Natural Features and Absolute Information. *Robotics and Autonomous Systems (RAS)*, 40(23):79–90, 2002.
35. J.-S. Gutmann, E. Eade, P. Fong, and M.E. Munich. Vector Field SLAM: Localization by Learning the Spatial Variation of Continuous Signals. *IEEE Transactions on Robotics (TRO)*, 28(3):650–667, June 2012.
36. P.E. Hart, N.J. Nilsson, and B. Raphael. A Formal Basis for the Heuristic Determination of Minimum Cost Paths. *IEEE Transactions on Systems Science and Cybernetics*, 4(2):100–107, July 1968.
37. A. Hornung, K.M. Wurm, M. Bennewitz, C. Stachniss, and W. Burgard. OctoMap: An Efficient Probabilistic 3D Mapping Framework Based on Octrees. *Autonomous Robots (AR)*, 34(3):189–206, April 2013.
38. G. Hu, S. Huang, and G. Dissanayake. Evaluation of Pose Only SLAM. In *Proceedings of the IEEE/RSJ International Conference on Intelligent Robots and Systems (IROS)*, pages 3732–3737, Taipei, Taiwan, October 2010.
39. S. Huang, Z. Wang, G. Dissanayake, and U. Frese. Iterated D-SLAM Map Joining : Evaluating its Performance in Terms of Cconsistency, Accuracy and Efficiency. *Autonomous Robots (AR)*, 27(4):409–429, November 2009.

40. IEEE. IEE 172-1983. `http://dx.doi.org/10.1109/IEEESTD.1983.82384`. Accessed: 2014-10-15.
41. ISO. ISO 8373:2012. `https://www.iso.org/obp/ui/#iso:std:iso:8373:ed-2:v1:en:en`. Accessed: 2014-10-15.
42. M.E. Jefferies and W. Yeap, editors. *Robotics and Cognitive Approaches to Spatial Mapping*, volume 38 of *Springer Tracts in Advanced Robotics*. Springer, Berlin, Germany, 2008.
43. M. Kaess, H. Johannsson, R. Roberts, V. Ila, J. Leonard, and F. Dellaert. iSAM2: Incremental Smoothing and Mapping with Fluid Relinearization and Incremental Variable Reordering. In *Proceedings of the IEEE International Conference on Robotics and Automation (ICRA)*, Shanghai, China, May 2011.
44. M. Kaess, A. Ranganathan, and F. Dellaert. iSAM: Incremental Smoothing and Mapping. *IEEE Transactions on Robotics (TRO)*, 24(6):1365–1378, December 2008.
45. Y. Kanayama, A. Nilipour, and C.A. Lelm. A Locomotion Control Method for Autonomous Vehicles. In *Proceedings of the IEEE International Conference on Robotics and Automation (ICRA)*, pages 1315–1317, Philadelphia, PA, USA, April 1988.
46. A. Kelly. *Mobile Robotics: Mathematics, Models, and Methods*. Cambridge University Press, Cambridge, UK, 2013.
47. G. Kitagawa. Monte Carlo Filter and Smoother for Non-Gaussian Nonlinear State Space Models. *Journal of Computational and Graphical Statistics*, 5(1):1–25, 1996.
48. S. Koenig and M. Likhachev. Fast Replanning for Navigation in Unknown Terrain. *IEEE Transactions on Robotics (TRO)*, 21(3):354–363, June 2005.
49. D. Koller and N. Friedman. *Probabilistic Graphical Models: Principles and Techniques*. MIT Press, Boston, MA, USA, 2009.
50. E.B. Kosmatopoulos and A. Kouvelas. Large Scale Nonlinear Control System Fine-tuning Through Learning. *Transactions on Neural Networks*, 20(6):1009–1023, June 2009.
51. R. Kümmerle, G. Grisetti, H. Strasdat, K. Konolige, and W. Burgard. g2o: A General Framework for Graph Optimization. In *Proceedings of the IEEE International Conference on Robotics and Automation (ICRA)*, pages 3607–3613, Shanghai, China, May 2011.
52. H.J. Kushner. Approximations to Optimal Nonlinear Filters. *IEEE Transactions on Automatic Control*, 12(5):546–556, October 1967.
53. H.J. Kushner. Dynamical Equations for Optimal Nonlinear Filtering. *Journal of Differential Equations*, 3(2):179 – 190, April 1967.
54. Y. Latif, C. Cadena, and J. Neira. Robust Loop Closing over Time for Pose Graph SLAM. *The International Journal of Robotics Research (IJRR)*, 32(14):1611–1626, December 2013.
55. Y. Latif, G. Huang, J. Leonard, and J. Neira. An Online Sparsity-Cognizant Loop-Closure Algorithm for Visual Navigation. In *Proceedings of Robotics: Science and Systems Conference (RSS)*, Berkeley, USA, July 2014.
56. J.C. Latombe. *Robot Motion Planning*. Kluwer Academic Publishers, Norwell, MA, USA, 1991.
57. S.M. LaValle. *Planning Algorithms*. Cambridge University Press, New York, NY, USA, 2006.
58. D. Lee. *The Map-building and Exploration Strategies of a Simple Sonar-equipped Mobile Robot*. Cambridge University Press, New York, NY, USA, 1996.
59. D.T. Lee and III R.L. Drysdale. Generalization of Voronoi Diagrams in the Plane. *SIAM Journal on Computing*, 10(1):73–87, 1981.

60. M. Likhachev. Search-Based Planning Library. https://github.com/sbpl/sbpl. Accessed: 2014-10-15.
61. M. Likhachev and D. Ferguson. Planning Long Dynamically Feasible Maneuvers for Autonomous Vehicles. *The International Journal of Robotics Research (IJRR)*, 28(8):933–945, August 2009.
62. M. Likhachev, D. Ferguson, G. Gordon, A. Stentz, and S. Thrun. Anytime Dynamic A*: An Anytime, Replanning Algorithm. In *Proceedings of the International Conference on Automated Planning and Scheduling (ICAPS)*, Monterey, CA, June 2005.
63. M. Likhachev, D. Ferguson, G. Gordon, A. Stentz, and S. Thrun. Anytime Search in Dynamic Graphs. *Artificial Intelligence*, 172(14):1613–1643, September 2008.
64. M. Likhachev, G. Gordon, and S. Thrun. ARA* : Anytime A* with Provable Bounds on Sub-Optimality. In S. Thrun, L.K. Saul, and B. Schölkopf, editors, *Advances in Neural Information Processing Systems 16*, pages 767–774. MIT Press, Cambridge, MA, 2004.
65. T. Lozano-Pérez and M.A. Wesley. An Algorithm for Planning Collision-free Paths Among Polyhedral Obstacles. *Communications of the ACM*, 22(10):560–570, October 1979.
66. F. Lu and E. Milios. Globally Consistent Range Scan Alignment for Environment Mapping. *Autonomous Robots (AR)*, 4:333–349, 1997.
67. D. Marinakis and G. Dudek. Pure Topological Mapping in Mobile Robotics. *IEEE Transactions on Robotics (TRO)*, 26(6):1051–1064, December 2010.
68. R. Martinez-Cantin and J.A. Castellanos. Unscented SLAM for Large-Scale Outdoor Environments. In *Proceedings of the IEEE/RSJ International Conference on Intelligent Robots and Systems (IROS)*, pages 328–333, Edmonton, Alberta, Canada, 2005.
69. M. Milford. Vision-based Place Recognition: How Low Can You Go? *The International Journal of Robotics Research (IJRR)*, 32(7):766–789, 2013.
70. M. Montemerlo and S. Thrun. *FastSLAM: A Scalable Method for the Simultaneous Localization and Mapping Problem in Robotics*. Springer Tracts in Advanced Robotics. Springer, Berlin, Heidelberg, Germany, 2007.
71. M. Montemerlo, S. Thrun, D. Koller, and B. Wegbreit. FastSLAM 2.0: An Improved Particle Filtering Algorithm for Simultaneous Localization and Mapping that Provably Converges. In *International Joint Conferences on Artificial Intelligence (IJCAI)*, Acapulco, Mexico, 2003.
72. H. Moravec. Sensor Fusion in Certainty Grids for Mobile Robots. *AI Magazine*, 9(2):61–74, July 1988.
73. H.P. Moravec and A Elfes. High Resolution Maps from Wide Angle Sonar. In *Proceedings of the IEEE International Conference on Robotics and Automation (ICRA)*, St. Louis, MO, pages 116–121, March 1985.
74. P. Moutarlier and R. Chatila. Stochastic Multisensory Data Fusion for Mobile Robot Location and Environment Modeling. In *5th International Symposium on Robotics Research*, volume 1. Tokyo, Japan, 1989.
75. P. Moutarlier and R. Chatila. An Experimental System for Incremental Environment Modelling by an Autonomous Mobile Robot. In *Experimental Robotics I*, V. Hayward, and O. Khatib (eds), pages 327–346. Springer, Berlin, Heidelberg, Germany, 1990.
76. K.P. Murphy. Bayesian Map Learning in Dynamic Environments. In S.A. Solla, T.K. Leen, and K. Müller, editors, *Advances in Neural Information Processing Systems 12*, pages 1015–1021. MIT Press, Cambridge, MA, 2000.
77. K.P. Murphy. *Machine Learning: A Probabilistic Perspective*. Adaptive computation and machine learning series. MIT Press, Cambridge, MA, USA, 2012.

78. R. Murphy. *Introduction to AI Robotics*. A Bradford book. MIT Press, Cambridge, MA, 2000.
79. U. Nehmzow. *Mobile Robotics: A Practical Introduction*. Applied computing. Springer, Berlin, Heidelberg, Germany, 2000.
80. J. Neira and J.D. Tardós. Data Association in Stochastic Mapping Using the Joint Compatibility Test. *IEEE Transactions on Robotics (TRO)*, 17(6):890–897, 2001.
81. V. Nguyen, S. Gächter, A. Martinelli, N. Tomatis, and R. Siegwart. A Comparison of Line Extraction Algorithms Using 2D Range Data for Indoor Mobile Robotics. *Autonomous Robots (AR)*, 23(2):97–111, August 2007.
82. E. Olson and P. Agarwal. Inference on Networks of Mixtures for Robust Robot Mapping. *The International Journal of Robotics Research (IJRR)*, 32(7):826–840, 2013.
83. A. Papoulis and S.U. Pillai. *Probability, Random Variables, and Stochastic Processes*. McGraw-Hill Electrical and Electronic Engineering Series. McGraw-Hill, New York City, NY, USA, 2002.
84. M. Pivtoraiko and A. Kelly. Generating Near Minimal Spanning Control Sets for Constrained Motion Planning in Discrete State Spaces. In *Proceedings of the IEEE/RSJ International Conference on Intelligent Robots and Systems (IROS)*, Edmonton, Canada, pages 3231–3237, August 2005.
85. M. Pivtoraiko, R.A. Knepper, and A. Kelly. Differentially Constrained Mobile Robot Motion Planning in State Lattices. *Journal of Field Robotics (JFR)*, 26(3):308–333, March 2009.
86. A. Pronobis, O.M. Mozos, B. Caputo, and P. Jensfelt. Multi-modal Semantic Place Classification. *The International Journal of Robotics Research (IJRR)*, 29(2–3):298–320, February 2010.
87. A. Ranganathan and F. Dellaert. Online Probabilistic Topological Mapping. *The International Journal of Robotics Research (IJRR)*, 30(6):755–771, 2011.
88. B. Siciliano and O. Khatib. *Springer Handbook of Robotics*. Springer-Verlag, Secaucus, NJ, USA, 2007.
89. B. Siciliano, L. Sciavicco, L. Villani, and G. Oriolo. *Robotics: Modelling, Planning and Control*. Springer-Verlag, Secaucus, NJ, USA, 1st edition, 2008.
90. R. Siegwart, I.R. Nourbakhsh, and D. Scaramuzza. *Introduction to Autonomous Mobile Robots*. MIT Press, Boston, MA, USA, 2nd edition, 2011.
91. R. Smith and P. Cheeseman. On the Representation and Estimation of Spatial Uncertainty. *The International Journal of Robotics Research (IJRR)*, 5(4):56–68, December 1986.
92. O.J. Sordalen and C.C. De Wit. Exponential Control Law for a Mobile Robot: Extension to Path Following. *IEEE Transactions on Robotics (TRO)*, 9(6):837–842, December 1993.
93. C. Stachniss. *Robotic Mapping and Exploration*, volume 55. Springer, Berlin, Germany, 2009.
94. H. Stark and J.W. Woods. *Probability and Random Processes with Applications to Signal Processing*. Prentice Hall, Upper Saddle River, New Jersey, 2002.
95. S. Thrun. Particle Filters in Robotics. In *Proceedings of Uncertainty in AI (UAI)*, Edmonton, Canada, 2002.
96. S. Thrun. Learning Occupancy Grid Maps with Forward Sensor Models. *Autonomous Robots (AR)*, 15(2):111–127, September 2003.
97. S. Thrun, W. Burgard, and D. Fox. *Probabilistic Robotics*. MIT Press, Boston, MA, USA, 2005.

98. S. Thrun and M. Montemerlo. The Graph SLAM Algorithm with Applications to Large-Scale Mapping of Urban Structures. *The International Journal of Robotics Research (IJRR)*, 25(5–6):403–429, May 2006.
99. B. Tovar, L. Muñoz-Gómez, R. Murrieta-Cid, M. Alencastre-Miranda, R. Monroy, and S. Hutchinson. Planning Exploration Strategies for Simultaneous Localization and Mapping. *Robotics and Autonomous Systems (RAS)*, 54(4):314–331, April 2006.
100. E. Vanmarcke. *Random Fields: Analysis and Synthesis.* World Scientific, Singapore, Singapore, 2010.
101. B. Williams, M. Cummins, J. Neira, P. Newman, I. Reid, and J. Tardós. A Comparison of Loop Closing Techniques in Monocular SLAM. *Robotics and Autonomous Systems (RAS)*, 57(12):1188–1197, 2009.

7 Path Planning and Collision Avoidance

Hubert Gattringer, Andreas Müller, and Klemens Springer

CONTENTS

7.1 STATE OF THE ART

The motion planning of robotic systems was for a long time restricted to *point-to-point* (PTP) or *continuous path* (CP) motions that are geometrically described by lines and circular arcs as defined by the ISO-6983 standard and were used to program numerical control (NC) units. However, the steady increase of the computation power of industrial controllers now allows the use of advanced trajectory planning algorithms that were developed in the last decade but were, for a long time, not applicable to standard industrial systems. The main innovative advantage of these methods is the optimization of the trajectories in response to the demands from automation industry for reduction of process time, energy consumption, and the overall costs in general. At present, several approaches exist for offline time-optimal trajectory planning along predefined continuous paths in space taking into account the physical constraints of robots [25,34,37], where all mechanical components are assumed to be ideally rigid. Only a very few approaches exist that attempt to determine optimal point-to-point trajectories, i.e., to find the geometric path as well as the motion in time along this path. The main challenge to this end is that it is very difficult to find the global optimum of the associated optimization problem. To tackle this problem, the geometric path and the actual motion are optimized either simultaneously or sequentially [12].

A fundamentally different approach to reduce energy consumption and process time is the constructive reduction of the robot mass. This consequently leads to lightweight manipulators and potentially, also to a minimization of production costs. Due to the decrease in structural stiffness, such robots tend to vibrate during motion, which impairs their positioning accuracy. Computing time-optimal trajectories for such flexible multibody systems requires incorporating additional constraints within the optimization problem.

This chapter provides a summary of classical path planning methods and an overview of state of the art methods for optimal trajectory planning, in particular such that they can be implemented in current industrial controller hardware. The main differences of the solutions proposed in the literature [35,36,39] as well as a specific strategy are discussed in detail. In addition to structurally compliant robotic systems, the motion planning for kinematically redundant serial robots is described. The latter systems are gaining importance in industry due to their increased dexterity and flexibility. Yet, only a few approaches to the time-optimal motion planning for such robots have been reported in the literature [11,21,31]. The last part of this chapter is dedicated to collision avoidance and methods to incorporate the existence of obstacles within the workspace into the optimization problem.[1]

Notation

Italic letters, e.g., a, are used for scalars and bold lowercase letters for vectors, e.g., $\mathbf{b}$, while matrices are written in bold uppercase letters, e.g., $\mathbf{C}$.

The vector of minimal coordinates, also called generalized coordinates, which is a minimal set of variables that uniquely describes the configuration of a robot, is denoted $\mathbf{q} = (q_1, \ldots, q_r)^T$, where r denotes the number of minimal coordinates, i.e., the degrees of freedom (DoF). The vector $\mathbf{q} \in \mathbb{R}^r$ is referred to as the configuration (or position) of the robot. The set of all possible configuration forms the configuration space of the robot, denoted by $\mathscr{Q} \subseteq \mathbb{R}^r$. The task space is the set $\mathscr{T} \subseteq \mathbb{R}^6$, which contains all possible positions.

The position of the robot end-effector (EE) is described by ${}_I\mathbf{z}_E^T = \left({}_I\mathbf{r}_E^T, \varphi_E^T\right) \in \mathbb{R}^6$, where ${}_I\mathbf{r}_E$ is the position coordinate vector represented in a world-fixed inertial frame

[1]This work has been supported by the "LCM – K2 Center for Symbiotic Mechatronics" within the framework of the Austrian COMET-K2 program.

(indicated by the leading subscript I), and φ_E contains three angles used to describe the EE orientation. The components of ${}_I\mathbf{z}_E$ are also called world coordinates.

For a description of subscripts, we distinguish different meanings of subscripts at position vectors, rotation matrices, and other vectors.

The leading subscript 1 of the position vector ${}_1\mathbf{r}_A$ indicates the frame in which this vector is represented. The subscript A indicates that the vector is pointing from the inertial frame to the point A. The rotation matrices that transform the coordinates of a vector represented in frame j to its coordinates when represented in frame i are denoted with $\mathbf{R}_{ij}$. That is, the subscripts must be read from right to left: j is transformed to i.

Zero matrices and vectors are denoted by 0, with the dimension depending on the context. Dependency on time is suppressed whenever possible. For brevity, $\dot{(\cdot)} = \frac{\mathrm{d}}{\mathrm{d}t}(\cdot)$ is used to denote derivative w.r.t. time t, and $(\cdot)' = \frac{\mathrm{d}}{\mathrm{d}s}(\cdot)$ represents the derivative with respect to (w.r.t.) a path coordinate s.

7.2 CURVE DESCRIPTIONS

Curves can be described in implicit or parametric form. A circle, for instance, is given implicitly as

$$\{(x,y) \in \mathbb{R}^2 | x^2 + y^2 = 1\} \tag{7.1}$$

but can be described in parametric form as

$$x(t) = \cos(t),\ t \in \mathbb{R} \tag{7.2a}$$

$$y(t) = \sin(t) \tag{7.2b}$$

where the coordinates x, y depend explicitly on the independent variable t. For path planning purposes, the parametric description is used.

7.2.1 POLYNOMIALS

A univariate polynomial in the variable t can be written as a (finite) sum of the form

$$x(t) = \sum_{i=0}^{n} d_i t^i \tag{7.3}$$

with (real) coefficients $d_i \in \mathbb{R}$ and $\forall i \in \mathbb{N} \cup \{0\}$. It can be written in matrix form as

$$x(t) = \begin{pmatrix} t^0 & t^1 & \cdots & t^n \end{pmatrix} \begin{pmatrix} d_0\ d_1\ \cdots\ d_n \end{pmatrix}^T \tag{7.4}$$

The polynomial $x(t)$ can be efficiently evaluated using the *Horner scheme* [1]. An accurate interpolation or approximation of a given curve usually requires using higher-order polynomials. The latter tend to oscillate, in particular at the domain boundaries. This is overcome using piecewise-defined polynomials, known as *splines*.

7.2.2 NON-UNIFORM RATIONAL BASIS SPLINES (NURBS)

The term *spline* originates from the field of naval architecture, where thin flexible wooden or metal laths (called splines) were used to define the shape of a ship. The first known reference to smooth, piecewise-defined polynomials as splines was in 1946 by I.J. Schoenberg [33]. Fundamental work on splines was carried out by de Casteljau, Bézier, de Boor, and others in the 1950s and 1960s. Comprehensive literature and standard references are found in de Boor [7], Piegl [27] and Piegl and Tiller [28]. In a survey paper of Piegl in 1991, the roots of NURBS were comprehensively reviewed:

> On the basis of theoretical works by Schoenberg, de Boor, Cox and Mansfield, Riesenfeld introduced B-Spline curves and surfaces .... Following Riesenfeld, Versprille extended B-Splines to rational B-Splines ([38]) ... the first written account of NURBS.
>
> [27]

A NURBS curve of degree m is defined by

$$x(t) = \sum_{i=0}^{n} R_{i,m}(t) d_i = \underbrace{\begin{pmatrix} R_{0,m}(t) & R_{1,m}(t) & \dots & R_{n,m}(t) \end{pmatrix}}_{\mathbf{r}(t)} \underbrace{\begin{pmatrix} d_0 \\ d_1 \\ \vdots \\ d_n \end{pmatrix}}_{\mathbf{d}} \tag{7.5}$$

where the $n+1$ rational basis functions of degree m

$$R_{i,m}(t) = \frac{N_{i,m}(t) w_i}{\sum_{j=0}^{n} N_{j,m}(t) w_j} \tag{7.6}$$

are defined by the weights w_i and the B-Spline basis functions $N_{i,m}(t)$, which are defined recursively via

$$N_{i,m}(t) = \frac{t - t_i}{t_{i+m} - t_i} N_{i,m-1}(t) + \frac{t_{i+m+1} - t}{t_{i+m+1} - t_{i+1}} N_{i+1,m-1}(t) \tag{7.7a}$$

$$N_{i,0}(t) = \begin{cases} 1 & t \in [t_i, t_{i+1}) \\ 0 & \text{otherwise} \end{cases} \tag{7.7b}$$

The knot vector $\mathbf{t} = (t_0, \dots, t_{n+m+1})$ comprises the monotonously increasing knots t_i. If a denominator is zero, the basis function is set to zero. The control points d_i are summarized in the so-called control vector $\mathbf{d}$.

The resulting curve $x(t)$ is $m - f$ times continuously differentiable, where f is the knot multiplicity. A very important property of NURBS is that the curve stays in the union of the local *convex hulls* formed by $m+1$ control points; see Rockafellar [32] for an explanation of convexity. Another important feature of NURBS is that the sum of basis functions $N_{i,m}(t)$ is unity for a particular value t, which is referred to as the *partition of unity*. Equations 7.7 show that a basis function $N_{i,m}(t)$ is only defined in a small interval and hence, is said to have a *local support*. This is a beneficial property of B-Splines and thus, of NURBS. Consequently, the change of a knot point only locally affects the NURBS curve $x(t)$, while the curve remains unchanged outside the local support of the corresponding basis function. This property is advantageous for manipulating trajectories within optimization procedures. An efficient algorithm for evaluating a NURBS curve is the *Cox–de Boor* algorithm [7,19].

If the weights $w_i = w_j = w$ in Equation 7.6 are constant, such that $R_{i,m}(t) = N_{i,m}(t)$, one obtains non-uniform non-rational B-Spline curves of the form

$$x(t) = \sum_{i=0}^{n} N_{i,m}(t) d_i = \mathbf{n}(t)\mathbf{d} \tag{7.8}$$

with $\mathbf{n} = (N_{0,m}, \ldots, N_{n,m})$.

7.3 GEOMETRIC PATH PLANNING

The parameterization just described allows adaptation of the curves so to achieve a motion with desired properties. The motion planning problem is frequently divided into the *geometric path planning* in task or configuration space and the *trajectory planning*. The former (also called static path planning) determines ${}_I\mathbf{z}_E(s)$ or $\mathbf{q}(s)$, respectively, as a function of a scalar coordinate s (the path parameter), while the latter (also called dynamic path planning) determines the actual motion along these paths in time, i.e., ${}_I\mathbf{z}_E(t)$ or $\mathbf{q}(t)$ [4]. Since most tasks are defined in task space, it seems advantageous to define the path in task space $\mathscr{T}$. However, deciding whether the path should be defined in task space $\mathscr{T}$ or in configuration space $\mathscr{Q}$ should take into account the aspects listed in Table 7.1.

In general, robot motion tasks fall into two categories: point-to-point (PTP) path and continuous path (CP) planning.

7.3.1 CONTINUOUS PATHS

Continuous Paths (CP) are generally used for most industrial tasks, e.g., welding, grinding, polishing, etc. In addition, obstacle and collision avoidance can be included in this path planning stage. In the following, only decoupled movements of position and orientation are considered. At first, the definition of the end-effector position ${}_I\mathbf{r}_E$ of the geometrical path using geometric primitives is shown. A more general strategy is to define the full path using polynomials or splines such as NURBS, as introduced in Section 7.2.2. In contrast to using geometric primitives, this allows arbitrary curves to be defined.

Table 7.1
Comparison of advantages and disadvantages when describing the path in task space or in joint space

	Joint Space $\mathscr{Q}$	Task Space $\mathscr{T}$
Requires inverse kinematics	no	yes
Prone to kinematic singularities	no	yes
Joint limits easily included	yes	no
Easy handling of obstacles	no[a]	yes[b]

[a]The configuration of the robot, and thus its geometric position, can be easily described in terms of the joint coordinates $\mathbf{q}$.
[b]But the detection of collisions with spatial obstacles leads to non-linear constraints in terms of $\mathbf{q}$. Since the EE position ${}_I\mathbf{z}_E$ is prescribed, detecting collisions of EE and obstacles is simple. But when collisions of all links of the robot need to be detected, the inverse kinematics must be solved, and the collision avoidance also leads to non-linear constraints in $\mathbf{q}$.

7.3.1.1 Geometric Primitives

One of the most straightforward ways is the concatenation of so-called elementary paths or path primitives in Cartesian space—e.g. lines and circular arcs:

$$\text{Line:} \qquad {}_I\mathbf{r}_E(s) = {}_I\mathbf{r}_s + s\left({}_I\mathbf{r}_e - {}_I\mathbf{r}_s\right), \qquad s \in [0,1] \tag{7.9a}$$

$$\text{Circular arc:} \qquad {}_I\mathbf{r}_E(s) = {}_I\mathbf{r}_s + \begin{pmatrix} r - r\cos(\frac{s}{r}) \\ r\sin(\frac{s}{r}) \\ 0 \end{pmatrix}, \qquad s \in [0, s_{arc}] \tag{7.9b}$$

with the radius r and arc length s_{arc}. Without loss of generality, the paths in Equations 7.9 are defined in the inertial frame I with the start and end position ${}_I\mathbf{r}_s, {}_I\mathbf{r}_e \in \mathbb{R}^3$. The main drawback of such a concatenation is the discontinuity of curvature at the points where the segments (line and circular arc) meet. In order to cope with this problem, one can use *Euler spirals* at the transition point (also known as clothoids).

7.3.1.2 Clothoids

The curvature κ of clothoid curves depends linearly on the curve length, and these were first introduced in 1744 by Leonhard Euler. For the combination of the primitive line and circular arc, see Figure 7.1; a short part of the arc is replaced by clothoids, so that the discontinuity is avoided.

Remark: Note that the resulting curve slightly differs from the direct combination of line and circular arc.

The derivation of the Euler spirals is conducted in the plane, since a subsequent spatial transformation allows a general representation. Clothoids are defined to have

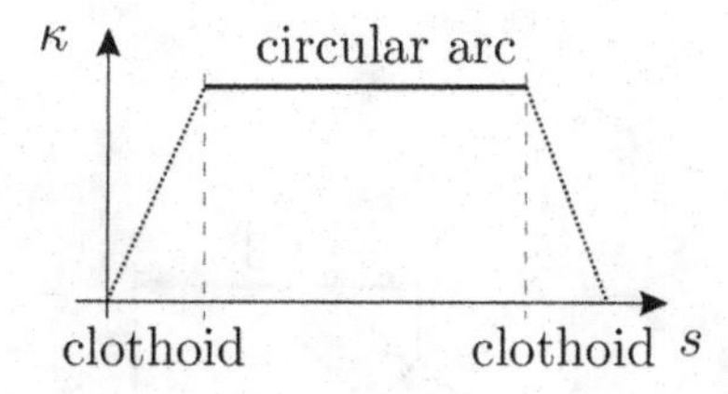

Figure 7.1 Continuous curvature when combining circular arc and clothoids.

a radius of curvature r_{cl} proportional to their length $s_{cl} = a^2/r_{cl}$ with the characteristic parameter a. Based on that, the curves are derived as a function of well-known Fresnel integrals

$$\begin{pmatrix} x_{cl}(s) \\ y_{cl}(s) \end{pmatrix} = a\sqrt{\pi}\begin{pmatrix} C(\frac{s}{a\sqrt{\pi}}) \\ S(\frac{s}{a\sqrt{\pi}}) \end{pmatrix} \tag{7.10}$$

with $s \in [0, s_{cl}]$ and

$$C(x) = \int_0^x \cos\left(\frac{\pi\xi^2}{2}\right) d\xi \tag{7.11}$$

$$S(x) = \int_0^x \sin\left(\frac{\pi\xi^2}{2}\right) d\xi \tag{7.12}$$

Figure 7.2 shows a combination of clothoid and circular arc with an arbitrary opening angle α. The deviation from the resulting curve to the reference circle strongly depends on the clothoid's opening angle ϕ_{cl}. Hence, a trade-off between a sufficient rate of change of curvature and the deviation of the curve has to be made.

By evaluating the projected lengths x_1, x_2, and x_3 in Figure 7.2, the unknown clothoid parameter a is obtained. It depends on the reference radius r, the clothoid's opening angle ϕ_{cl}, as well as the overall opening angle α. The length of the combination *clothoid–circular arc–clothoid* is composed of

$$s_{comb} = 2\underbrace{\left(a\sqrt{2\phi_{cl}}\right)}_{s_{cl}} + \underbrace{r_{arc}(\alpha - 2\phi_{cl})}_{s_{arc}} \tag{7.13}$$

7.3.1.3 Orientation Parametrization

The complete description of the EE motion ${}_I\mathbf{z}_E(s)$ further requires describing the orientation. The rotation matrix that transforms from the end to the start configuration of the EE when rotating about a constant axis, for instance, can be described using the angle-axis parameterization as $\mathbf{R}_{se}(s) = \mathbf{R}_{se}(\mathbf{u}, s\alpha)$. Therein, $\mathbf{u}$ is a unit vector along the axis of rotation, α is the corresponding rotation angle, and $s \in [0,1]$. For a general spatial rotation, the axis and angle can be described by motion primitives or NURBS. It is important to ensure continuity for the angular velocity and possibly, its time derivatives [22].

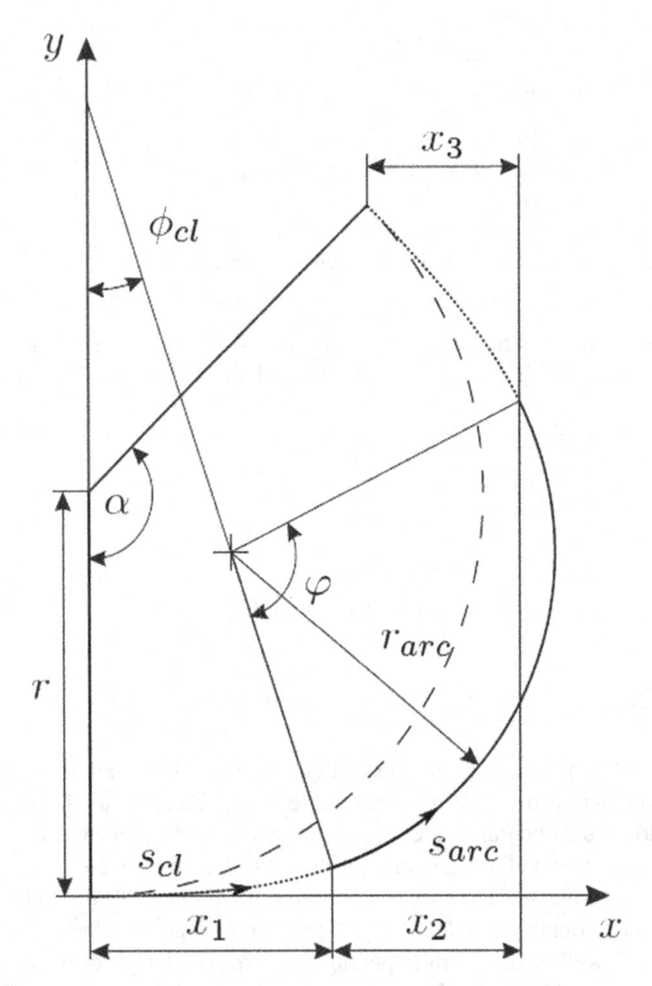

Figure 7.2 An arbitrary combination of clothoid and circular arc. Dashed line: reference curve.

In summary, the complete EE motion is parametrized and can be used for the geometric path planning. A comprehensive introduction to the use of motion primitives can be found in Biagiotti and Melchiorri [4].

7.3.2 POINT-TO-POINT PATHS

As the most general formulation, PTP motion is only defined by a start and end position in configuration or task space: $\mathbf{q}_s$ and $\mathbf{q}_e$ or $\mathbf{z}_s$ and $\mathbf{z}_e$, respectively. Hence, the path between the two points is not defined a priori. This problem can be solved:

- By directly moving in joint space without a defined intermediate path.
- By using geometrical paths to connect two positions, e.g., a line.

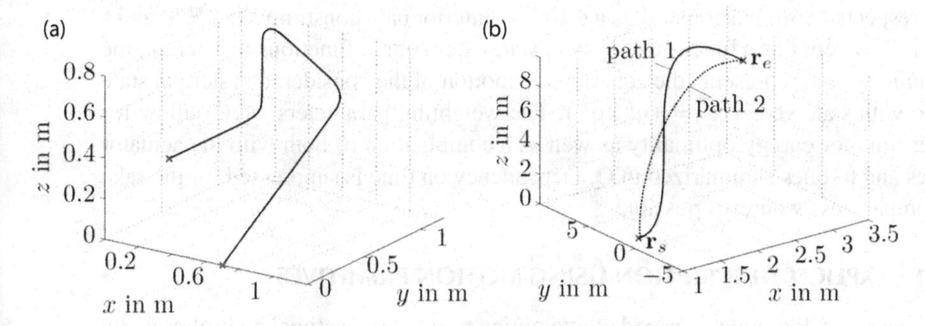

Figure 7.3 Comparison between CPs and PTP paths. (a) Example of a CP motion composed of line, circular arc, and clothoid primitives. (b) Example of two possible PTP paths with arbitrary curves between start and end.

- By solving the path implicitly within an optimal trajectory planning problem based on some optimality measure and constraints. This may be minimum-length, collision avoidance, etc.

Figure 7.3 shows a comparison between CP and PTP motions.

7.4 DYNAMIC PATH PLANNING

In the second stage of the motion planning problem, a *motion law*, which relates a geometric path with time information, is sought to obtain the actual trajectory. As described in Section 7.3 for CP motions, the movement on a geometric path can be described by a single path coordinate s. Hence, in this case, the motion law is represented by an appropriate time-dependent function $s = s(t) \in \mathbb{R}, t \in [0, t_e]$, which implicitly defines the trajectory ${}_I\mathbf{z}_E(s(t))$ or $\mathbf{q}(s(t))$. In contrast, for PTP motions, one has to provide several functions of time ${}_I\mathbf{z}_E(t) \in \mathbb{R}^6, t \in [0, t_e]$ or $\mathbf{q}(t) \in \mathbb{R}^r, t \in [0, t_e]$ and properly ensure synchronization.

In order to obtain a trajectory that is optimal in a certain sense, optimization theory is applied to finding (an) appropriate function(s) solving the optimal control problem. For details on the theory of optimal control problems, the reader is referred to Nocedal and Wright [23] and Betts [3]. In robotic applications, the objective is usually to achieve time- or energy-optimal trajectories that comply with technological restrictions. In the subsequent sections, *continuous-path time-optimal motions* (CPTOM) and *point-to-point time-optimal motions* (PTPTOM) will be addressed. Generally, such an optimal control problem is formulated as

$$\min_{t_e \in \mathbb{R}, \mathbf{u}(\cdot)} \quad \int_0^{t_e} \left[w_1 + w_2 \mathbf{Q}(\mathbf{x}(t), \mathbf{u}(t))^T \mathbf{Q}(\mathbf{x}(t), \mathbf{u}(t)) \right] \mathrm{d}t \tag{7.14a}$$

$$\text{s.t.} \quad \dot{\mathbf{x}}(t) = \mathbf{f}(\mathbf{x}(t), \mathbf{u}(t)), \; \mathbf{x}(0) = \mathbf{x}_s \tag{7.14b}$$

$$\mathbf{g}(\mathbf{x}(t_e)) = 0 \tag{7.14c}$$

$$\mathbf{h}(\mathbf{x}(t), \mathbf{u}(t)) \leq 0 \tag{7.14d}$$

$$\forall t \in [0, t_e] \tag{7.14e}$$

with respect to terminal constraints $\mathbf{g} \in \mathbb{R}^{2r}$, n_c interior path constraints $\mathbf{h} \in \mathbb{R}^{n_c}$, and a dynamic system $\mathbf{f}$ as a function of system states $\mathbf{x}$ and input functions $\mathbf{u}$. Therein, the dynamic system represents the equations of motion of the considered system in state space with state vector $\mathbf{x}^T = \left(\mathbf{q}^T, \dot{\mathbf{q}}^T\right)$. The weighting parameters w_1, w_2 allow for either time- or energy-optimality as well as a combination of both with the actuator forces and torques summarized in $\mathbf{Q}$. Dependency on time is suppressed for the sake of compactness wherever possible.

7.4.1 EXPLICIT DESCRIPTION USING MOTION PRIMITIVES

In the case of CP motions, instead of attempting to solve the optimal control problem (Equation 7.14), the most straightforward way is to define a motion law

$$s(t) = \beta t \tag{7.15}$$

with a constant time scaling factor β. In general, a more appropriate function in regard to the requirement of a given task as well as of different characteristics is chosen, such as joint (velocity/acceleration/jerk) constraints, trajectory duration, or continuity. Triangular and trapezoidal profiles as well as $\sin^2(t)$ functions for velocity or acceleration are common choices. More advanced laws are based on polynomials or splines.

In the case of PTP movements (e.g., $\mathbf{q}(t) = \mathbf{q}_s + (\mathbf{q}_e - \mathbf{q}_s)s(t)$), the motion law describes the trajectory, e.g., on acceleration level

$$\ddot{s}_i(t) = \begin{cases} \ddot{s}_{i,max} \sin^2(\omega_i t) & t \in [0, t_e/2[\\ -\ddot{s}_{i,max} \sin^2(\omega_i t) & t \in [t_e/2, t_e] \\ 0 & \text{otherwise} \end{cases}, \qquad i = 1, \ldots, r \tag{7.16}$$

with amplitudes and frequencies $\ddot{s}_{i,max}$ and $\omega_i(t_e)$, respectively.

It must be observed that an increase of the degree of continuity (as required for elastic robots) leads to an increase of the trajectory duration t_e. A comparison of different trajectories concerning end time and continuity but equal maximum acceleration is shown in Figure 7.4. For utilizing all required characteristics in the best possible way, optimal trajectories are discussed next.

7.4.2 CONTINUOUS PATH TIME-OPTIMAL MOTION

For CPTOM, the optimization problem can be formulated in different ways. Nevertheless, the path is always described in either $\mathscr{Q}$ or $\mathscr{T}$ as a function of the coordinate s, and an optimal solution $s^*(t)$ is sought [5,6,25,34,37].

7.4.2.1 Dynamic Programming and Numerical Integration

Objective function: The integration limit in the objective function in Equation 7.14a depends on the unknown end time t_e:

$$\int_0^{t_e} \left[w_1 + w_2 \mathbf{Q}^T \mathbf{Q}\right] \mathrm{d}t \tag{7.17}$$

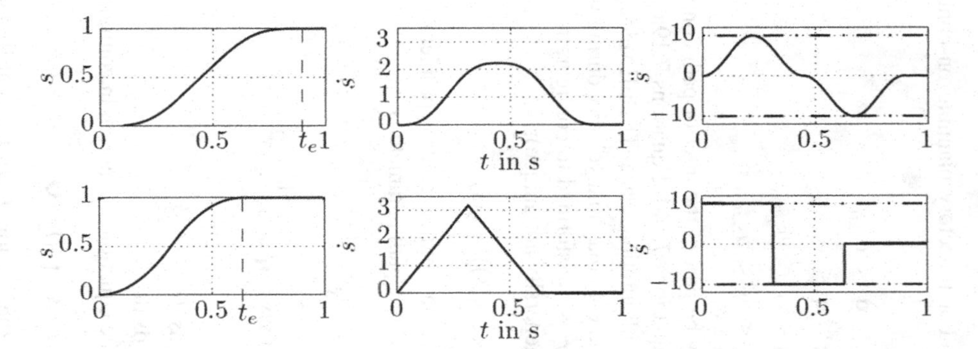

Figure 7.4 Comparison between a trajectory generated with a $\sin^2(t)$ function for acceleration (a) and a trajectory generated with a triangular profile for velocity (b) with end time t_e. Dotdashed lines: maximum acceleration.

For solving the optimization problem, it is advantageous to have a fixed integration limit. This is obtained by a transcription of time to the path coordinate $\dot{s} = \mathrm{d}s/\mathrm{d}t$, and thus $\mathrm{d}t = 1/\dot{s}\,\mathrm{d}s$, yielding

$$\int_0^{s_e} \frac{1}{\dot{s}}\left[w_1 + w_2\mathbf{Q}^T\mathbf{Q}\right]\mathrm{d}s \tag{7.18}$$

Thus, the end time is computed as a result of the optimization.

Velocity/Acceleration Constraints: For trajectory planning on continuous paths, velocity and acceleration constraints

$$\dot{\mathbf{q}}_{min} \leq \dot{\mathbf{q}} \leq \dot{\mathbf{q}}_{max} \tag{7.19a}$$

$$\ddot{\mathbf{q}}_{min} \leq \ddot{\mathbf{q}} \leq \ddot{\mathbf{q}}_{max} \tag{7.19b}$$

$$v_{E,min} \leq v_E \leq v_{E,max} \tag{7.19c}$$

$$a_{E,min} \leq a_E \leq a_{E,max} \tag{7.19d}$$

are of particular importance and may be part of the interior path constraints $\mathbf{h}(\mathbf{s}, u)$ in the optimal control problem in Equations 7.14d. Equations 7.19a and 7.19b denote drive-dependent quantities in configuration space. The EE velocity v_E and EE acceleration a_E are process dependent, e.g., maximal velocity during welding.

The path in world coordinates ${}_I\mathbf{z}_E$ is transformed to configuration space via inverse kinematics (notice that this is generally not unique):

$$\mathbf{q} = \mathbf{f}_{IK}(\mathbf{z}_E(s)) = \mathbf{q}(s) \tag{7.20}$$

In order to include the constraints in an optimal control problem such as Equation 7.14, they are rewritten in terms of the path parameter s:

$$\dot{\mathbf{q}} = \frac{\mathrm{d}\mathbf{q}}{\mathrm{d}s}\dot{s} = \mathbf{q}'\dot{s} \tag{7.21}$$

$$\ddot{\mathbf{q}} = \frac{\mathrm{d}}{\mathrm{d}s}(\mathbf{q}'\dot{s})\dot{s} = \mathbf{q}''\dot{s}^2 + \frac{1}{2}\mathbf{q}'(\dot{s}^2)' \tag{7.22}$$

using the identity

$$\ddot{s} = \frac{\mathrm{d}\dot{s}}{\mathrm{d}s}\frac{\mathrm{d}s}{\mathrm{d}t} = \frac{(\dot{s}^2)'}{2} \tag{7.23}$$

Torque Constraints: Starting from the equations of motion in configuration space,

$$\mathbf{M}(\mathbf{q})\ddot{\mathbf{q}} + \mathbf{g}(\mathbf{q}, \dot{\mathbf{q}}) = \mathbf{Q} \tag{7.24}$$

with the mass matrix $\mathbf{M} \in \mathbb{R}^{r \times r}$ and the friction, Coriolis, centrifugal, and gravity terms in $\mathbf{g} \in \mathbb{R}^r$, we rewrite them in terms of the path parameter s as

$$\mathbf{A}(s)(\dot{s}^2)' + \mathbf{B}(s)(\dot{s}^2) + \mathbf{C}(s) = \mathbf{Q} \tag{7.25}$$

The right-hand side vector $\mathbf{Q}$ denotes the actuator torques. Assuming a fully actuated system, the equation for the kth coordinate becomes with $z = (\dot{s}^2)$

$$A_k z' + B_k z + C_k = Q_k, \quad k = 1, \ldots, r \tag{7.26}$$

There are various advantages of this reformulation in terms of the new coordinate z, such as:

- The equations of motion are linear in $\dot{s}^2$. Hence, a geometric evaluation in a modified phase plane $[(\dot{s}^2)', \dot{s}^2]$ and in the space $[(\dot{s}^2)', \dot{s}^2, s]$ is possible.
- The structure of Equation 7.26 is perfectly suited for optimization, since it is a one-dimensional problem in $z = \dot{s}^2(s)$.

Feasible Robot Motions: For the calculation of possible robot motions respecting the constraints given by the optimization problem, we start with the idea of evaluating Equation 7.26 as a discretized linear equation in the plane $(\dot{s}^2)' = f(\dot{s}^2)$. Therefore, the coefficients $A_k(s)$, $B_k(s)$, and $C_k(s)$ are obtained for the entire path $s(t), t \in [0, t_e]$. Subsequently, the kth constraint

$$-Q_{k,max} \leq A_k(s)z' + B_k(s)z + C_k(s) \leq Q_{k,max}, \quad k = 1, \ldots, r \tag{7.27}$$

represents two parallel straight lines in the $z - z'$ plane. Each torque constraint limits the feasible area in this plane to another stripe between such two straight lines. Adding velocity constraints

$$z_{vc} = \dot{s}_{vc}^2 = \min_k \left\{ \frac{\dot{q}_{k,max}^2}{q_k'^2}, \frac{v_{max}^2}{|\mathbf{r}_E'|^2} \middle| k = 1, \ldots, r \right\} \tag{7.28}$$

by squaring and combining of $-\dot{q}_{k,max} \leq q_k' \dot{s} \leq \dot{q}_{k,max}$ and $-v_{max} \leq |\mathbf{r}_E'| \dot{s} \leq v_{max}$ results in a polytope, describing the feasible motion in the z'-z plane. An exemplary feasible polytope is depicted in Figure 7.5. For each point in time $s(t), t \in [0, t_e]$, a different polytope exists, such that in the $z' - z - s$ space, a polyhedron of feasible robot motions results (Figure 7.6a). The projection of the maximum velocity z_{max} onto the z–s plane yields the limit curve z_{limit} (Figure 7.6b).

Remark: For the derivation of the limit curve, we assumed a linear dependence of z. If viscous friction is not neglected, this is not the case any longer. However, feasible robot motions can still be calculated, but they do not lead to a polyhedron in general.

Dynamic Programming: With dynamic programming, a powerful optimization method was developed in the mid-20th century by R.E. Bellman [2] for solving optimal control problems. The methodology is based on the *principle of optimality* postulated by Bellman, which states that [2]

> An optimal policy has the property that whatever the initial state and initial decision are, the remaining decisions must constitute an optimal policy with regard to the state resulting from the first decision.

This means that the optimal solution starting from a specific point only depends on decisions from this point to the end. This fact justifies pursuing a reverse optimization starting at the end of the path. The limit curve z_{limit} in Figure 7.6b constitutes the base

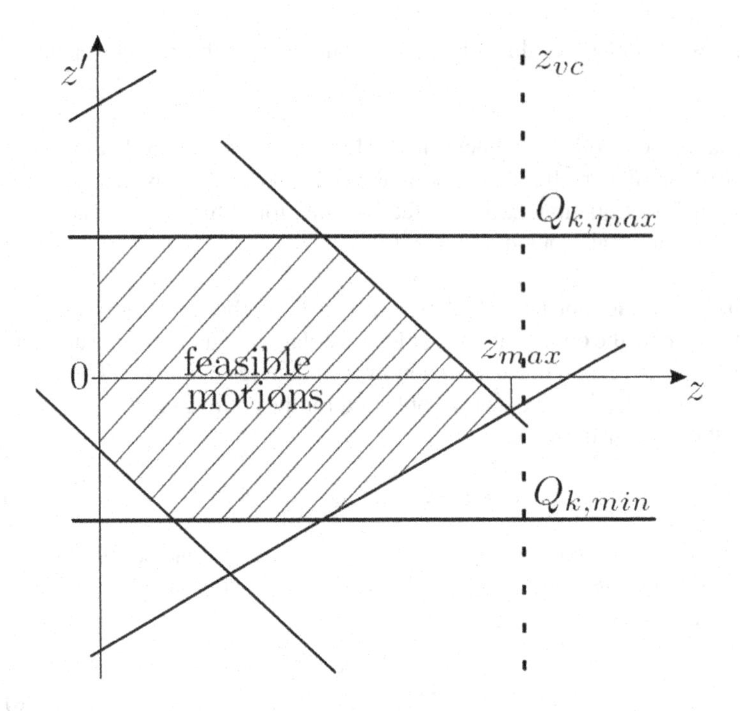

Figure 7.5 Area of feasible robot motions.

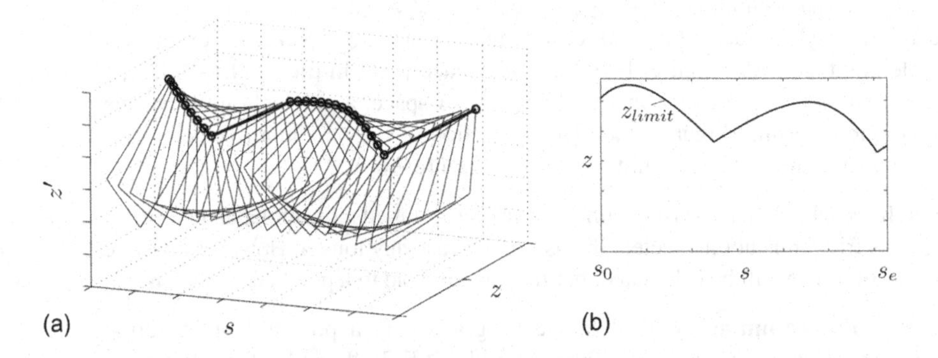

Figure 7.6 (a) Polyhedron of feasible robot motions and (b) typical limit curve.

for the optimization. For more details on solving optimal control problems using dynamic programming, the reader is referred to Pfeiffer [24].

Numerical Integration: A particular technique for obtaining time-optimal motions only is the method of numerical integration [5,26], which is based on the fact that at each point of a time-optimal trajectory, the velocities have to be at their maximum with respect to the given constraints. Hence, the resulting trajectory goes along curves with minimum or maximum slope—called extremals—or at the limit curve z_{limit}. The field of extremals is found by evaluating Equation 7.26 numerically for different initial conditions for s_0 and Q_k, e.g., using *Euler* integration. Based on that, the time-optimal trajectory for the path parameter is obtained by finding optimal switching points for acceleration and deceleration in the phase plane (forward and backward integration).

7.4.2.2 Numerical Solution of the Optimal Control Problem

For higher-order dynamic systems and a high number of interior path constraints n_c, these methods rapidly increase in complexity and suffer from long computation times. Hence, direct or indirect methods may be better suited for solving such optimal control problems. Most often, direct optimization methods are used, which first discretize the optimal control problem and subsequently solve the finite dimensional optimization problem using, e.g., Multiple Shooting [20] in connection with sequential quadratic programming (SQP) based on Interior-Point or Active Set techniques for inequality constraint handling. Some powerful optimization software packages for optimal control problems are:

- ACADO: **A**utomatic **C**ontrol **A**nd **D**ynamic **O**ptimization
  `www.acadotoolkit.org`
- MUSCOD-II: **MU**ltiple **S**hooting **COD**e for Optimization II
  `www.proxy.iwr.uni-heidelberg.de/~agbock/RESEARCH/muscod.php`
- IPOPT: **I**nterior **P**oint **OPT**imizer
  `www.projects.coin-or.org/Ipopt`

It is important to mention that the first two toolkits generate the required derivatives based on *algorithmic differentiation* (AD). AD tools decompose the function of interest into elementary operations and consequently, can also deliver derivative information [15]. For AD, there exist two different methods of implementation: *operator overloading* (see, e.g., the tools ADOL-C (utilized by MUSCOD-II) or CppAD) and *source code transformation* (see, e.g., the tools ADIC or CasADi).

Unfortunately, the problem in Equation 7.14 cannot be expressed as a convex optimization problem, which makes it difficult to obtain the global time-optimal solution. Additionally, the degree of continuity cannot be directly chosen. Therefore, an integrator chain of order l is used as the dynamic system in the minimum-time problem in Equation 7.14, which ensures continuity of degree $l-1$:

$$\dot{\overline{\mathbf{s}}} = \frac{\mathrm{d}}{\mathrm{d}t} \underbrace{\begin{pmatrix} s \\ \vdots \\ s^{(l-1)} \\ s^{(l)} \end{pmatrix}}_{\overline{\mathbf{s}}} = \underbrace{\begin{pmatrix} 0 & 1 & & 0 \\ \vdots & \ddots & \ddots & \\ 0 & \dots & 0 & 1 \\ 0 & \dots & \dots & 0 \end{pmatrix} \overline{\mathbf{s}} + \begin{pmatrix} 0 \\ \vdots \\ 0 \\ 1 \end{pmatrix} u}_{\mathbf{f}(\overline{\mathbf{s}},u)} \tag{7.29}$$

In order to solve the non-convex optimization problem and to determine a locally optimal solution, it is separated into two nested subproblems. The first problem,

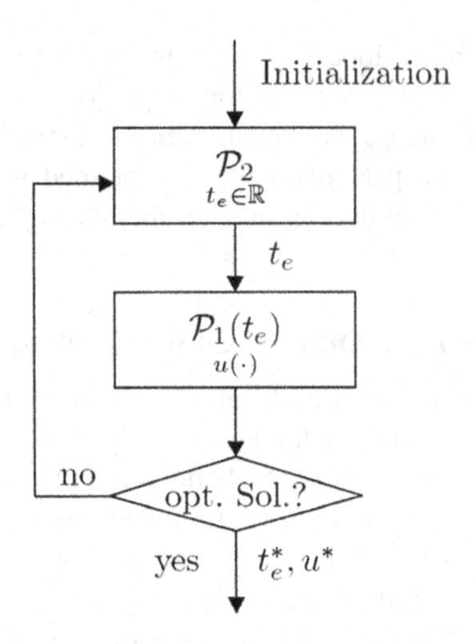

Figure 7.7 Cascaded optimization procedure for CPTOM.

denoted $\mathscr{P}_1$, is to determine the optimal trajectory for a fixed end time t_e, and the second problem, denoted $\mathscr{P}_2$, is to find the shortest end time such that there exists a solution of $\mathscr{P}_1$. The nested problem is depicted in Figure 7.7. The optimal control problem (for fixed end time t_e) $\mathscr{P}_1$ is summarized as

$$\arg\min_{u(\cdot)} \int_0^{t_e} 1\mathrm{dt} \tag{7.30a}$$

$$\text{s.t.} \quad \dot{\bar{\mathbf{s}}} = \mathbf{f}(\bar{\mathbf{s}}, u) \ , \ \bar{\mathbf{s}}(0) = \bar{\mathbf{s}}_0 \tag{7.30b}$$

$$\mathbf{g}(\bar{\mathbf{s}}(t_e)) = 0 \tag{7.30c}$$

$$\mathbf{h}(\bar{\mathbf{s}}, u) \leq 0 \tag{7.30d}$$

$$\forall t \in [0, t_e] \tag{7.30e}$$

and the second problem $\mathscr{P}_2$ as

$$\min_{t_e \in \mathbb{R}} \ t_e \tag{7.31a}$$

$$\text{s.t.} \quad t_{e,min} \leq t_e \leq t_{e,max} \tag{7.31b}$$

$$\exists u \text{ solving } \mathscr{P}_1 \tag{7.31c}$$

Solving $\mathscr{P}_2$ requires evaluation of the optimal control problem $\mathscr{P}_1$ in Equations 7.30 for a given end time t_e. Since the optimal control problem is non-convex, possibly n local solutions u_i^*, $i = 1 \ldots n$ of $\mathscr{P}_1$ are obtained. By continuously decreasing t_e, an approximation of the global solution for the t_e^* and u^* is found. To find the solution of Problem Equation 7.31, bisection algorithms or iterative techniques can be used [9].

7.4.3 CONTINOUS PATH TIME-OPTIMAL MOTION FOR REDUNDANT ROBOTS

Due to the demand for enlarged workspaces, the acceptance of robots with more than 6 DoF is increasing in industry. If a robot comprises more DoF than required for reaching a certain pose ${}_I\mathbf{z}_E$ in task space $\mathscr{T}$, we call it a redundant robot. Consequently, also for this type of robot, time-optimal motions parametrized in task space are required to minimize process duration. Redundant robots reveal completely new capabilities concerning the utilization of null space movement for TOMs. Note that due to the possibility of a reformulation of PTPTOM problems to cascaded CPTOM formulations (see Section 7.4.4), we focus only on the latter.

Basically, only a few authors have discussed the problem of time-optimal trajectory planning for redundant robots. In Galicki [11], the EE paths are restricted to certain classes, and the resulting optimal control problem is solved using indirect methods. In general, well-known formulations [5,6,25,34,37]) can be used when parametrizing the path in task space due to the fact that the null space utilization is shifted to inverse kinematics—called *implicit approaches*. This gives rise to CPTOM as in Sections 7.4.2 and 7.4.4. There is a different approach by Ma and Watanabe in 2000 [21], that separates inverse kinematics into a redundant and non-redundant part and solves the optimization problem using phase plane analysis and linear programming—called an *explicit approach*. This explicit approach is extended using a B-Spline parametrization of the path coordinate and the redundant DoF in [31]. Thereby, arbitrary continuity of the input and state trajectories is guaranteed.

The explicit approach splits the minimal coordinates into a set of n_{red} redundant joint coordinates $\mathbf{q}_{red}$ and a set of n_{nred} non-redundant joint coordinates $\mathbf{q}_{nred}$, such that the minimal coordinates are $\mathbf{q}^T = \left(\mathbf{q}_{nred}^T, \mathbf{q}_{red}^T\right)$. The direct kinematics is then written up to acceleration level as

$$
{}_I\mathbf{z}_E = \mathbf{f}_K(\mathbf{q}) = \mathbf{f}_K(\mathbf{q}_{nred}, \mathbf{q}_{red}) \tag{7.32a}
$$

$$
{}_I\dot{\mathbf{z}}_E = \mathbf{J}\dot{\mathbf{q}} = \mathbf{J}_{nred}\dot{\mathbf{q}}_{nred} + \mathbf{J}_{red}\dot{\mathbf{q}}_{red} \tag{7.32b}
$$

$$
\begin{aligned}
{}_I\ddot{\mathbf{z}}_E &= \mathbf{J}\ddot{\mathbf{q}} + \dot{\mathbf{J}}\dot{\mathbf{q}} \\
&= \mathbf{J}_{nred}\ddot{\mathbf{q}}_{nred} + \mathbf{J}_{red}\ddot{\mathbf{q}}_{red} + \dot{\mathbf{J}}_{nred}\dot{\mathbf{q}}_{nred} + \dot{\mathbf{J}}_{red}\dot{\mathbf{q}}_{red}
\end{aligned} \tag{7.32c}
$$

and the redundant DoFs are accessed *explicitly*. Next, a transformation to the scalar path coordinate ${}_I\mathbf{z}_E(s(t))$ is carried out, so that ${}_I\dot{\mathbf{z}}_E = {}_I\mathbf{z}_E'\dot{s}, {}_I\ddot{\mathbf{z}}_E = {}_I\mathbf{z}_E''\dot{s}^2 + {}_I\mathbf{z}_E'\ddot{s}$. Hence, inverse kinematics of the non-redundant joints follows as

$$
\mathbf{q}_{nred} = \mathbf{f}_{IK}({}_I\mathbf{z}_E(s), \mathbf{q}_{red}) \tag{7.33a}
$$

$$
\dot{\mathbf{q}}_{nred} = \mathbf{J}_{nred}^{-1}\left({}_I\dot{\mathbf{z}}_E - \mathbf{J}_{red}\dot{\mathbf{q}}_{red}\right) = \mathbf{J}_{nred}^{-1}{}_I\mathbf{z}_E'\dot{s} - \mathbf{J}_{nred}^{-1}\mathbf{J}_{red}\dot{\mathbf{q}}_{red} \tag{7.33b}
$$

$$
\begin{aligned}
\ddot{\mathbf{q}}_{nred} &= \mathbf{J}_{nred}^{-1}\left({}_I\ddot{\mathbf{z}}_E - \dot{\mathbf{J}}_{nred}\dot{\mathbf{q}}_{nred} - \mathbf{J}_{red}\ddot{\mathbf{q}}_{red} - \dot{\mathbf{J}}_{red}\dot{\mathbf{q}}_{red}\right) \\
&= \mathbf{J}_{nred}^{-1}\left({}_I\mathbf{z}_E''\dot{s}^2 + {}_I\mathbf{z}_E'\ddot{s} - \dot{\mathbf{J}}_{nred}\left(\mathbf{J}_{nred}^{-1}{}_I\mathbf{z}_E'\dot{s} - \mathbf{J}_{nred}^{-1}\mathbf{J}_{red}\dot{\mathbf{q}}_{red}\right)\right) \\
&\quad + \mathbf{J}_{nred}^{-1}\left(-\mathbf{J}_{red}\ddot{\mathbf{q}}_{red} - \dot{\mathbf{J}}_{red}\dot{\mathbf{q}}_{red}\right)
\end{aligned} \tag{7.33c}
$$

For the consideration of torque constraints, the equations of motion in minimal description in Equation 7.24 are reformulated using $\mathbf{q}^T = \left(\mathbf{q}_{nred}^T, \mathbf{q}_{red}^T\right)$ to

$$\mathbf{Q} = \mathbf{M}_{nred}\ddot{\mathbf{q}}_{nred} + \mathbf{M}_{red}\ddot{\mathbf{q}}_{red} + \mathbf{g}(\mathbf{q}_{nred}, \mathbf{q}_{red}, \dot{\mathbf{q}}_{nred}, \dot{\mathbf{q}}_{red}) \tag{7.34}$$

with the actuator torques $\mathbf{Q}$. Substituting Expressions 7.33 into Equation 7.34 yields a separation into terms dependent on the path coordinate s and the redundant coordinates $\mathbf{q}_{red}$:

$$\mathbf{Q} = \mathbf{A}(s, \mathbf{q}_{red})\ddot{s} + \mathbf{B}(s, \mathbf{q}_{red})\ddot{\mathbf{q}}_{red} + \mathbf{g}(s, \mathbf{q}_{red}, \dot{s}, \dot{\mathbf{q}}_{red}) \tag{7.35}$$

Then, the path coordinate $s(t)$ as well as the redundant DoFs $\mathbf{q}_{red}(t)$ are parametrized using B-Splines from Equation 7.8, so that

$$s(t) = \mathbf{n}(t)\mathbf{d}_{nred} \tag{7.36a}$$

$$q_{red,i}(t) = \mathbf{n}(t)\mathbf{d}_{red,i}, \quad i \in \{1 \dots n_{red}\} \tag{7.36b}$$

Based on this, continuity for the splines can be chosen arbitrarily and is defined to degree $m = 3$ in this case. The optimal control problem is discretized and solved in parameter space

$$\min_{\xi \in \mathbb{R}^{(1+n_{red})(n+1)}} \quad \int_0^{t_e} 1 \mathrm{d}t \tag{7.37a}$$

$$\text{s.t.} \quad \mathbf{g}(s(0), \dot{s}(0), \ddot{s}(0), s(t_e), \dot{s}(t_e), \ddot{s}(t_e)$$
$$\mathbf{q}_{red}(0), \dot{\mathbf{q}}_{red}(0), \ddot{\mathbf{q}}_{red}(0), \mathbf{q}_{red}(t_e), \dot{\mathbf{q}}_{red}(t_e), \ddot{\mathbf{q}}_{red}(t_e)) = 0 \tag{7.37b}$$

$$\mathbf{h}(s(\xi), \dot{s}(\xi), \ddot{s}(\xi), \mathbf{q}_{red}(\xi), \dot{\mathbf{q}}_{red}(\xi), \ddot{\mathbf{q}}_{red}(\xi)) \leq 0 \tag{7.37c}$$

$$\forall t \in [0, t_e] \tag{7.37d}$$

with $n+1$ control points for each spline and $\xi = (\mathbf{d}_{nred}^T, \mathbf{d}_{red,1}^T, \dots, \mathbf{d}_{red,n_{red}}^T)^T$. The inequality constraints are chosen to limit joint torques exemplarily:

$$\mathbf{h}(s(\xi), \dot{s}(\xi), \ddot{s}(\xi), \mathbf{q}_{red}(\xi), \dot{\mathbf{q}}_{red}(\xi), \ddot{\mathbf{q}}_{red}(\xi)) = \begin{pmatrix} \mathbf{Q}(s, \dot{s}, \ddot{s}, \mathbf{q}_{red}, \dot{\mathbf{q}}_{red}, \ddot{\mathbf{q}}_{red}) - \mathbf{Q}_{ub} \\ \mathbf{Q}_{lb} - \mathbf{Q}(s, \dot{s}, \ddot{s}, \mathbf{q}_{red}, \dot{\mathbf{q}}_{red}, \ddot{\mathbf{q}}_{red}) \end{pmatrix} \tag{7.38}$$

Note that this optimization is part of the cascaded CPTOM formulation in Equation 7.30 with sequential constant end time t_e. The solution is only optimal with respect to the chosen separation of kinematics $\mathbf{q}^T = \left(\mathbf{q}_{nred}^T, \mathbf{q}_{red}^T\right)$ and the selection of the B-Spline parameters.

The problem in Equation 7.37 is optimized using Active Set SQP approaches and compared with the results presented in Ma and Watanabe [21]. Therein, a planar articulated robot with 3 DoF and uniform thin rods of equal length (l = 1 m) and mass (m = 10 kg) is used with neglected gravity. To introduce redundancy, the EE position is specified with ${}_I\mathbf{z}_E^T = ({}_Ix_E, {}_Iy_E)$. The geometric path is defined by a straight line of 1 m length going from the start position

$${}_I\mathbf{z}_s = \begin{pmatrix} x_s \\ y_s \end{pmatrix} = \begin{pmatrix} 1.7321 \\ 0.0 \end{pmatrix} \text{m} \tag{7.39}$$

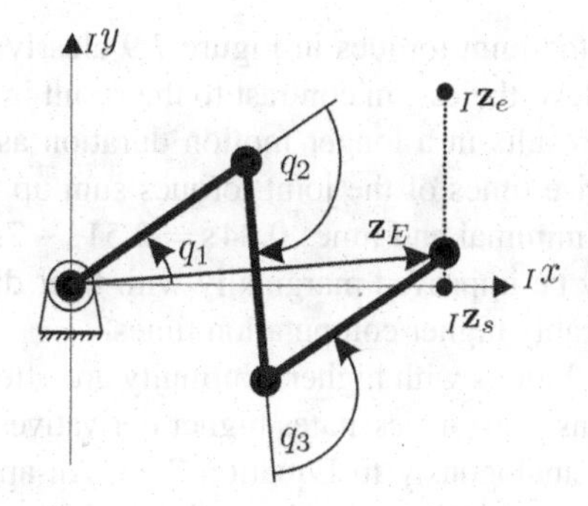

Figure 7.8 Redundant planar robot. Dashed line desired path.

to an end position $_I\mathbf{z}_e$ (see Figure 7.8). The solution is computed for every possibility of the redundant coordinate $q_{red} = q_i,\ i \in \{1,2,3\}$. The optimization problem in Equations 7.37 is solved w.r.t. to joint torque constraints $\mathbf{Q}_{ub} = -\mathbf{Q}_{lb} = (10,10,10)^T$ Nm. For the choice $q_{red} = q_3$ for the redundant coordinate, the joint torques normalized to ± 1 are shown in Figure7.9.

Remark: The graphs are evaluated at the discretization points only, where the inequality constraints from Equation 7.37c have to be satisfied. Between these points, the constraints may not be satisfied.

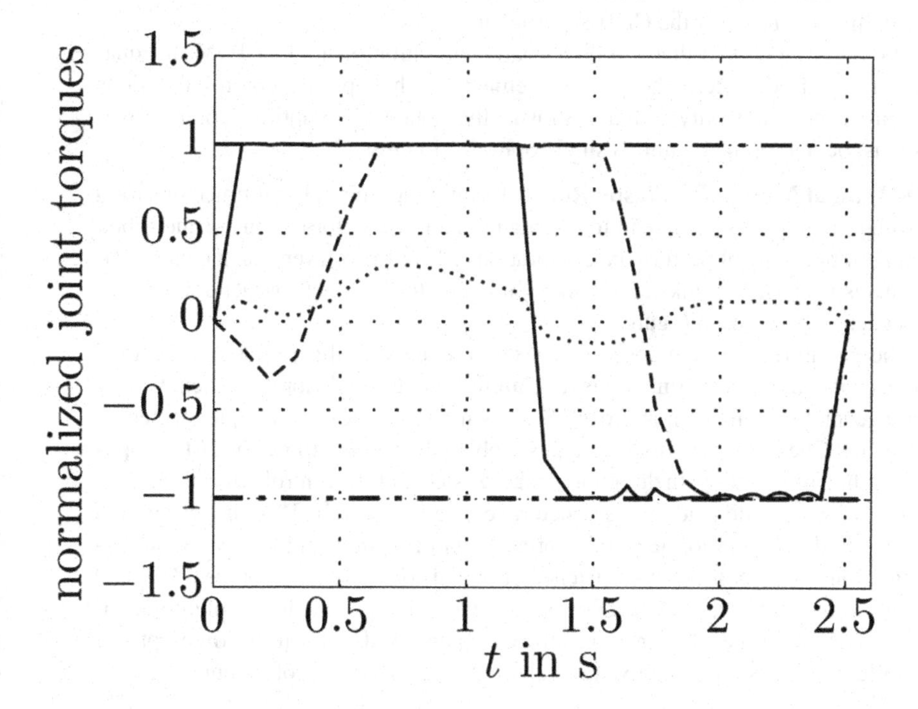

Figure 7.9 Optimized trajectory for the end time $t_e = 2.51$ s.

The min-max behavior of the joint torques in Figure 7.9 clearly indicates an almost time-optimal solution. Nevertheless, in contrast to the result in Ma and Watanabe [21], with $t_e = 2.17\,\mathrm{s}$, it results in a longer motion duration as a consequence of increased continuity. The rise times of the joint torques sum up to $0.33\,\mathrm{s}$, which explains the difference of the minimal end times $0.34\,\mathrm{s} = 2.51\,\mathrm{s} - 2.17\,\mathrm{s}$ for this geometric path. The results may be improved marginally with finer discretization but with the drawback of considerably higher computation times.

In practice, time-optimal solutions with higher continuity are often required, e.g., to avoid excitation of vibrations. This necessitates higher derivatives of forward and inverse kinematics solutions, analogously to Equation 7.33. An approach to time-optimal motion planning of kinematically redundant robots with higher-order continuity has been reported [29,30].

7.4.4 POINT-TO-POINT TIME-OPTIMAL MOTION

For solving the time-optimal PTP, sequential or direct optimization in $\mathcal{Q}$ or $\mathcal{T}$, respectively, is distinguished. As presented in Gattringer et al. [12], it is convenient to reformulate this as a sequential CPTOM problem and to optimize the geometric and dynamic paths independently. To this end, denote with $\mathcal{P}_3$ the problem of finding the optimal geometric path, e.g., $\mathbf{q}(\xi, s) \in \mathcal{Q}$, for given end time t_e. Then, the actual CPTOM optimization, denoted with $\mathcal{P}_4$, determines the minimal end time t_e for which $\mathcal{P}_3$ possesses a solution. NURBS (see Section 7.2.2) can be used to parameterize the path, so that the variables $\xi = \mathbf{d}$ are the optimization variables. Solving $\mathcal{P}_4$ yields a local solution t_e^* and u^* of the CPTOM problem.

It should be remarked that if a direct parametrization of the PTPTOM in Equations 7.14 is used, where the path is defined by the optimal control functions $\mathbf{q}(t) = \mathbf{u}(t)$, the complexity increases drastically, since then, r optimal control functions must be determined within a single optimization run.

Time-Optimal Motion for Elastic Robot: Planning a time-optimal trajectory for a lightweight manipulator can lead to considerable elastic deformations of the robot, resulting in bad control performance, vibrations at the EE or even mechanical damage. This is because the links are often flexible due to the robot's structural elasticities; see, e.g., the robot in Figure 7.10.

In the literature, different approaches exist to cope with this problem. A method that does not claim time-optimality is to minimize the residual vibration excitation by using a genetic optimization approach [17]. A more advanced technique, presented by Zhao and Chen [39], tries to handle this problem by formulating a multistage optimization that minimizes path duration in the first step and the control effort in the second step, which should lead to less structural excitation as well. This energy-optimal objective used for vibration-less trajectories is, unfortunately, at the expense of motion duration. The simultaneous existence of bounds on generalized forces (GF) and generalized force derivatives (GFD) was, to the author's knowledge, explicitly mentioned for the first time by Shin and McKay [35] and is also useful for time-optimal trajectories with less vibration excitation but at the expense of motion duration.

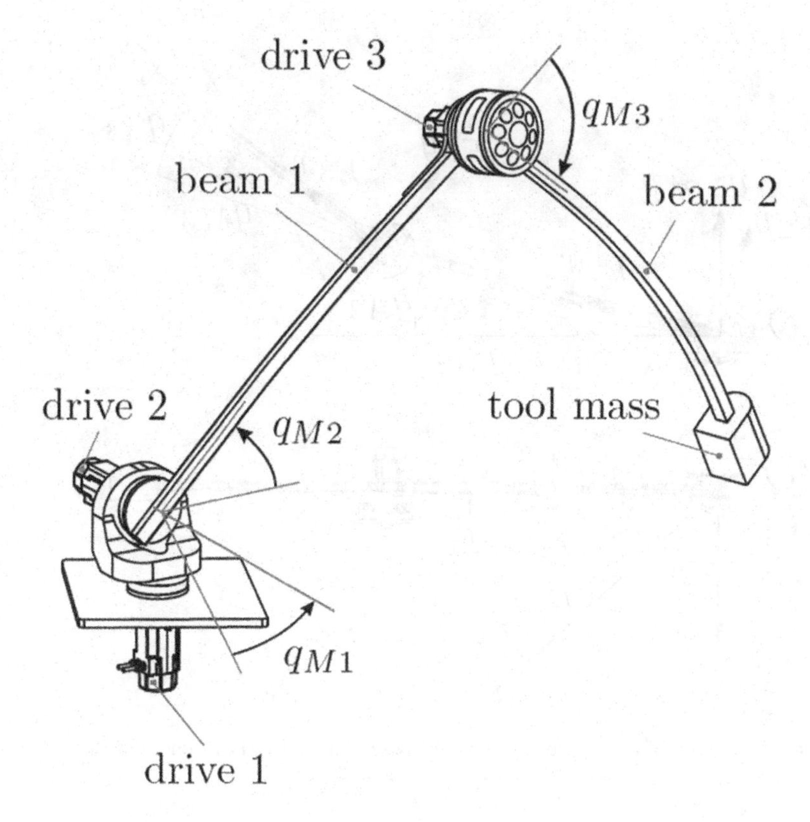

Figure 7.10 Elastic robot with three drives, built at the JKU Linz.

In this part, a novel technique is presented, first mentioned in Springer et al. [36], that overcomes the oscillations by utilizing additional constraints based on limiting only those parts of the GFs and GFDs that depend on the elastic deflections in order to minimize motion duration. The limiting term, resulting from the elastic potential of the robot, affects the vibration excitation of the flexible links and is the primary restriction in the resulting problem formulation. The technique is applied to the elastic articulated robot in Figure 7.10.

Therefore, the robot is modeled as a rigid structure with all elasticities condensed to one elasticity for each driving unit, represented by equivalent springs with the stiffness parameters c_i , $i = 1,2,3$. This model is called a lumped element model (LEM); see Figure 7.11 for a schematic illustration. Therein, $\mathbf{q}_M = (q_{M1}, q_{M2}, q_{M3})^T$ denote the vector of motor and $\mathbf{q}_A = (q_{A1}, q_{A2}, q_{A3})^T$ the virtual link angles that lead to the set of six minimal coordinates:

$$\mathbf{q} = \begin{pmatrix} \mathbf{q}_M \\ \mathbf{q}_A \end{pmatrix} \in \mathbb{R}^6 \tag{7.40}$$

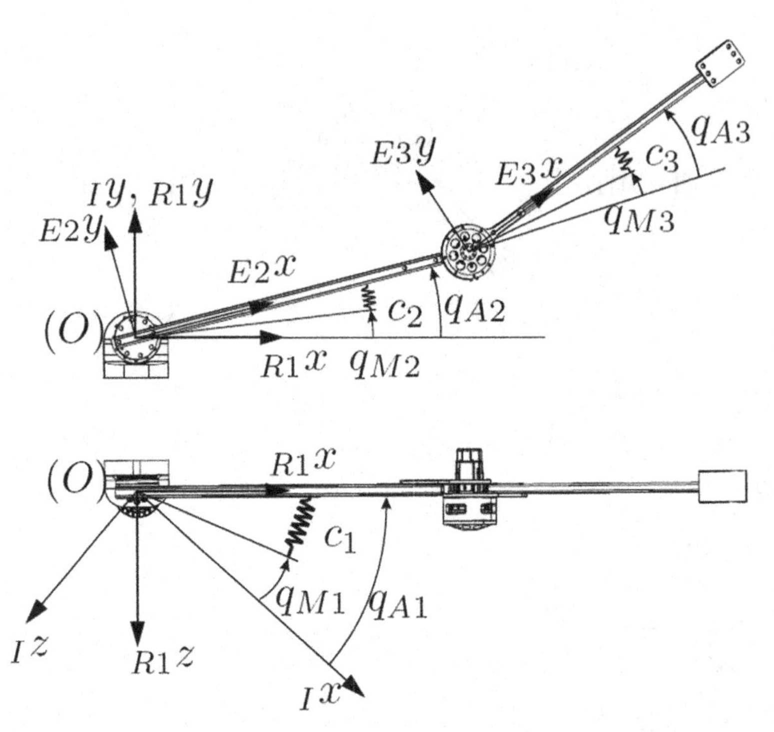

Figure 7.11 Illustration of the elastic deformations based on the LEM in front view (a) and top view (b).

In order to understand the influence of the elastic structure of the robot during trajectory tracking, we write the generalized forces due to the elastic potential V_{el}:

$$\mathbf{Q}_A = \left(\frac{\partial V_{el}}{\partial \mathbf{q}_M}\right)^T = \left(\frac{\partial}{\partial \mathbf{q}_M}\left(\sum_{i=1}^{3}\frac{1}{2}c_i(q_{Ai} - q_{Mi})^2\right)\right)^T = -\left(\frac{\partial V_{el}}{\partial \mathbf{q}_A}\right)^T \tag{7.41a}$$

$$= \mathbf{K}(\mathbf{q}_M - \mathbf{q}_A), \quad \mathbf{K} = \operatorname{diag}(c_1,\ c_2,\ c_3). \tag{7.41b}$$

Besides constraining the required physical limitations concerning the actuator torques $\mathbf{Q}$, motor angular rates $\dot{\mathbf{q}}_M$, as well as the jerk of the link angles $\overset{(3)}{\mathbf{q}_A}$, it is useful to include limitations on the term (Equation 7.41). This will lead to slower motion, on the one hand, but induces much less vibration at the EE, on the other hand. Similarly, the GFDs are physically bounded as well. In the case of electric actuators, for example, the available supply voltage necessarily limits the GFs' variation rate. Analogously, only the terms concerning the elastic potential resulting from Equation 7.41 are considered, and therefore, $\frac{\mathrm{d}}{\mathrm{d}t}\mathbf{Q}_A$ is constrained as well. Hence, the optimization problem $\mathscr{P}_1$ to determine a time-optimal trajectory on a path (cf. Equation 7.30) is given with the dynamic system in Equation 7.29 ($l = 3$) as

$$\arg\min_{u(\cdot)} \int_0^{t_e} 1 \mathrm{d}t \tag{7.42a}$$

$$\text{s.t.} \quad \dot{\bar{\mathbf{s}}} = \mathbf{f}(\bar{\mathbf{s}}, u) \tag{7.42b}$$

$$\dot{\mathbf{q}}_{M,lb} \leq \dot{\mathbf{q}}_M(\bar{\mathbf{s}}, u) \leq \dot{\mathbf{q}}_{M,ub} \tag{7.42c}$$

$$\overset{(3)}{\mathbf{q}}_{A,lb} \leq \overset{(3)}{\mathbf{q}}_A(\bar{\mathbf{s}}, u) \leq \overset{(3)}{\mathbf{q}}_{A,ub} \tag{7.42d}$$

$$\mathbf{Q}_{lb} \leq \mathbf{Q}(\bar{\mathbf{s}}, u) \leq \mathbf{Q}_{ub} \tag{7.42e}$$

$$\mathbf{Q}_{A,lb} \leq \mathbf{Q}_A(\bar{\mathbf{s}}, u) \leq \mathbf{Q}_{A,ub} \tag{7.42f}$$

$$\dot{\mathbf{Q}}_{A,lb} \leq \dot{\mathbf{Q}}_A(\bar{\mathbf{s}}, u) \leq \dot{\mathbf{Q}}_{A,ub} \tag{7.42g}$$

$$s(0) = 0 \,, \; s(t_e) = 1 \tag{7.42h}$$

$$\dot{s}(0) = 0 \,, \; \dot{s}(t_e) = 0 \tag{7.42i}$$

$$\dot{s} \geq 0 \,, \; \forall t \in [0, t_e] \tag{7.42j}$$

with the lower and upper bounds $\mathbf{Q}_{lb}$ and $\mathbf{Q}_{ub}$, $\mathbf{Q}_{A,lb}$ and $\mathbf{Q}_{A,ub}$, $\dot{\mathbf{Q}}_{A,lb}$ and $\dot{\mathbf{Q}}_{A,ub}$, $\dot{\mathbf{q}}_{M,lb}$ and $\dot{\mathbf{q}}_{M,ub}$, as well as $\overset{(3)}{\mathbf{q}}_{A,lb}$ and $\overset{(3)}{\mathbf{q}}_{A,ub}$. In Figure 7.12a, one can clearly see, based on the CPTOM optimization without additional constraints, that the bending stresses σ_{b1}, σ_{b2}, and σ_{b3} exceed the maximum admissible stress σ_{max}. In contrast, Figure 7.12b shows that σ_{max} is satisfied when optimizing with additional constraints.

7.5 COLLISION AVOIDANCE

Up to now, we considered time-optimal motions with no restricted workspace. In general, and in particular in an industrial environment, different objects may be present in the robot's workspace that have to be taken into account during trajectory planning. Therefore, we approximate the robot as a union of n_r convex polyhedra

$$R = \bigcup_{i=1}^{n_r} R_i, \quad \text{with} \quad R_i = \left\{ \mathbf{x} \in \mathbb{R}^3 \,|\, \mathbf{A}_i \mathbf{x} \leq \mathbf{b}_i \right\} \tag{7.43}$$

with the number of faces n_{rf}, so that $\mathbf{A}_i \in \mathbb{R}^{n_{rf} \times 3}, \mathbf{b}_i \in \mathbb{R}^{n_{rf}}$. Analogously, we define obstacles

$$O = \bigcup_{j=1}^{n_o} O_j, \quad \text{with} \quad O_j = \left\{ \mathbf{x} \in \mathbb{R}^3 \,|\, \mathbf{C}_j \mathbf{x} \leq \mathbf{d}_j \right\} \tag{7.44}$$

with the number of polyhedra n_o and the number of its faces n_{of}, so that $\mathbf{C}_j \in \mathbb{R}^{n_{of} \times 3}, \mathbf{d}_j \in \mathbb{R}^{n_{of}}$. Since the union of $R \cap O$ has to be the empty set for collision avoidance, the following equation must not hold for any point $\mathbf{x} \in \mathbb{R}^3$

$$\begin{bmatrix} \mathbf{A}_i \\ \mathbf{C}_j \end{bmatrix} \mathbf{x} \leq \begin{pmatrix} \mathbf{b}_i \\ \mathbf{d}_j \end{pmatrix}, \quad i = 1, \ldots, n_r, \; j = 1, \ldots, n_o \tag{7.45}$$

for each possible combination (i, j). Using Farkas' Lemma [10], we know that the linear system in Equation 7.45 has no solution if there exists a vector $\mathbf{v} \geq 0 \in \mathbb{R}^{n_{rf}+n_{of}}$

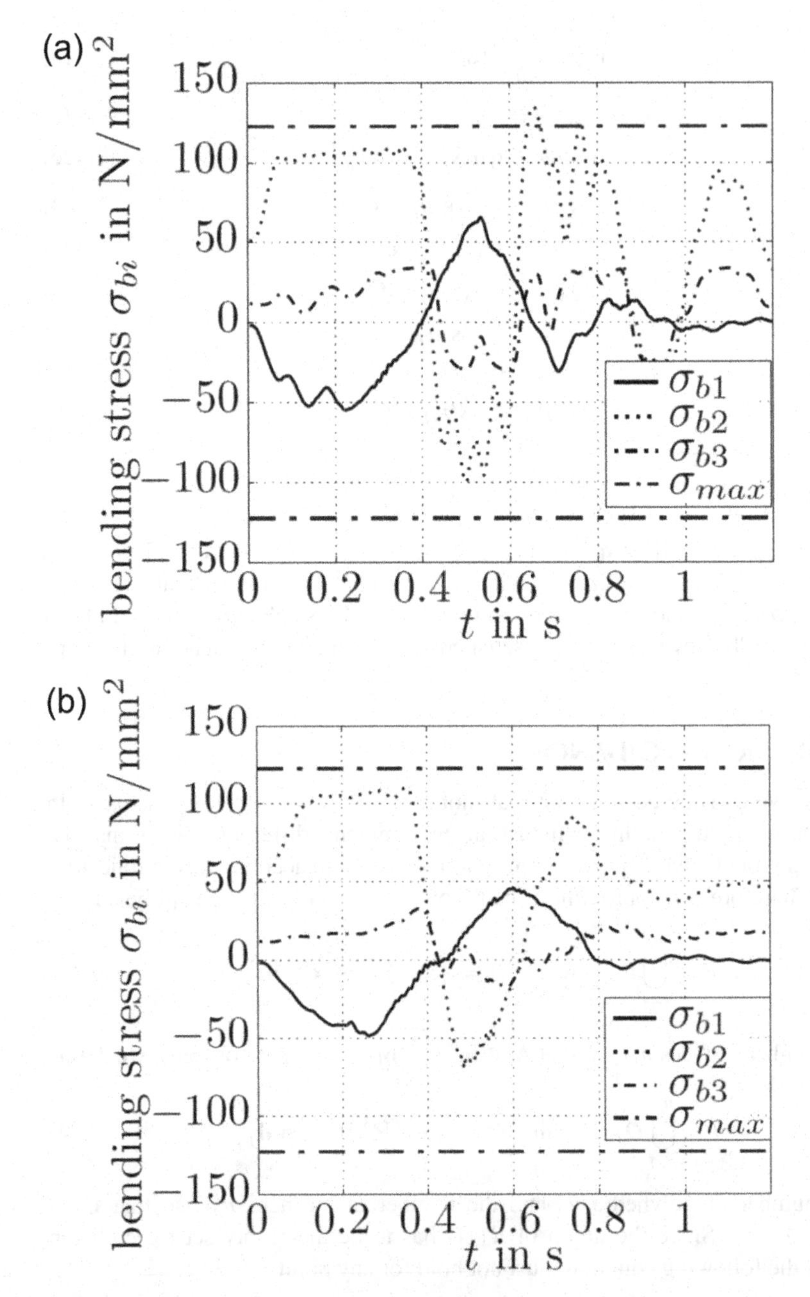

Figure 7.12 Bending stresses occurring for the CPTOM with and without additional constraints for optimization. (a) Bending stresses occurring for the CPTOM without additional constraints. (b) Bending stresses occurring for the CPTOM with additional constraints.

satisfying the equations

$$\begin{bmatrix}\mathbf{A}_i\\ \mathbf{C}_j\end{bmatrix}^T \mathbf{v}=0\,,\ \begin{pmatrix}\mathbf{b}_i\\ \mathbf{d}_j\end{pmatrix}^T \mathbf{v}<0,\quad i=1,\ldots,n_r,\ j=1,\ldots,n_o$$

Consequently, for obtaining a collision-free solution $\mathscr{P}_1(t_e)$ of the optimal control problem in Equation 7.30, the constraints

$$\mathbf{N}_l(s)=\begin{bmatrix}\mathbf{A}_i(s)\\ \mathbf{C}_j\end{bmatrix},\ \mathbf{n}_l=\begin{pmatrix}\mathbf{b}_i(s)\\ \mathbf{d}_j\end{pmatrix} \tag{7.46}$$

$$l=1,\ldots,n_r n_o,\quad i=1,\ldots,n_r,\quad j=1,\ldots,n_o \tag{7.47}$$

with $n_r \cdot n_o$ possible combinations (i,j) are introduced. Thereby, the robot approximation R is obtained as a function of the path parameter s.

Remark: Note that the matrices and vectors in Equation 7.46 are functions of time.

The interior path constraints in Equation 7.30 are extended with the non-convex constraints

$$\mathbf{h}(\bar{\mathbf{s}},u)\begin{cases}\mathbf{N}_l^T(s)\mathbf{v}_l=0\\ \mathbf{n}_l^T(s)\mathbf{v}_l\le-\tau\\ \mathbf{v}_l\ge 0\end{cases},\quad l=1,\ldots,n_r n_o \tag{7.48}$$

and the additional optimization variables $\mathbf{v}_l=(v_{l,1},\ldots,v_{l,n_{rf}+n_{of}})^T$. The parameter $\tau>0$ can be used as relaxation parameter for conservative constraint consideration [8,18].

With the description of the robot geometry as function of the path parameter, these relations can now be incorporated in the optimal control problem [13,14,16].

REFERENCES

1. E.J. Barbeau. *Polynomials*, volume 32. Springer Verlag, 1989.
2. R.E. Bellman. *Dynamic Programming*. Dover Publications, 1957.
3. J.T. Betts. *Practical Methods for Optimal Control Using Nonlinear Programming*. Society for Industrial and Applied Mathematics, 2001.
4. L. Biagiotti and C. Melchiorri. *Trajectory Planning for Automatic Machines and Robots*. Springer Verlag, 1st edition, 2008.
5. J.E. Bobrow, S. Dubowsky, and J.S. Gibson. Time-Optimal Control of Robotic Manipulators Along Specified Paths. *International Journal of Robotics Research*, 4:3–17, 1985.
6. D. Constantinescu and E.A. Croft. Smooth and Time-Optimal Trajectory Planning for Industrial Manipulators along Specified Paths. *Journal of Robotic Systems*, 17(5):233–249, 2000.
7. C. de Boor. *A Practical Guide to Splines*. Springer Verlag, 1978.
8. F. Debrouwere, W. Van Loock, G. Pipeleers, M. Diehl, J. De Schutter, and J. Swevers. Time-Optimal Path Following for Robots with Object Collision Avoidance using Lagrangian Duality. In *Robot Motion and Control (RoMoCo), 2013 9th Workshop on*, pages 186–191. IEEE, 2013.

9. L. Van den Broeck, M. Diehl, and J. Swevers. Time Optimal MPC for Mechatronic Applications. In *Proceedings of the 48th IEEE Conference on Decision and Control*, pages 8040–8045, 2009.
10. D. Gale, H.W. Kuhn, and A.W. Tucker. Linear Programming and the Theory of Games. *Activity Analysis of Production and Allocation*, 13:317–335, 1951.
11. M. Galicki. Time-Optimal Controls of Kinematically Redundant Manipulators with Geometric Constraints. *IEEE Transactions on Robotics and Automation*, 16(1):89–93, 2000.
12. H. Gattringer, M. Oberherber, and K. Springer. Extending Continuous Path Trajectories to Point-to-Point Trajectories by Varying Intermediate Points. *International Journal of Mechanics and Control*, 15(01):35–43, 2014.
13. E.G. Gilbert and D.W. Johnson. Distance Functions and Their Application to Robot Path Planning in the Presence of Obstacles. *IEEE Journal on Robotics and Automation*, 1(1):21–30, 1986.
14. E.G. Gilbert, D.W. Johnson, and S.S. Keerthi. A Fast Procedure for Computing the Distance Between Complex Objects in Three-Dimensional Space. *IEEE Journal on Robotics and Automation*, 4(2):193–203, 1988.
15. A. Griewank and A. Walther. *Evaluating Derivatives: Principles and Techniques of Algorithmic Differentiation.* Other Titles in Applied Mathematics. SIAM, 2nd edition, 2008.
16. D. Kaserer, H. Gattringer, and A. Müller. Optimal Point-to-Point Trajectory Planning with Collision Avoidance for Dual Arm Setup. In *ECCOMAS Thematic Conference on Multibody Dynamics*, 2017.
17. H. Kojima and T. Kibe. Optimal Trajectory Planning of a Two-Link Flexible Robot Arm Based on Genetic Algorithm for Residual Vibration Reduction. In *Proceedings of the 2001 IEEE/RSJ International Conference on Intelligent Robots and Systems*, pages 2276–2281, 2001.
18. C. Landry, M. Gerdts, R. Henrion, and D. Hömberg. Path-Planning with Collision Avoidance in Automotive Industry. In *System Modeling and Optimization - 25th IFIP TC 7 Conference*, volume 391 of *IFIP Advances in Information and Communication Technology*, pages 102–111. Springer, 2011.
19. E.T.Y. Lee. A Simplified B-Spline Computation Routine. *Computing*, 29(4):365–371, 1982.
20. D.B. Leineweber, I. Bauer, H.G. Bock, and J.P. Schlöder. An Efficient Multiple Shooting Based Reduced SQP Strategy for Large-Scale Dynamic Process Optimization - Part I: Theoretical Aspects. *Computers & Chemical Engineering*, 27:157–166, 2003.
21. S. Ma and M. Watanabe. Minimum Time Path-Tracking Control of Redundant Manipulators. In *Proceedings of the IEEE/RSJ International Conference on Intelligent Robots and Systems*, volume 1, pages 27–32, 2000.
22. M. Neubauer and A. Müller. Smooth Orientation Path Planning with Quaternions Using B-Splines. In *Proceedings of the IEEE/RSJ International Conference on Intelligent Robots and Systems*, pages 2087–2092, 2015.
23. J. Nocedal and S.J. Wright. *Numerical Optimization.* Springer Verlag, 2nd edition, 2006.
24. F. Pfeiffer. *Mechanical System Dynamics*. Springer Verlag, 2008.
25. F. Pfeiffer and R. Johanni. A Concept for Manipulator Trajectory Planning. *IEEE Journal of Robotics and Automation*, 3(2):115 –123, 1987.
26. Q.C. Pham. A General, Fast, and Robust Implementation of the Time-Optimal Path Parameterization Algorithm. *IEEE Transactions on Robotics*, 30:1533–1540, 2014.

27. L. Piegl. On NURBS: A Survey. *IEEE Computer Graphics and Applications*, 11(1):55–71, 1991.
28. L. Piegl and W. Tiller. *The NURBS Book*. Springer Verlag, 2nd edition, 1997.
29. A. Reiter, A. Müller, and H. Gattringer. Inverse Kinematics in Minimum-Time Trajectory Planning for Kinematically Redundant Manipulators. In *IECON 2016-42nd Annual Conference of the IEEE Industrial Electronics Society*, pages 6873–6878, 2016.
30. A. Reiter, A. Müller, and H. Gattringer. On Higher-Order Inverse Kinematics Methods in Time-Optimal Trajectory Planning for Kinematically Redundant Manipulators. *IEEE Transactions on Industrial Informatics*, 14(4):1681–1690, 2018.
31. A. Reiter, K. Springer, H. Gattringer, and A. Müller. An Explicit Approach for Time-Optimal Trajectory Planning for Kinematically Redundant Robots. In *Proceedings in Applied Mathematics and Mechanics (PAMM)*, pages 67–68. WILEY-VCH Verlag GmbH & Co. KGaA, 2015.
32. R.T. Rockafellar. *Convex Analysis*. Princeton University Press, 1970.
33. I.J. Schoenberg. Contributions to the Problem of Approximation of Equidistant Data by Analytic Functions. Part A: On the Problem of Smoothing of Graduation. A First Class of Analytic Approximation Formulae. *Quarterly of Applied Mathematics*, 4:45–88, 1946.
34. Z. Shiller. Time-Energy Optimal Control of Articulated Systems With Geometric Path Constraints. In *Proceedings of the IEEE International Conference on Robotics and Automation (ICRA)*, volume 4, pages 2680–2685, 1994.
35. K. Shin and N. McKay. A Dynamic Programming Approach to Trajectory Planning of Robotic Manipulators. *IEEE Transactions on Automatic Control*, 31(6):491–500, 1986.
36. K. Springer, H. Gattringer, and P. Staufer. On Time-Optimal Trajectory Planning for a Flexible Link Robot. *Proceedings of the Institution of Mechanical Engineers, Part I: Journal of Systems and Control Engineering*, 227(10):752–763, 2013.
37. D. Verscheure, B. Demeulenaere, J. Swevers, J. De Schutter, and M. Diehl. Time-Energy Optimal Path Tracking for Robots: A Numerically Efficient Optimization Approach. In *Proceedings of the 10th IEEE International Workshop on Advanced Motion Control*, pages 727–732, 2008.
38. K.J. Versprille. *Computer-Aided Design Applications of the Rational B-spline Approximation Form*. PhD thesis, Syracuse University, 1975.
39. H. Zhao and D. Chen. Optimal Motion Planning for Flexible Space Robots. In *Proceedings of the IEEE International Conference on Robotics and Automation (ICRA)*, volume 1, pages 393–398, 1996.

8 Robot Programming

Christian Schlegel, Dennis Stampfer, Alex Lotz, and Matthias Lutz

CONTENTS

Instructing a robot by demonstration is considered *robot programming*, as is the implementation of a software library for robotics. This already shows that robot programming comprises a huge variety of diverse activities carried out by an even greater diversity of experts. They cover different roles in different phases of a robot's life-cycle. All are decisive in building, integrating, deploying, operating, adapting, and maintaining modern robotic systems. Advanced robotic systems are expected to interact, collaborate, and cooperate with persons in different environments (production assistant, robots co-working with persons, cognitive factory, domestic services, service robots, etc.), and they need to be able to perform a multitude of different tasks (multi-purpose systems). Advanced robotic systems are composite systems where one always has to deal with a huge body of heterogeneous resources. Due to open-ended environments, advanced robotic systems need to be configurable and adaptable even at run-time, which adds another complexity to their software systems. Thus, robot programming is nowadays just one aspect of the much broader topic of software systems engineering for robotics.

8.1 HISTORY AND STATE-OF-THE-ART

As vital functions of advanced robotic systems are provided by software, the increasing requirements and capabilities of such robots are reflected in a still growing dominance and an ever increasing complexity of a robot's software. Mastering the software complexity has already become pivotal towards exploiting the capabilities of advanced robotic algorithms and robotic components. As long as the software part implies risks and efforts that are difficult to manage (see the EFFIROB-Study [14]), the software challenge will very likely become a show-stopper towards the next level of robotics innovations. This has also been identified by the *European SPARC Robotics initiative* in its *Strategic Research Agenda (SRA)* [37]. Related to software systems engineering in robotics, it states: Investment in these technologies is critical to the timely development of products and services and a key enabling factor in the stimulation of a viable robot industry.

In earlier times, robot programming was more of an art than a systematic engineering process. Robot programming was often related to bare-metal programming and could be done only by highly skilled programming experts. This approach neither scaled with respect to the complexity challenge nor allowed its application domains to easily get access to and make use of robotics technology. In short, despite its tremendous role in advanced robotics, robot programming lacked all the structures and processes that consolidate a path towards commercial innovations.

Nowadays, robot programming is a completely different story. Robotics software has seen tremendous changes and shifts up to now. In the past, it has been left to the reader of a paper to try to implement a described algorithm on his own. He never knew how close he came to the original implementation, and it was nearly impossible to reproduce experiments. Nowadays, it is quite standard that papers are accompanied by freely accessible (often open source and at least prototypical) implementations. Means to organize world-wide repositories of software artifacts resulted in a shift from proprietary lab experiments to world-wide exchange of open-source libraries with a perspective of experiment reproduction. Thus, access to software plays a significant role in managing scientific standards in robotics like reproduction of experiments and third-party evaluation of results.

In general, there is a long-standing fruitful relationship with mutual benefits between robotics and other related domains. For example, middleware systems are researched and pushed forward by a community of their own, and robotics makes use of their outcomes. On the other hand, the middleware community is stimulated by insights and demanding needs from robotics. For example, robotics has a need to address quality and resource properties that are just abstracted away for many other application domains. This interplay of robotics with other domains is one driver of the robotics software technology cycles. However, there is another important aspect not to be underestimated. Since roboticists are typically not software engineers, all the advanced software systems and software design patterns need to be made accessible via abstraction layers and presentation layers as well as according to usage guidelines. For instance, modern general-purpose middleware systems have a hard time in making their way into robotics, because they do not match the needs and expectations of roboticists. Instead, roboticists then prefer to stay with software technologies they are familiar with. In consequence, one builds most complex robotic software systems without gaining from the latest software technologies and software systems. As result, the overall performance of robotic systems and robotic systems development in terms of, e.g., time-to-market, dependability, and cost-efficiency falls behind expectations as raised by the maturity levels of the available building blocks.

A comprehensive overview on the rich history of different robotics software frameworks is impossible. Of course, each framework favors different needs and thus, fills a particular niche. However, they all can only be understood in their time, since they have been heavily influenced by at least (i) the technology available and accessible at that time, (ii) the kinds of robots and domains put into focus (mobile robots, manipulation, flying robots, etc.), (iii) whether the focus has been more on functionalities (typically driven by academia) or on (application) scenarios (typically driven by industry and also by competitions like Robocup), and even (iv) the (non)availability and (non)affordability of standard robot platforms (*Pioneer* mobile robot, *NAO* humanoid, *Care-O-bot* mobile robot with manipulator, the *KUKA youBot*, the *FESTO Robotino*, the *Universal Robots UR-Arm* and many others), cheap sensors (laser rangers, RGB-D cameras, tablets for user interfaces, etc.), ubiquitous and always available internet connections as well as enough and cheap computing power.

Robot software has been notoriously specific to the hardware and target task, and as both change quickly, robotics engineers always dreamed of separating concerns in order to avoid starting from scratch over and over again. This has become more and more feasible with constantly emerging hardware/software technologies.

One influential software framework has been *Player* with its simulation backends *Stage* and *Gazebo*. It consisted of a large set of device drivers for robots and sensors and came with a simple client/server communication mechanism. Its wide acceptance in academia resulted from (at least to the degree possible at that time) factoring out source code of algorithms from communication and from details of devices. *Gazebo* has been continuously maintained and is still one of the major simulators used in robotics. However, there are powerful alternatives, like *V-REP*, *MORSE*, and *Webots*.

Microsoft Robotics Developer Studio (MRDS) was a freely available .NET-based programming environment for building robotic applications, for professional as well as non-professional developers and even for hobbyists. It included a light-weight REST-style service-oriented run-time, a set of visual programming and simulation tools, as well as lots of tutorials. However, due to the vendor lock-in, it never gained widespread acceptance within the robotics research community.

The open-source *Mobile Robot Programming Toolkit (MRPT)* provides developers with portable and well-tested applications and libraries covering data structures and algorithms employed in common robotics research areas. The open-source *OROCOS Real-Time Toolkit* provides a framework targeting the implementation of (realtime and non-realtime) robotic control systems (applications with a focus on industrial robotics and manipulators). *YARP (Yet Another Robot Platform)* is a thin middleware mostly used with humanoid robots.

Robot Technology (RT) Middleware [1] is based on the *OMG Robot Technology Component (RTC)* standard. The interesting part here is that an implementation-independent single description of its structures exists in the form of a standard, which comes with many different implementations. The reference implementation *OpenRTM-aist* includes an Integrated Development Environment (IDE) based on *Eclipse*. *RT-Middleware* was one of the leading modeling initiatives in robotics around a decade ago, with at that time, a significant impact on robotics software development. Compared with recent approaches, *RT-Middleware* has a rather simplified data flow–like communication model, and the Eclipse-based tooling lacks state-of-the-art modeling techniques for structured system integration and deployment.

ROS (Robot Operating System) gained tremendous acceptance within the robotics community. It is a collection of tools, libraries, and conventions that aim to simplify the task of creating complex and robust robot behavior across a wide variety of robotic platforms. It has been the first time in robotics software that so many have adopted a particular framework. However, this should not distract from its deficiencies, e.g., the lack of prescribed patterns for building and structuring systems and the lack of features like life-cycle management and configurations for deployment. This led to the still ongoing design of *ROS 2* [12]. The goal of the *ROS 2 project* is to leverage what is great about *ROS 1* and improve what is not.

SmartSoft/SmartMDSD has already very early brought into focus structured communication patterns [34,35], a sound software component model [26], and modeling of its key concepts agnostic to implementation technologies [27]. This finally evolved into an open-source *Eclipse*-based model-driven IDE [41]. Its core concepts proved to be stable, and as outlined in more detail in Section 8.4, it became an essential part of the consolidated body-of-knowledge of well-accepted best practices in model-driven software engineering for robotics that we explicate in the EU H2020 project *RobMoSys—Better Models, Better Tools, Better Systems* (www.robmosys.eu).

PROTEUS is a French initiative that also introduced formal modeling early into robotics software development. It is also based on *Eclipse* but builds on top of the *Papyrus* modeling tool. It introduced *RobotML* (Robot Modeling Language), which is a domain-specific language to create abstract views of robotic problems and solutions [10]. In its objectives and also in some of its approaches, it is similar to *SmartSoft/SmartMDSD*. *Papyrus* has been extended within *RobMoSys* to *Papyrus4Robotics*.

The need for systematic tooling has also been identified by the *BRICS* project and its integrated development environment *(BRIDE)*, which is based on the *BRICS Component Model (BCM)* [8]. One of the core motivations behind *BRIDE* and *BCM* has been to harmonize (i.e., to find a common denominator for) common robotic frameworks. This, however, inevitably resulted in a simplified component model (abstracting away differences between component models) and thus, consequently resulted in tooling that leaves important aspects unaddressed. For example, the communication between components cannot be sufficiently defined. In consequence, at the level of a component implementation, communication characteristics can be modified beyond the model, which breaks reuse and integration via composition. In alignment to *Software Product Lines (SPLs)*, *Hyperflex* (an extension of *BRIDE*) defines application-specific reference architectures based on feature models [13].

The *Unified Component Model (UCM)* is an interesting initiative of the Object Management Group (OMG), which might also have an impact on robotics. *Domain Specific Languages (DSLs)* will also gain more and more momentum in robotics, as they allow domain-specific views to be offered, based on domain-specific vocabularies. However, right now, they still lack integration, and too many just coexist [24].

Recently, more and more tools are showing up that allow a graphical arrangement of configurable skills (see, e.g., the Institute of Electrical and Electronics Engineers Industrial Electronics Society Emerging Technologies and Factory Automation [IEEE IES ETFA] workshop series *Skill-Based Systems Engineering*). Unexperienced users can easily parameterize and arrange reusable function blocks or even reuse complete application templates to solve their robotics use cases. This so-called skill-based robot programming does not require deep robotics or programming knowledge anymore as long as the given set of skills can cover your application needs. In particular, such tools overcome the vendor-specific programming languages, as the very same skills can be used with different robots. Examples are the *ArtiMind Robot Programming Suite* and the *drag & bot* approach and tooling.

The latest developments now link cloud platforms and robotic systems. Examples are the *Microsoft Azure Cognitive Services* and the *Amazon Web Services RoboMaker*. Cloud services have the potential to boost object recognition and learning algorithms as well as inference algorithms and add these to the variety of skills of robots. Cloud services have inherent advantages when it comes to cheap processing power, sharing of data sets for learning, and sharing of experiences made by a robot with others. This then enables taking advantage of others' experience when facing a new situation that other robots have dealt with already. A slightly different approach is followed by *NVIDIA* with its *ISAAC GEMS* and the *ISAC SIM*, which finally run on their cheap, powerful, and energy-efficient hardware to be used onboard a robot for demanding robotic algorithms.

The *euRobotics AISBL Topic Group on Software Engineering, System Integration, System Engineering* shapes the European road-mapping in software systems engineering for robotics and closely interacts with the *Topic Group on Standardization* with respect to software issues. The broadest coverage of concepts, tools, implementations, and applications of model-driven software engineering in robotics is represented by the EU H2020 project *RobMoSys* and can be found in the *RobMoSys Wiki* (https://robmosys.eu/wiki). It is continuously updated with broadest community involvement based on discussions in the relevant forum (https://discourse.robmosys.eu/).

The community also includes the *Technical Committee on Software Engineering for Robotics and Automation (IEEE RAS TC-Soft)*. The workshop series *DSLRob (Domain-Specific Languages and Models for Robotic Systems)* and *MORSE (Model-Driven Robot Software Engineering)*, the *International Conference on Simulation, Modeling, and Programming for Autonomous Robots (SIMPAR)*, the *International Conference on Model Driven Engineering Languages and Systems (MODELS)*, as well as the *Journal of Software Engineering for Robotics (JOSER)* are major community services. An overview on software engineering for experimental robotics in its early stages is given in Brugali [4]. Relevant software and modeling activities with an impact on robotics are also driven by the *Reference Architecture Model Industrie 4.0 (RAMI)*, the *Asset Administration Shell*, and the related *OPC UA companion specifications*.

8.2 SOFTWARE ENGINEERING IN ROBOTICS: GOALS, OBJECTIVES, CHALLENGES

Although a large number of open-source software packages for robotic systems have become available worldwide, available software integration frameworks have still not achieved the status of robotics commodities. There is still a tremendous lack of systematic software engineering processes in robotics (indicative of the lack of overall software architectural models and methods and the large fragmentation of robotics software) [37].

Robotics as a science of integration depends on structures that guide the overall system design and the system integration and that even support run-time adaptation according to the executed task, current context, and available resources. Such structures are called architectural patterns (https://robmosys.eu/wiki/general_principles:architectural_patterns:start), and these should result in, e.g., flexibility, easy (re-)configurability, support for quick prototyping, and reliable operation of complete systems. Architectural patterns are considered good if they provide partitioning schemes that organize different and various views on a robotics system, such that one can effectively and efficiently cope with the complexity of the whole life-cycle of such a system (design, composition, configuration, deployment, operation, evolution, adaptation, etc.). Good architectural patterns result in an appropriate ratio of *size of change* and *effort in change* when reengineering of systems is required. Similar requirements should be achievable by requiring only minor effort (based on already available solutions: either modifying them or composing them in a new way).

Separation of concerns is one of the most fundamental partitioning schemes in software engineering and in robotics [33]. It aims at identifying orthogonal concerns in order to address complexity by decoupling. In most engineering approaches as well as in robotics, at least the following are dominant dimensions of concerns that should be distinguished [3,25]: (i) *Computation* provides the functionality of an entity. Computation activities require communication to access input data and to provide computed results to other entities; (ii) *Communication* exchanges data between entities; (iii) *Configuration* comprises the binding of configurable parameters, e.g., connections between entities; and (iv) *Coordination* determines how the activities of all entities in a system should work together. It relates to orchestration and resource management. There exist variants that split *configuration* into *connection* and *configuration* [7] or that treat *configuration* and *coordination* in the same way [2]. With respect to robotics software systems, e.g., functional libraries like *OpenCV* should be clearly separated from a software component model and should provide stable interfaces independently of the underlying middleware system.

Separation of roles [33] is another fundamental partitioning scheme. Roboticists, application experts, professional users as well as private end-users, and others all fill different roles with different responsibilities, needs, and duties. They all need to be supported in filling their role without being required to become an expert in what is covered by other roles. In other words, architectural patterns for robotics organize modularity (decomposability, composability, understandability, etc.) on a technical level and on an organizational level.

Further important principles are *composability*, *compositionality*, and *composition*. *Composability* is the ability to (re-)combine as-is building blocks into different systems for different purposes. It requires that properties of sub-systems are invariant ("remain satisfied") under composition. *Splittability* is the inverse relationship of composability. *Compositionality* requires that the behavior of a system is predictable from its sub-systems and from that of the composition operators. *System composition* is the activity of putting together a set of existing building blocks to match system needs with a focus on flexible (re-)combination (just like putting Lego bricks together). In contrast, *system integration* is the activity that requires effort to combine components, requiring modifications or additional actions to make them work together. Once integrated, it is difficult to put them apart again. For example, *resource shares* are composable, as already assigned resource shares are not affected when the remaining resource shares get assigned. *Constraints* are also composable.

Separation of concerns, *separation of roles*, and *composability* proved to be the prerequisites towards a *business ecosystem* [23]. In a business ecosystem, a large number of loosely interconnected participants depend on each other for their mutual effectiveness and survival. Customers, lead producers, competitors, and other stakeholders are nodes in a constantly evolving network of relationships. A successful business ecosystem benefits from a symbiotic coexistence of its participants and their dedicated expertise (robotics expert, application domain expert, system integrator, developers of frameworks and tools, professional users, consumers, and the robot itself). The power of a business ecosystem shows up best when the different participants can act independently of each other. Instead of tightly managing all the details of the collaboration of the different roles involved in a project, a business ecosystem gets structure by managed interfaces between different roles and between different building blocks. A *business ecosystem for robotics software* [29,40,42] would share software efforts and lower software risks, would reduce costs, development time, and time-to-market, would take advantage of specialized and second source suppliers, and would increase the robustness and quality of products and services.

As an example, the personal computer industry has already shown very well the advantages of a business ecosystem based on separation of roles, separation of concerns, and composition. The personal computer market is based on stable interfaces that change only slowly but allow for parts changing rapidly, since the way parts interact can last longer than the parts themselves, and there is a huge number of cooperating and competing players involved. This resulted in a tremendous offer of systems and components, all with their specifics, ranging from cost-efficient office systems to high-end gaming systems, from portable laptops to desktop systems, from non-modifiable systems to extendable systems.

The central challenge is to identify the sweet spot between *freedom of choice* and *freedom from choice* [16,21]. Freedom of choice wants to support as many different schemes as possible and then leaves it to the user to decide which one best fits individual needs. However, this requires huge expertise and discipline at the user-side in order to avoid mixing non-interoperable schemes. Typically, academia tends towards preferring freedom of choice, since it seems to be as open and flexible as possible. However, the price to pay is high, since there is no guidance with respect to composability and system-level conformance. In contrast thereto, freedom from choice gives clear guidance with respect to architectural schemes and can ensure composability and system-level conformance. As long as a component developer follows these schemes, he can be sure that others can use his component in their system designs. Well-thought-out restrictions are not a universal negative, and freedom from choice gives guidance and assurance of properties beyond one's responsibilities in order to, e.g., ensure system-level conformance [16]. However, there is a high responsibility in selecting the proper schemes such that progress and future designs are not blocked.

An appropriate sweet spot between freedom of choice and freedom from choice can only be found by first agreeing on paramount objectives. In our view, these are the need for *separation of roles* and *composability*. These drive the proper clustering of subsets of concerns as a reasonable building block surviving in a market is always based on the combination of different concerns. Again, proper clustering of different concerns contributes to good architectural patterns.

Good architectural patterns should leave as much freedom as possible at a local level and should impose structure to a local level only to the extent that otherwise, global-level structures could not be fulfilled. Typically, industry tends towards freedom from choice by agreeing on (de facto) standards that represent the minimum set of structures to establish a business ecosystem. With respect to robotics software engineering, one has to identify which architectural patterns could organize separation of roles and achieve composability in the most appropriate way for establishing a robotics software business ecosystem.

This is related to different granularities of building blocks (libraries, components, architectures, system of systems, etc.), different schemes to manage variability (plugins into a framework, variants of a product line architecture, composition based on loosely-coupled services, etc.), and also the support for different roles in binding variability along the life-cycle of the robotics system (design time versus run-time binding, stepwise refinement, binding more and more variability within the development workflow by different stakeholders, etc.). It is also about means to support the different stakeholders such that they can most easily use such structures, adhere to them, and focus on their dedicated role (separation of development and use in time and space).

An example for freedom of choice is *ROS*. What its founders mean by "we do not wrap your main" [9] is, among other things, that they do not want to enforce any architectural design decisions for developers. In consequence, each developer uses his own personally preferred architecture, which is then very likely to conflict with those defined by others. Everyone first needs to understand the architectural decisions of each individual component before being able to reuse them in their own system. Their proposed solution is just to extensively document each *ROS component* on the *ROS portal*. However, this does not circumvent the need to extensively analyze and understand the source code in order to adjust it or to implement workarounds in order to somehow make components compatible and reusable. It is obvious that *ROS* (in line with its overall design philosophy) does not yet give enough structure in an appropriate format in order to better support separation of roles and separation of concerns. The minimally required structures are a sound software component model, which has to be formalized for use in model-driven tools in order to support separation of concerns (e.g., to maintain semantics independently of the OS/middleware mapping), to assist the different roles in conforming to structures like component life-cycles, and to reduce exposed complexity by systematic and computer-assisted management of variation points. As already mentioned, that led to the ongoing development of *ROS 2*, which introduces more structures.

8.3 SOFTWARE ENGINEERING TECHNOLOGIES BENEFICIAL IN ROBOTICS

In order to achieve separation of roles, we need to support a black-box view of the software building blocks. A black box view comes with explicated interfaces, properties, and variation points. That is needed for system composition, proper configuration, and hand-over to another role without requiring knowledge of inside details of the black box as long as you are not responsible for that building block. The following explanations partly reuse our previous descriptions [33].

Component-based software engineering (CBSE) [15] is an approach that has arisen in the software engineering community in the last decade. It shifts the emphasis in system-building from traditional requirements analysis, system design, and implementation to composing software systems from a mixture of reusable off-the-shelf and custom-built components. Software components explicitly consider reusable pieces of software, including notions of independence and late composition [44]. Composition can take place during different stages of the life-cycle of components; that is, during the design phase (design and implementation), the deployment phase (system integration), and even the run-time phase (dynamic wiring of data flow according to situation and context). *CBSE* is based on the explication of all relevant information of a component to make it usable by other software elements without the need to get in contact with the component provider. The key properties of *encapsulation* and *composability* result in the following seven criteria that make a good component: (i) may be used by other software elements (clients), (ii) may be used by clients without the intervention of the component's developers, (iii) includes a specification of all dependencies (hardware and software platform, versions, other components), (iv) includes a precise specification of the functionalities it offers, (v) is usable on the sole basis of that specification, (vi) is composable with other components, and (vii) can be integrated into a system quickly and smoothly [22]. An overview on *CBSE in robotics* is given in Brugali and Scandurra [5] and Brugali and Shakhimardanov [6].

Service-oriented architectures (SOA) are the policies, practices, and frameworks that enable application functionality to be provided and consumed as sets of services published at a granularity relevant to the service consumers. Services can be invoked, published, discovered, and abstracted away from the implementation using a single, standards-based form of interface [38]. In component models, services are represented as ports. This view puts the focus on the question of a proper level of abstraction of offered functionalities. Services combine information and behavior, hide the internal workings from outside intrusion, and present a relatively simple interface to the rest of the program [38].

Services are the key entities performing communication between providers and consumers. *SOA* is all about style (policy, practice, and frameworks), which makes process matters an essential consideration. A *SOA* has to ensure that services are not get reduced to the status of interfaces; rather, they have an identity of their own. With *SOA*, it is critical to implement processes ensuring that there are at least two different and separate processes – for providers and consumers. According to Sprott and Wilkes [38], principles of good service design enabled by characteristics of *SOA* are: (i) *reusable*: use of service, not reuse by copying of code/implementation, (ii) *abstracted*: service is abstracted from the implementation, (iii) *published*: precise, published specification functionality of service interface, not implementation, (iv) *formal*: formal contract between endpoints places obligations on provider and consumer, and (v) *relevant*: functionality is presented at a granularity recognized by the user as a meaningful service.

Model-driven software development (MDSD) is a technology that introduces significant efficiencies and rigor to the theory and practice of software development. *MDSD* is much more than code generation for different platforms to address the technology change problem and to make development more efficient by automatically generating repetitive code. The benefits of *MDSD* are manifold [39,45]: (i) models are free of implementation artifacts and directly represent reusable domain knowledge, including best practices, (ii) domain experts can play a direct role and are not requested to translate their knowledge into software representations, (iii) design patterns, sophisticated and optimized software structures, and approved software solutions can be made available to domain experts and enforced by embedding them in templates for use by highly optimized code generators such that even novices can immediately take advantage from an immense experience, and (iv) parameters and properties of components required for system-level composition and the adaptation to different target systems are explicated and can be modified within a model-driven toolchain.

A *domain-specific language (DSL)* is a programming or modeling language dedicated to a particular problem domain that offers specific notations and abstractions, which, at the same time, decrease the coding complexity and increase programmer productivity within that domain. *DSLs* based on good architectural patterns offer a high-level way for domain users to specify the functionality of their system at the right level of abstraction. *DSLs* and models have historically been used for programming complex systems and have become a separate field of study.

CBSE separates the component development process from the system integration process and aims at component reusability. *MDSD* separates domain knowledge (formally specified by domain experts) from how it is being implemented (defined by software experts using model transformations). *SOA* is about the right level of granularity for offering functionality and strictly separates service providers and service consumers. A *DSL* is a programming or modeling language specialized to a particular application domain. It aims at easiness for domain experts.

8.4 THE STEP CHANGE TO MODEL-DRIVEN APPROACHES IN ROBOTICS

Over the last decade, these software engineering technologies and the accompanying tooling workbenches have reached a maturity level that enables them to be tailored even to such complex domains as robotics. The most advanced spearhead is *SmartSoft/SmartMDSD*, which has already put together all these technologies to systematically address the software challenges in robotics. It provided the foundation for the next step change as moderated by the EU H2020 project *RobMoSys—Better Models, Better Tools, Better Systems*.

RobMoSys is the first approach of a community-wide consolidation process for model-driven software engineering in robotics with the broadest possible community involvement being agnostic to implementation technologies. *RobMoSys* developed and provides a methodology to organize the *robotics body of knowledge*. It aims for composable models and software for robotic systems. It provides consistent, agreed, and resilient reference structures for making robotics accessible and usable. Its aim is to shape an *EU digital industrial platform for robotics* together with the EU H2020 project *ROSIN*. *RobMoSys* moderates a community-wide consolidation process with the broadest possible community involvement. Due to its model-driven approach, it has the potential to explicate the relationships between the different, so far fragmented, approaches and to make understood their individual strengths but also their limitations. It thus helps in selecting the proper approach for the structure and size of the problem faced, being just as complex as needed to address the problem in a sound way. Most importantly, the model-driven approach has the potential to overcome fragmentation, link between the different worlds, and express stable structures and solutions independently of implementation technologies.

The *SmartSoft World* [41] has been one of the earliest approaches to systematically address the above-mentioned software engineering challenges in robotics based on the above-mentioned software technologies. Its contributions cover not only a service-oriented software component model with life-cycle management [29,30], meta-models and model-driven tooling [28,32,33], but also task-level coordination [42,43] as well as variability management [19,21,31].

Its concepts, tools, implementations, and applications form one of the pillars—together with other pillars outlined in Section 8.1—that prepared the ground for the *RobMoSys* project. The core concepts of the *SmartSoft World* have been particularly influential in shaping *RobMoSys* due to their maturity, illustrated in their industrial user community [11]. They became part of the consolidated body-of-knowledge of well-accepted best practices in model-driven software engineering for robotics that we explicated so far in *RobMoSys*. Within *RobMoSys*, we continuously further develop and consolidate these concepts to be consistent in a much bigger context. Naturally, at the same time, the *SmartSoft World* evolved into one of the fully conformant references for *RobMoSys* with a very high coverage of the *RobMoSys* content. This latest version of the *SmartSoft World* is refered to in Section 8.6. It comprises the extensions described in Lotz [17] and Stampfer [40].

8.5 COMPOSABLE MODELS AND SOFTWARE FOR ROBOTIC SYSTEMS

The following subsections now go through some selected key concepts of how models, model-driven software development, component-based software engineering, and service-oriented architectures are used to achieve *separation of roles* and *composability* for robotics software. These key concepts are part of the core of the body of knowledge compiled, consolidated, and explicated by *RobMoSys*. The full body of knowledge, which is continuously being extended, is accessible via the *RobMoSys Wiki* and related discussions via the *RobMoSys Forum*. Many of the following key concepts have already been part of the *SmartSoft World* but are now extended and well-edited along the much broader systematic context we established within *RobMoSys*.

8.5.1 ECOSYSTEM TIERS, BLOCKS, PORTS, CONNECTORS

An ecosystem is typically structured along three different tiers, as illustrated in Figure 8.1 and also explained in the *RobMoSys Wiki*. Tier 1 structures the ecosystem in general for robotics. It is shaped by the drivers of the ecosystem, which define an overall structure that enables composition and to which the lower tiers conform. Tier 1 is shaped by few representative experts, in this case the *core members of the RobMoSys project* as kick-starters. Tier 2 conforms to these foundations, structuring the particular domains within robotics. Each domain at Tier 2 (e.g., mobile platforms, manipulation, vision systems, simultaneous localization and mapping [SLAM], etc.) is shaped by representative domain experts. Tier 3 conforms to the domain-structures of Tier 2. Here are the main users of the ecosystem, for example, component suppliers and system builders. The number of users and contributors is significantly larger than the number on the earlier tiers, as everyone contributing or using a building block is located at Tier 3.

Another example of such ecosystem tiers is the *OPC UA ecosystem*. The *OPC Unified Architecture (OPC UA)* with all its elements for modeling machine data, its communication stack to access and transport them, and the related predefined application programming interfaces (APIs) forms the overall frame and thus, is located at Tier 1. The ecosystem driver at Tier 1 managing and keeping things consistent is the *OPC Foundation*. However, this is just the basic frame, which still gives too much freedom, as everybody can invent their own information model for a device. At the

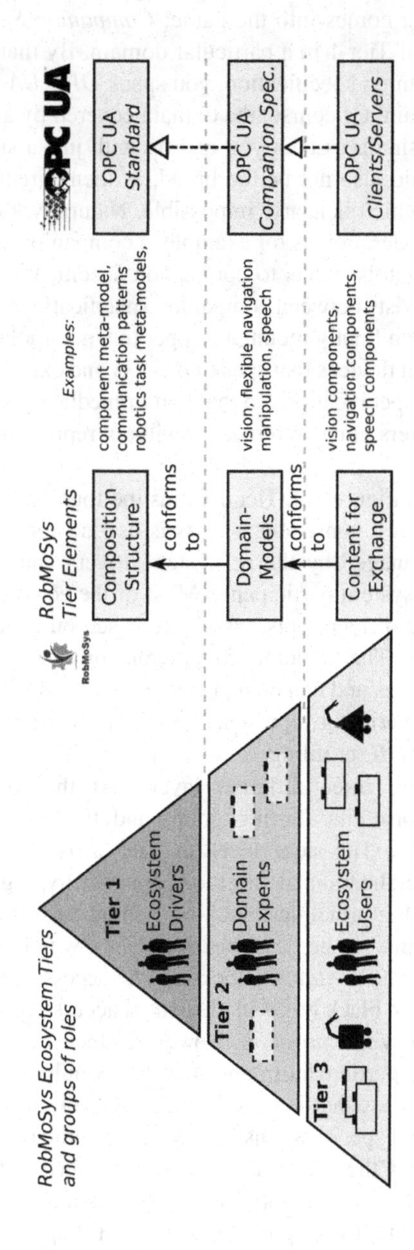

Figure 8.1 Tiers organizing a business ecosystem.

end, all device interfaces are based on *OPC UA*, but devices of the same category from different vendors all use different information models. In consequence, one can still not replace one device by another one without adjusting the software that speaks to a device. This is where Tier 2 comes into the game: *Companion Specifications* define how to use the elements of Tier 1 in a particular domain. By that, as long as you conform to a specific companion specification, you speak *OPC UA* in the same way as the others do in that domain. Of course, the domain covered by a companion specification should not be too small, as then you end up with just a single manufacturer with its devices. It should also not be too broad, as then agreement across domains needs to be achieved, which is nearly impossible. Naturally, that is a continuous process with constant moderation as for example, a companion specification in robotics has links via coordinate systems to companion specifications of vision systems. The same kind of link exists between companion specifications for manipulators and those for grippers as you finally mount a gripper to a manipulator. Finally, at Tier 3, there are all the different devices that speak *OPC UA* and express their conformance to specific companion specifications so that you immediately know how to use that device and by which others you can replace it without reprogramming your software to talk to it.

Details of Tier 1 are shown in Figure 8.2. Tier 1 forms the foundation to express the content of the lower layers in a consistent way as is needed for model-driven tooling. These structures are managed by the ecosystem drivers and are thus not directly seen by most of the ecosystem participants. Most of the ecosystem participants just use role-specific model-driven tools, which are based on these structures and thus, allow consistent results. The modeling foundations in Tier 1 are independent of specific tooling technologies and can be represented in different ways within concrete toolings. For instance, Variant 1 represents Tier 1 via *Ecore* meta-models, while variant 2 goes for *UML/SysML* profiling.

Tier 1 itself is split up into three different layers. At the topmost level of Tier 1, *hypergraphs* [36] form the starting point, and they are the anchor for all the other formalizations. The next level in Tier 1 is the *block-port-connector* level which is a specialization of the more abstract hypergraph model (https://robmosys.eu/wiki/modeling:principles:block-port-connector). Basically, the structure of most complex systems can be described by *blocks*, which are arranged via different relations like *has-a* and *contains*. Blocks can be accessed from outside only via *ports*. Thus, blocks form a black box with dedicated access points hiding all the internals. Ports can be linked via *connectors*. However, blocks, ports, and connectors are so generic that every participant in an ecosystem would again come up with its own semantics. By the way, that is exactly the drawback of, e.g., *SysML*, which also requires further refined specifications in order to become really useful. It is key not only to precisely specify the structure of a concept in terms of blocks and ports, but also to precisely specify the semantics of the blocks and the ports. This then allows for late binding of technologies producing exactly the specified behavior (that is, decoupling the body-of-knowledge from the much faster pace of technologies, e.g., becoming middleware-agnostic). Otherwise, blocks and ports are just so generic that everybody can fit their individual interpretation with arbitrary behavior.

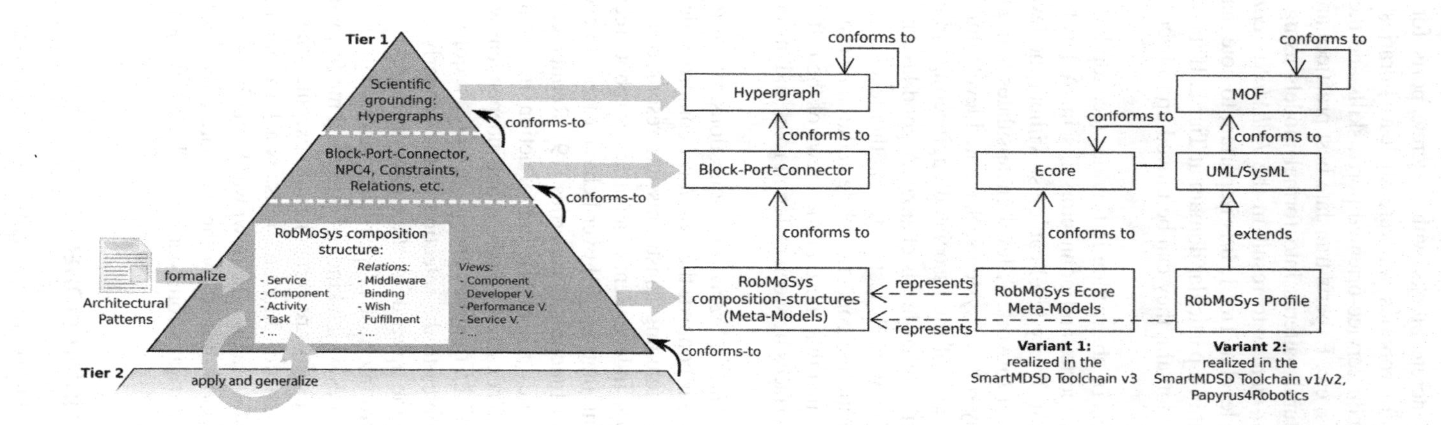

Figure 8.2 Details of Tier 1 and different options to map its structures into tooling technologies.

Thus, one needs to specify concrete blocks with concrete ports for concepts needed in a business ecosystem for robotics software such as libraries and their APIs, software components and their service-oriented ports, skills and their way to interact with software components, etc. Experts translate best practices and lessons learned as described by architectural patterns into formal models using the *RobMoSys block-port-connector meta-models* to result in the *RobMoSys composition structures*, which form the lowest level at Tier 1. It is important to note that composition structures can also arise within a particular domain at Tier 2. If these prove to be of use beyond just a particular domain, they can be moved up to become part of the Tier 1 composition structures.

Figure 8.3 shows different levels of abstraction in robotics (see top left) and how the universal approach of blocks, ports, and connectors fits in. A level is on top of another because it depends on it. Every level can exist without the levels above it, but it requires the levels below it to work. A level encapsulates and addresses a different part of the needs of many robotic systems. Separated levels thus decouple different parts, reduce the complexity of the associated engineering solutions, and allow reuse of a level even when the other levels change. A good layering goes for abstraction levels. Otherwise, different levels just go for another level of indirection. The challenge is to adhere to separation of concerns while at the same time, package, different concerns into structures such that these fit the views of the different roles. Thus, it is about managing the interfaces between different roles and different levels of abstraction.

Typically, the levels *task plot*, *skill*, and *service* rise questions, as these terms are used in many different flavors. In our context, the *task level* comprises descriptions of actions and their arrangement, including constraints like resource requirements and also qualities. The vocabulary is independent of a particular robot. Technologies can range from (dynamic) state automatons [42] over behavior trees to even hierarchical task nets [43] including non-functional properties [19]. Refinements of tasks end up at *skills*, which are the smallest reasonable unit of action one wants to reuse in task plots. *Upwards*, a skill is thus described in the robot-independent way needed for generic task plots but also with its qualities and its resource requirements. The latter two are decisive when it comes to the selection of the properly fitting skills when executing task plots. While the upward interface to tasks represents a generic description of a skill, the downward interface is specific to individual software components. *Downwards*, a skill describes which kinds of operating modes, wirings, and resource shares it requires at which software component to execute the skill. In that way, different components can be part of different skills, as a 1:1 mapping of component and skill proved not to be favorable. The downward interface of a skill talks to software components via the ports offered by the software components. In our system architecture, all ports of software components are based on *services*, as is explained in Section 8.5.2.

8.5.2 THE SOFTWARE COMPONENT MODEL

A *software component* is one concrete example of a block with ports that needs to be specified with its structure and its semantics at the lowest level of Tier 1 for shaping

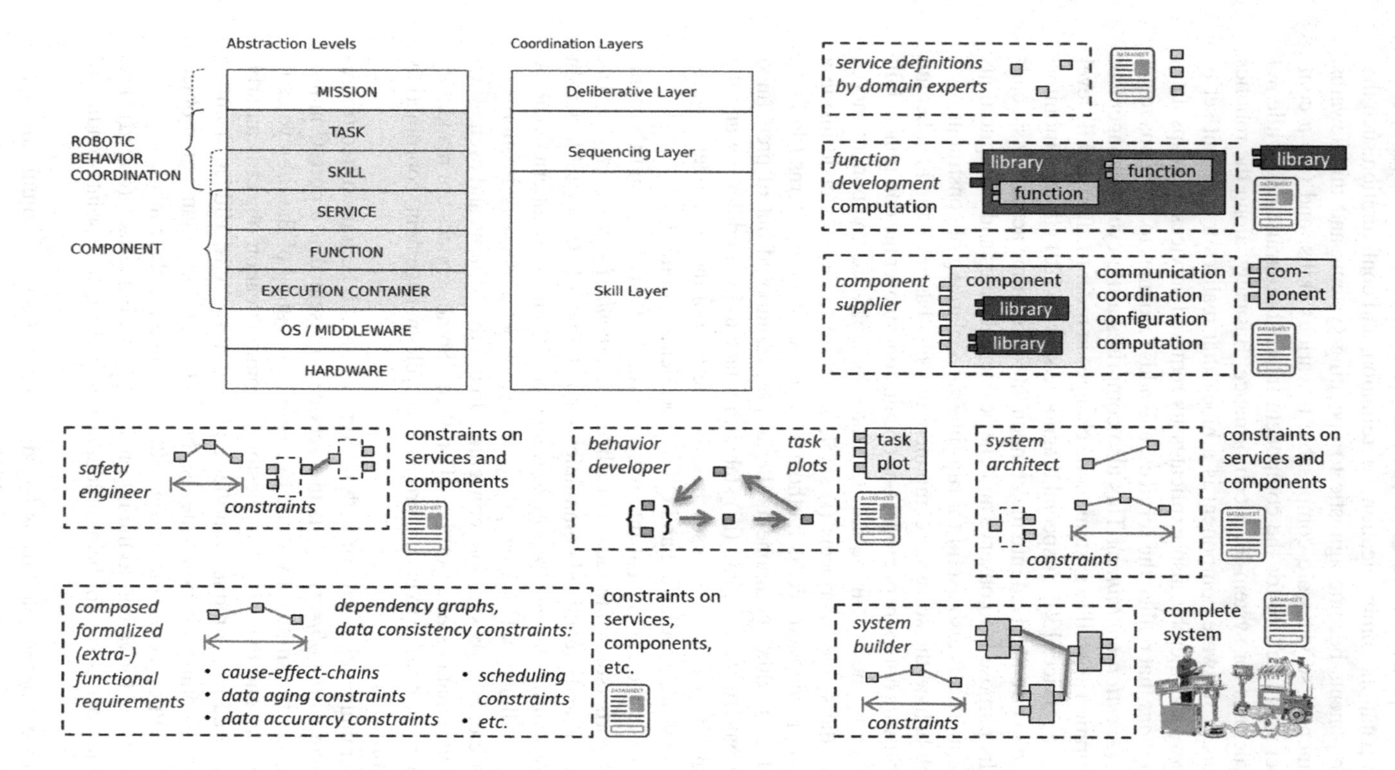

Figure 8.3 Levels of abstraction in robotics, blocks-ports-connectors, different roles, and data sheets.

composability and separation of roles. A software component model combines the concerns of computation, communication, coordination, and configuration such that (i) software components become agnostic to operating systems and middlewares, (ii) their implemented services are composable to various skills, and (iii) a component developer gets freedom for the component internal programming while not breaking the structures for system-level conformance. Basically, along the robotics abstraction levels, a software component is a block that realizes a coherent set of provided *services* (and thereby also explicates its required services), encapsulates the (private) *functions* that realize the services, and adds resources to functions to get them executed (*execution container*). This also comprises a life-cycle management covering configuration as well as system mode changes in order to allow a skill-level composition of the services [30]. A special diagnostic service [20] allows a monitoring infrastructure to be embedded into the component for run-time access by specialized monitors. Its purpose is to observe and check internal states and configurations of that component. This is also useful for run-time safe-guarding of constraints.

The approach behind the software component model—based on the idea behind the *SmartSoft* software component model—is to gain control over the component hull of the software components. Figure 8.4 illustrates the software component model and how it links the stable interfaces: (i) stable interface to the user code inside a component due to predefined access methods of communication patterns (dashed), (ii) stable interface to other components due to ports composed out of predefined communication patterns (solid), and (iii) stable interface to the middleware and the operating system (hidden from the user, internal interface of the component execution container to ensure middleware and OS independence) (dotted).

The link between the component internal ports (dashed interface) and the component external ports (solid interface) is realized via *communication patterns*. The small set of generic and predefined communication patterns is listed in Table 8.1. In general, a *publish/subscribe* and a *request/response* communication pattern would be sufficient to cover all needs. However, going for a small set of dedicated patterns on top of the generic ones makes it more convenient to use the patterns and even easier to understand the semantics of a port based on them. Some services are mandatory for every component in order to ensure proper life-cycle management, coordination, and configuration.

Binding a communication pattern with the type of data to be transmitted (represented as *communication object*) results in a service represented as a port. Communication objects are transmitted by-value to ensure decoupling of the life-cycles of the service-provider and the service-requestor. A communication object is an arbitrary object enriched by a unique identifier and get/set-methods. Hidden from the user and inside the communication patterns, the content of a communication object provided via the API of a communication pattern (dashed interface inside a component) is extracted and forwarded to the internal middleware interface (dotted). This hidden dotted interface is key for being middleware-agnostic and for late-binding of middlewares.

Since each externally visible port is based on one of these communication patterns, the port semantics is defined by the type of communication pattern used and

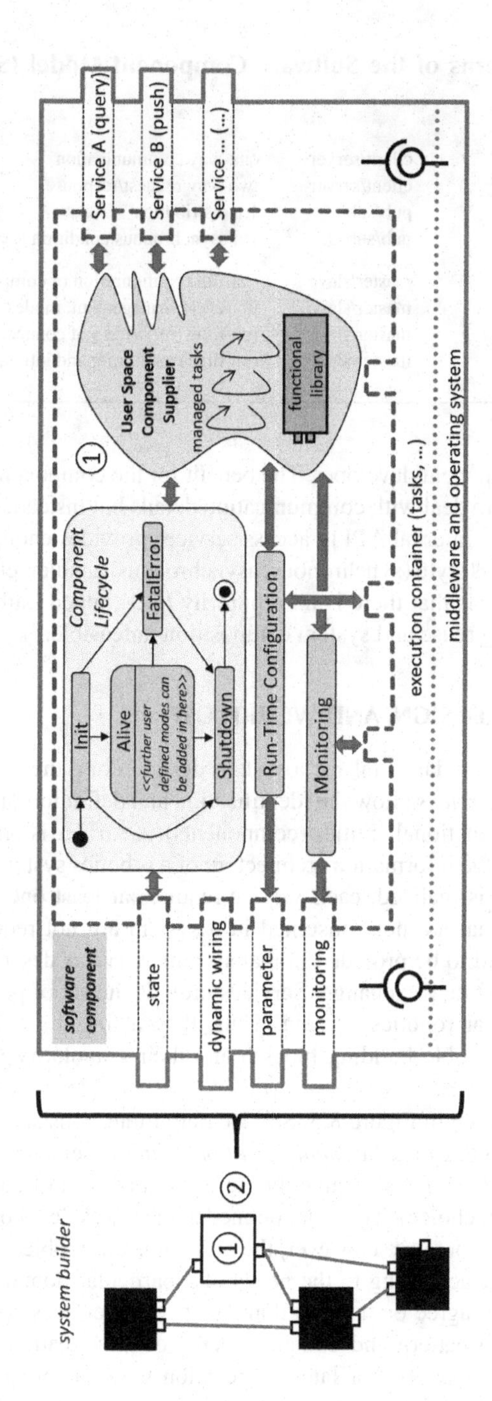

Figure 8.4 The software component model with the inside view of the component supplier ① and the outside view of the system builder ②.

Table 8.1
The Communication Patterns of the Software Component Model (See the *RobMoSys* Wiki)

Communication	*Send*	client/server	one-way communication
	Query	client/server	two-way request/response
	Push	pub/sub	1:n distribution
	Event	pub/sub	1:n asynchronous condition notification
Coordination	*Parameter*	master/slave	run-time configuration of components
	State	master/slave	life-cycle management, modes
	Dynamic Wiring	master/slave	run-time (re-)wiring of connections
	Monitoring	master/slave	run-time monitoring and introspection

cannot be modified by a component developer. The benefit for the component developer is that he does not have to deal with communication details but instead, can just program against services. The internal API to access services provides enough freedom to use the preferred access style (synchronous, asynchronous, handler, etc.). The benefit for the system builder is that there is no arbitrarily fine-grained multiplicity of port characteristics making reuse and system composition unfeasible.

8.5.3 SERVICES, SYSTEM DESIGN, AND WORKFLOW

The definition of *services* is fundamental for a system design along the principles consolidated in *RobMoSys*. *Services* allow the identification and definition of boundaries between encapsulated functional entities (components). A *service* is one of the visible entities that structure the information architecture of a robotics system. When the information architecture is realized, each *service* requires at least one component providing it. *Services* that are not consumed in a system are not required in that system. They do not need to be provided in that system. *Services* describe *how* components interact and *what* they exchange. Services provide the set of ports used by components of a particular robotics domain. Thus, they define the borders of software components (composable building blocks) and their granularity within a domain.

Some more details are shown in Figure 8.5. Service definitions consist of a *communication object* (data structure) and a *communication pattern* (semantics of the data structure transfer); see ①. A service can only use one out of a set of communication patterns (freedom from choice). These are defined at the lowest level of Tier 1, and they are the same for all domains. However, the communication objects can be defined individually at Tier 2 according to the needs of a particular domain. Thus, at Tier 2, domain experts can agree on a set of domain-specific services consisting of a standard communication pattern and domain-specific communication objects. Service definitions are the foundation for later aggregation to components (which implement the functionality behind services) [21].

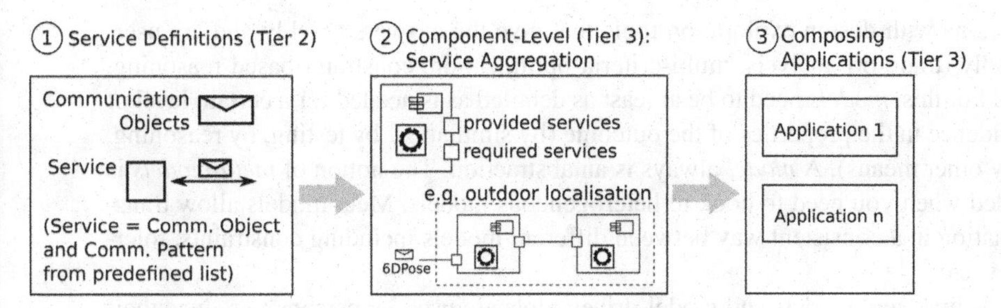

Figure 8.5 From service definitions over their use in information models of system architectures to applications based on components providing the required services.

Figure 8.3 introduces some roles in a business ecosystem for robotics software, such as a *function developer*, a *component developer*, a *system builder*, a *behavior developer*, and others like a *safety engineer* and a *service designer*. However, as a business ecosystem is based on managed interfaces between artifacts and roles (*composition* instead of *integration*), all the different roles coexist, work independently of each other and concurrently, and form ad hoc collaboration networks with added value when needed. For example, component developers can produce and offer components whenever they see a niche for a new one. They offer a component as-is, and there is no need to tightly work together with them for using that component.

In that setting, a *system architect* talks to customers that are seeking a robotics solution for their use case. The system architect translates the requirements into a set of services with their properties, like accuracy, etc. He now seeks for components or systems matching these services. In the ideal case, the *system builder* then finds an already existing solution that can be reconfigured to the new requirements. A bit more effort is needed when due to service requirements, some of the components need to be replaced. It might also be the case that some required services are not yet covered by any component on the market and thus, require the development of a new one. However, due to composition, the different building blocks—from devices over software components to task plots—can be reused and rearranged as far as needed. Services are valuable already for the design time, as they provide a checklist for early identification of white spots (services that are not yet provided/required by others).

Part ② in Figure 8.5 shows how an information model of a robotics system based on services is aligned with components that provide and require these services. Selecting the components fullfilling the service needs of the information model and configuring them accordingly corresponds to instantiating the information model. It is the composition of the application out of configurable as-is components by the *system builder*; see ③.

8.5.4 MODELS, DATA SHEETS, AND DEPENDENCY GRAPHS

You cannot manually go through all combinations of all parameter values in order to know about the properties of your system. You need to be able to answer *what-if* questions with design exploration tools that give the answers quickly and are user-friendly (trade-off analysis, multi-criteria-optimization, constraint-based reasoning, etc.). For this, *models* need to be at least as detailed as is needed for a certain level of confidence in the properties of the outcome (by simulation, by testing, by reasoning, or by other means). A *model* always is an abstraction. The notion of *meta-models* is needed when you need to hook to heterogeneous models. Meta-models allow transformation in a consistent way between different models including constraints, tolerances, etc.

In a first step, models and model-driven tools are used by persons to reduce their cognitive load. Adherence to the structures ensuring composability is then no longer by discipline but by tooling support. In a next step, robots themselves use models at run-time to monitor and to explain what they are doing. That can be followed by a third step where robots adapt models and improve them based on experiences made.

Figure 8.3 already annotated each block with a *data sheet* (https://wiki.servicerobotik-ulm.de/how-tos:documentation-datasheet:start). A data sheet is a particular flavor of a model. It is *the* means in a business ecosystem to make artifacts usable by others. It just comprises all the information you need to know in order to check how an artifact will behave in your setting. A data sheet explains (i) how to interact with the artifact via its ports, (ii) what are the foreseen variation points you can exploit for its configuration, and (iii) what are the functionalities with their extra-functional properties you get under which environmental conditions. A data sheet thus allows you to check beforehand whether a particular artifact fulfills your needs in your context and thus, then allows you to select the most adequate artifact. A data sheet respects intellectual property and does *not* contain details of the internals of an artifact. A data sheet is also not rich enough for synthesizing the artifact. However, it is a consistent abstraction of the artifact that gives just the information you need to know as a user with an outside view on the artifact. The composition of data sheets allows you to predict what you get without actually putting together the artifacts. However, when you then put together the artifacts described by the data sheets, you get as a real-world system what you expected.

It is decisive to understand one particularly important use of composable models in an ecosystem (Figure 8.6). We do not just model everything first or just compose all kinds of models first so that we then press the button to get the implementation of that model as a real-world artifact. One of the rare examples where this works is three-dimensional (3-D) printing where models are rich enough to allow a 3-D printer to synthesize the artifact. Instead, we use model-driven tools to guide a particular role in keeping conformance to composition structures. For example, the software component model comes with structures one needs to adhere to in order to gain from composability. Nevertheless, there is a lot of freedom in how to fill the internals of a software component, either just by programming or, of course, by again using model-driven tools.

As illustrated in Figure 8.7, a data sheet consists of two parts, a *formal part* (technical, used for composition) and a *descriptive part* (non-technical, used for finding and selecting the asset on a market platform). The formal part of the data sheet is derived from the models used in the model-driven tooling. In the case of a software

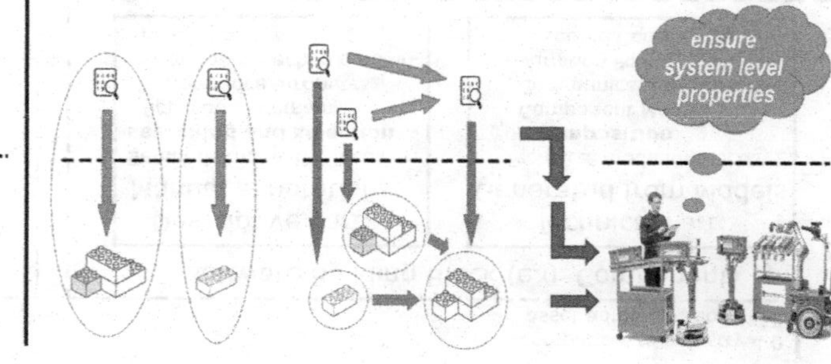
Composing different models for a full-fledged model for synthesis as the last step in the workflow so far only works in selected use-cases of 3D-printing.
from models to models
enrich,combine, analyze,predict, ...
from models to models
enrich,combine, analyze,predict, ...
Modeling without means to make models act is not a solution in robotics as robots finally need to act.
Data sheets (models of artefacts that act) represent components, sub-systems, task-plots etc. Suitability, traceability, simulation, etc. of system properties all via composed data sheets. When all is fine, then compose (put together and accordingly configure) the real artefacts to get the real system with properties as expected.
ensure system level properties

Figure 8.6 From models to robots that act: Composition of models and of artifacts.

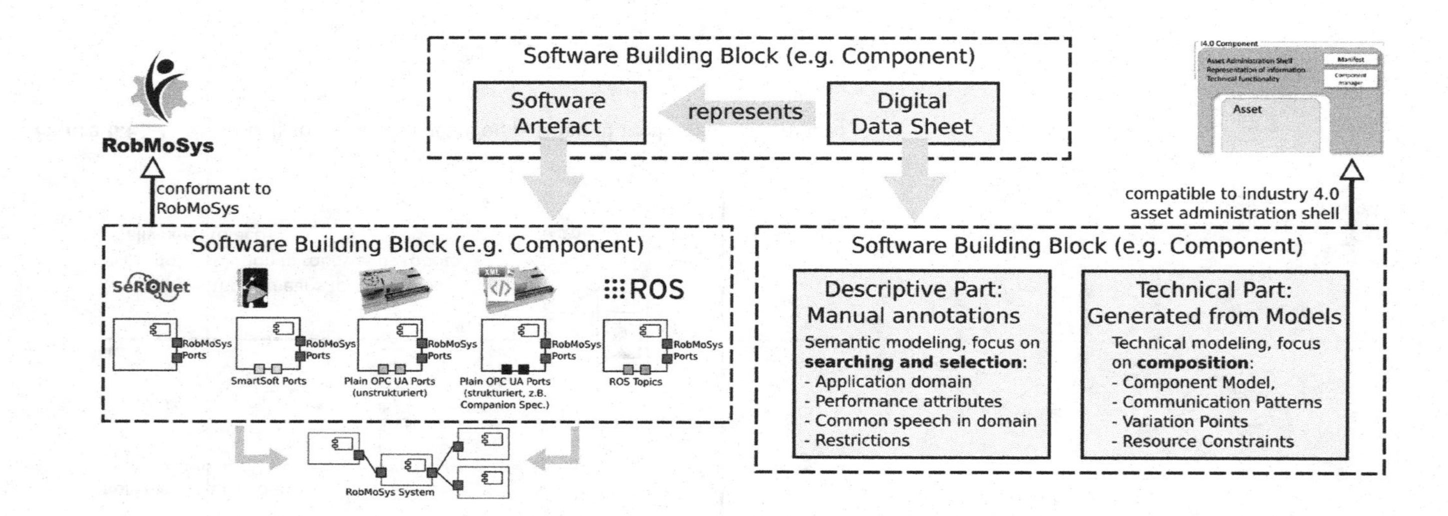

Figure 8.7 Models, structure of data sheets, and sub-models in the Industry 4.0 Asset Administration Shell.

component, these are, for example, the modeled ports including the variation points for configuring the component, its required resource shares, etc. That part of the data sheet can be used for what-if analysis, for predicting properties in your context, for a proper configuration along your system-level requirements, and for much more. The descriptive part is annotated manually and uses colloquial terms that are common in a particular domain. One has to think in descriptions for a market place. For example, one might use the term *collaborative robot* here. Such a term is then linked to relevant external references such as standards to define its semantics.

Dependency Graphs are a means to express relationships between different entities and/or properties. Basically, they can be used to check requirements. For example, a dependency graph can explicate along which chain of services raw sensor data ends up in a map service. One can then check whether the selected sensor is a safety device, as this might be required for the safety-critical map service. One can also check whether the cycle rates of the involved components ensure a timely update rate and thus match freshness requirements of the map service. Further examples are error propagation chains along services, so that one can check whether the variance of a selected sensor is good enough to match the variance constraints of the outcome of the processing chain. Dependency graphs can be brought up by, e.g., a system architect or a safety engineer, and they can cover in the same way functional requirements as well as extra-functional requirements. Their fulfillment can be traced through the complete life-cycle, ranging from the component selection process (working on data sheets) over the system composition process (working on data sheets and using analysis tools to find suitable configurations of components) to finally, the run-time binding of left-open variability (selecting and configuring skills, e.g., constraining the maximum speed in order to avoid blurring of sensor data). Another instance of dependency graphs is *cause-effect-chains* [18].

While *RobMoSys* provides the modeling foundations for coming up with a data sheet approach, there is still the need for agreeing on terms used in the *descriptive part*. While the formal structure of a data sheet is part of Tier 1, the finally used set of terms and their semantics is specific for a particular domain and is thus located at Tier 2. A suitable starting point might be *eCl@ss*. The data sheet, and in particular its descriptive part, is pushed within the *BWMi PAiCE SeRoNet* project. It conforms to *RobMoSys* but puts its focus onto the *OPC UA* technology and on ramping up a market place and brokerage platform for robotics components, systems, and applications. The concept of the data sheet is also an example of a submodel in the asset administration shell. The data sheet approach thus provides a link between model-driven approaches in robotics and those of Industry 4.0.

8.5.5 THE MIXED PORT COMPONENT AS A MIGRATION PATH

It is of paramount importance to provide a smooth migration path from current systems to the full-fledged composition structures offered by *RobMoSys*. The standard situation is that you have to cope with *brownfield settings*, and extremely rarely, you can go for a *greenfield project*. Thus, once you want to add your new parts along the full-fledged *RobMoSys* composition structures, you still need to be able to interact with your legacy parts. This is required anyway, as robotic systems need to interact with environments following other structures, such as cloud-based services or Industry 4.0 settings.

A *mixed port component* consists of two different sides of ports, as is illustrated in Figure 8.8. Side 1 presents ports following the *RobMoSys* composition structures. Side 2 consists of ports connecting to another world, for example, to *ROS topics*, *YARP*, *OPC UA devices*, and many others. From Side 1, the component looks like any other *RobMoSys* component. It comes with configurations, resource requirements, etc. From this side, the ports of Side 2 are not visible, and the systems to which the component connects are fully hidden behind the component. Thus, their resource management does not conflict with the *RobMoSys* way of managing resources. This is why a mixed port component connecting to *OPC UA* has the potential to fully replace all annoying proprietary device drivers in a robotics system by a standardized way to access devices just via *OPC UA*. The support of *OPC UA* is pushed in the scope of the *BMWi PAiCE SeRoNet* project.

It is important *not* to confuse mixed port components with the property of *RobMoSys* ports being middleware-agnostic. While a mixed port component talks at Side 2 via the original ports of that other world, it talks at Side 1 with *RobMoSys* ports adhering to their fixed semantics. Every *RobMoSys* port is middleware-agnostic and allows for a late binding of the underlying middleware (such as *ACE*, *DDS*, or even *OPC UA*) without changing the semantics or the properties of the ports (see Table 8.1). One can even select different middlewares for different *RobMoSys* ports of the same component. Of course, ports can interoperate technically only when they are mapped to the same middleware. However, due to the late binding of the middleware (which is, e.g., supported by the *SmartMDSD* Toolchain), the system builder can easily select the same one for interacting ports.

8.5.6 INTEROPERABILITY, COVERAGE, AND CONFORMANCE

Structures shaping an ecosystem and technologies filling it are continuously evolving and changing. As *RobMoSys* provides a methodology to organize the body-of-knowledge in robotics, coping with continuous change is an inherent part of this methodology. The tiers already provide a partitioning such that different domains can coexist with their specifics while still adhering to superordinated common and consolidated structures. There is a mechanism by which structures from Tier 2 can make it into Tier 1. Nevertheless, one also needs a means to express the relationships between different "islands" in that ecosystem. Key concepts here are *coverage* and *conformance*. Being able to express for any type of artifact (whether it is models, tools, software components, or whatever) which parts of the consolidated ecosystem structures it covers, and what is the level of conformance of that coverage, allows non-conforming contributions to coexist. This is key to being inclusive for so far "outside worlds" until their concepts are consistently included into the body-of-knowledge. This is also key for allowing new approaches to evolve within that body-of-knowledge but without confusing the participants about what you get from which artifacts and how they fit in where. In Figure 8.9b, this is illustrated by

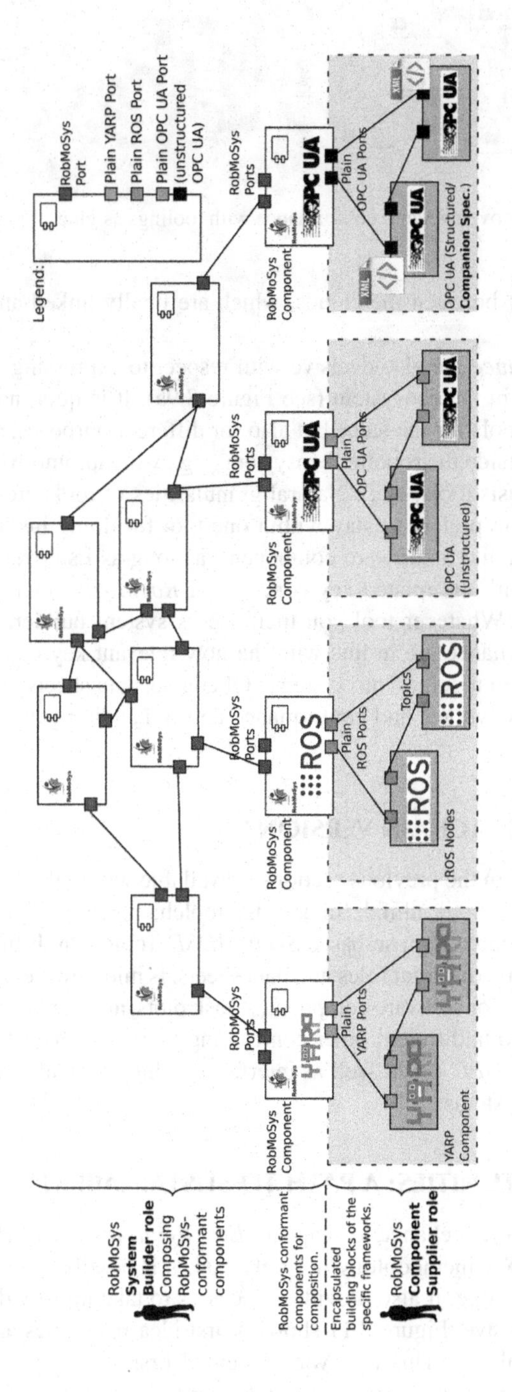

Figure 8.8 Middleware-agnostic software components and mixed port components as migration path.

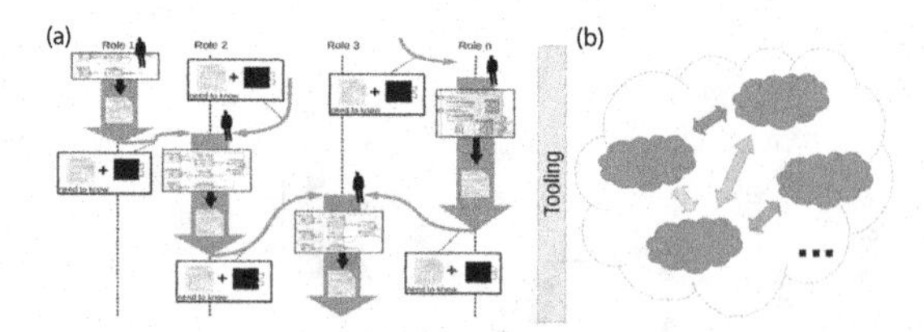

Figure 8.9 Interoperability, coverage, and conformance with toolings as glue.

arrows of varying intensity between the clouds, which are finally linked and made accessible via tooling.

Coverage and *conformance* are also decisive with respect to expressing interoperability of different tools in the ecosystem (see Figure 8.9a). It is quite natural to have various (coexisting) tools for the same but also for different purposes, and that these interact with tools outside the robotics ecosystem (e.g., with automotive timing analysis tools, safety analysis tools, etc.). Natural granularities of tools are defined by *roles*, as a particular role prefers to stay within one tool for doing his/her task. Whatever tool you use to build a software component, as long as its coverage and conformance are in line with the related key concepts of *RobMoSys*, then you get a composable component. Whatever tool you then use as system builder, as long as its coverage and conformance are in line with the now relevant key concepts of *RobMoSys*, you can put them together into systems. Of course, data sheets are a central concept in expressing coverage and conformance and in handing over building blocks.

8.6 SMARTMDSD TOOLCHAIN VERSION 3

All that has been described in the previous sections is available within the *SmartSoft World* (https://wiki.servicerobotik-ulm.de/smartmdsd-toolchain:start). Figure 8.10 illustrates the latest open-source Eclipse-based *SmartMDSD* Toolchain. It fully conforms to *RobMoSys*, supports different roles and data sheets, is middleware-agnostic, and comes with repositories for software components, task plots, and application systems. Via plug-ins, it supports different other conforming worlds such as *SeRoNet*, which includes support for *OPC UA* as well as interfaces to interact with brokerage platforms of business ecosystems.

8.7 FUTURE OPPORTUNITIES: A PATH TO SUSTAINABILITY

There is for the first time a real chance to shape an *EU Digital Industrial Platform for Robotics*. The *RobMoSys* methodology not only makes it possible to structure the robotics body-of-knowledge; it also provides processes to take up new developments and to be most inclusive. Figure 8.11 shows a first idea for bodies and roles needed to achieve sustainability. This idea was presented first at the workshop of the *Topic Group on Software Engineering, System Integration, System Engineering*

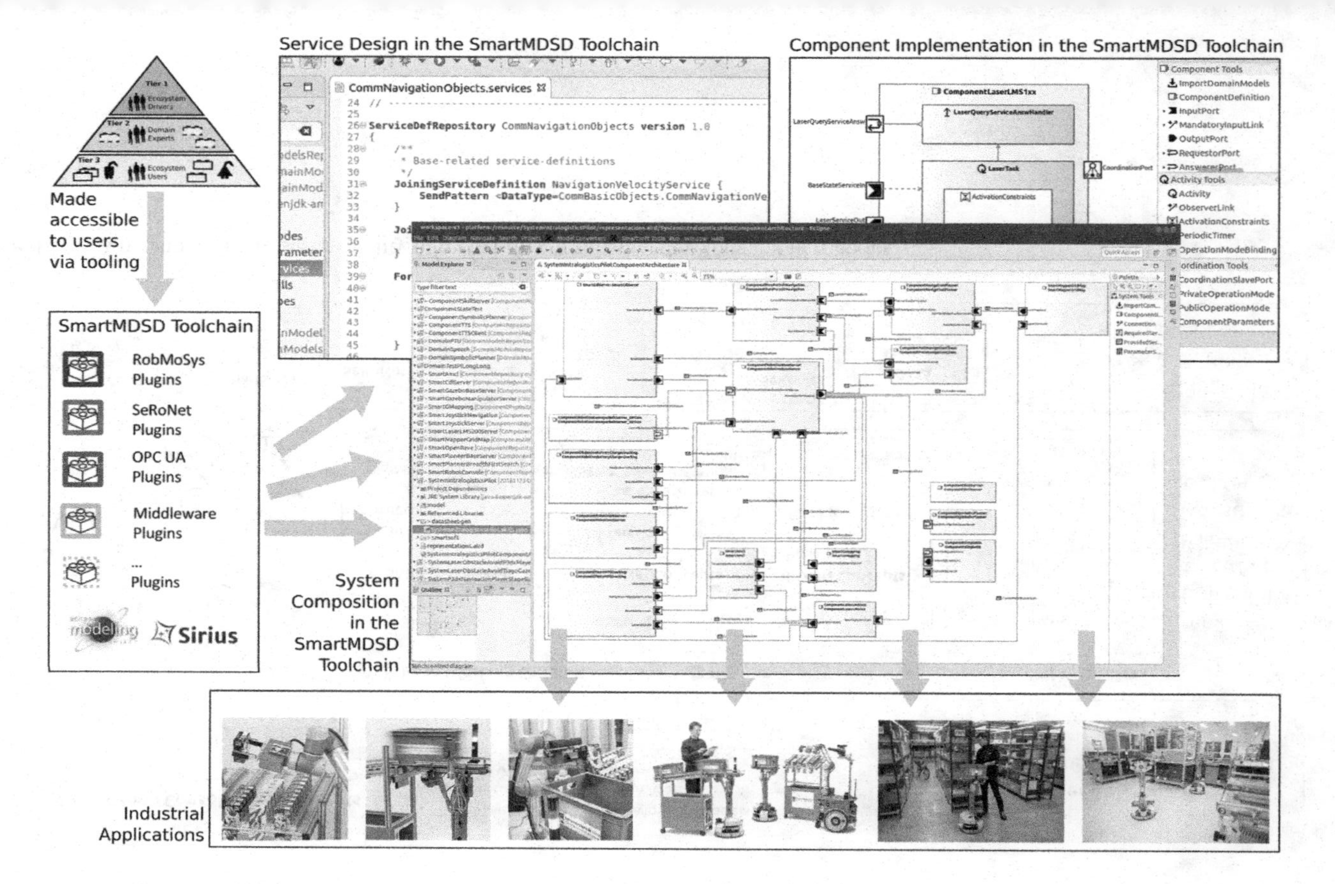

Figure 8.10 The *SmartSoft World*: *SmartMDSD* Toolchain, *SmartSoft* implementation, repositories.

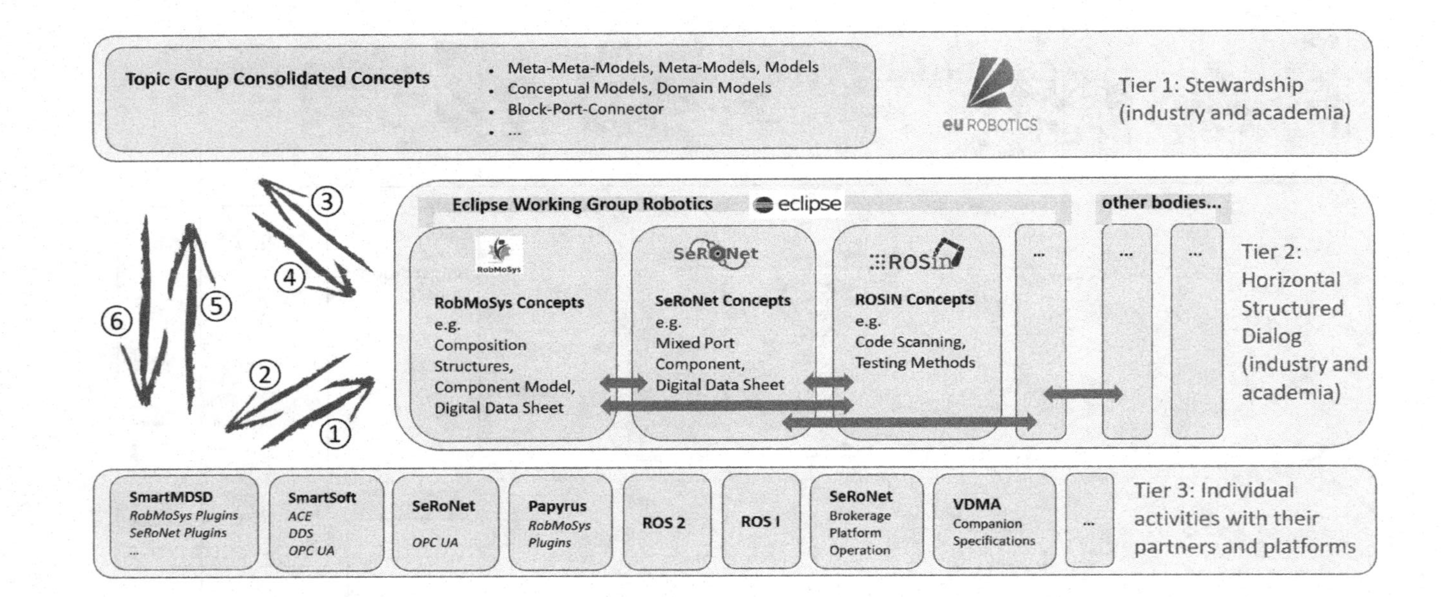

Figure 8.11 A potential path to sustainability of model-driven software development in robotics.

at the *European Robotics Forum (ERF)* in 2019. This first idea was received very well. It is not at all carved in stone yet, but it forms the starting point for further discussions, improvements, refinements, and concrete activities. The cornerstones are again three different tiers: (i) the role of a trustworthy *stewardship* (Tier 1) for established structures (e.g., under the roof of the *euRobotics aisbl*, which already gathers the community), (ii) an *institutionalized setting* (Tier 2) for a horizontal structured dialogue for coming up with aligned proposals for extending Tier 1, and (iii) finally, the individual activities with their platforms and partners. Obviously, the glue can be explicated again via *coverage* and *conformance*.

In short, individual projects provide new concepts and bring them into Tier 2 ①. Projects can align with horizontal technical discussions at Tier 2 and they have implementations of these concepts ②. Tier 2 proposes consolidated extensions to Tier 1 ③, while Tier 1 gives hints on what is needed and what one might want to harmonize ④. Tier 3 projects can also directly bring in consolidated extensions to Tier 1 ⑤. Projects use the consolidated concepts of Tier 1 and express their coverage and conformance ⑥.

To summarize, there is now for the first time the chance that we as the robotics community can all together make the shift from handcrafted robotic software systems to an *EU Digital Industrial Platform for Robotics*.

REFERENCES

1. N. Ando, T. Suehiro, K. Kitagaki, T. Kotoku, and Woo-Keun Yoon. RT-Component Object Model in RT-Middleware Distributed Component Middleware for RT (Robot Technology). In *IEEE International Symposium on Computational Intelligence in Robotics and Automation (CIRA '05)*, Espoo, Finland, pages 457–462, June 2005.
2. L. Andrade, J. L. Fiadeiro, J. Gouveia, and G. Koutsoukos. Separating Computation, Coordination and Configuration. *Journal of Software Maintenance and Evolution: Research and Practice*, 14(5):353–369, 2002.
3. A. Björkelund, L. Edström, M. Haage, J. Malec, K. Nilsson, P. Nugues, S. Robertz, D. Störkle, A. Blomdell, R. Johansson, M. Linderoth, A. Nilsson, A. Robertsson, A. Stolt, and H. Bruyninckx. On the Integration of Skilled Robot Motions for Productivity in Manufacturing. In *IEEE International Symposium on Assembly and Manufacturing (ISAM)*, Tampere, Finland, pages 1–9, 2011.
4. D. Brugali. *Software Engineering for Experimental Robotics (Springer Tracts in Advanced Robotics)*. Springer-Verlag, Berlin, Heidelberg, 2007.
5. D. Brugali and P. Scandurra. Component-based Robotic Engineering (Part I). *IEEE Robotics & Automation Magazine*, 16(4):84–96, December 2009.
6. D. Brugali and A. Shakhimardanov. Component-Based Robotic Engineering (Part II). *IEEE Robotics & Automation Magazine*, 17(1):100–112, March 2010.
7. H. Bruyninckx. Separation of Concerns: The 5Cs - Levels of Complexity, 2011. Lecture Notes, Embedded Control Systems, Spring.
8. H. Bruyninckx, M. Klotzbücher, N. Hochgeschwender, G. Kraetzschmar, L. Gherardi, and D. Brugali. The BRICS Component Model: A Model-Based Development Paradigm for Complex Robotics Software Systems. In *Proceedings of the 28th Annual ACM Symposium on Applied Computing*, SAC '13, New York, NY, USA, 2013. ACM.

9. S. Cousins, B. Gerkey, K. Conley, and W. Garage. Sharing Software with ROS [ROS Topics]. *IEEE Robotics Automation Magazine*, 17(2):12–14, June 2010.
10. S. Dhouib, S. Kchir, S. Stinckwich, T. Ziadi, and M. Ziane. RobotML, a Domain-Specific Language to Design, Simulate and Deploy Robotic Applications. In I. Noda, N. Ando, D. Brugali, and J. J. Kuffner, editors, *3rd International Conference on Simulation, Modeling and Programming for Autonomous Robots, SIMPAR*, Lecture Notes in Computer Science. Springer Berlin Heidelberg, November 2012.
11. FESTO Openrobotino Wiki. http://wiki.openrobotino.org/index.php?title=Smartsoft.
12. B. Gerkey. http://design.ros2.org/articles/why_ros2.html, 2014. Open Source Robotics Foundation.
13. L. Gherardi and D. Brugali. Modeling and Reusing Robotic Software Architectures: The Hyperflex Toolchain. In *IEEE International Conference on Robotics and Automation (ICRA)*, Hong Kong, China, May 2014.
14. M. Hägele, N. Blümlein, and O. Kleine. EFFIROB - Wirtschaftlichkeitsanalysen neuartiger Servicerobotik-Anwendungen und ihre Bedeutung für die Robotik-Entwicklung, 2011. Fraunhofer-Studie.
15. G. T. Heineman and W. T. Councill, editors. *Component-Based Software Engineering: Putting the Pieces Together*. Boston, MA, Addison-Wesley, June 2001.
16. E. A. Lee and S. A. Seshia. *Introduction to Embedded Systems, A Cyber-Physical Systems Approach*. Cambridge, MA, MIT Press, second edition, 2017.
17. A. Lotz. *Managing Non-Functional Communication Aspects in the Entire Life-Cycle of a Component-Based Robotic Software System*. Dissertation, Technische Universität München, München, 2018.
18. A. Lotz, A. Hamann, R. Lange, C. Heinzemann, J. Staschulat, V. Kesel, D. Stampfer, M. Lutz, and C. Schlegel. Combining Robotics Component-Based Model-Driven Development with a Model-Based Performance Analysis. In *Proceedings IEEE International Conference on Simulation, Modeling, and Programming for Autonomous Robots (SIMPAR)*, San Francisco, CA, pages 170–176, December 2016.
19. A. Lotz, J. F. Inglés-Romero, D. Stampfer, M. Lutz, C. Vicente-Chicote, and C. Schlegel. Towards a Stepwise Variability Management Process for Complex Systems: A Robotics Perspective. In *International Journal of Information System Modeling and Design (IJISMD)*, Hershey, PA, volume 5, pages 55–74. IGI Global, 2014.
20. A. Lotz, A. Steck, and C. Schlegel. Runtime Monitoring of Robotics Software Components: Increasing Robustness of Service Robotic Systems. In *15th International Conference on Advanced Robotics (ICAR)*, Tallinn (Estonia), June 2011.
21. M. Lutz, D. Stampfer, A. Lotz, and C. Schlegel. Service Robot Control Architectures for Flexible and Robust Real-World Task Execution: Best Practices and Patterns. In *Informatik 2014, Workshop Roboter-Kontrollarchitekturen*. Stuttgart, Germany, Springer LNI der GI, September 2014. ISBN 978-3-88579-626-8.
22. B. Meyer. What to Compose. *Software Development*, 8(3):59, 71, 74–75, 2000.
23. J. F. Moore. Predators and Prey: A New Ecology of Competition. *Harward Business Review*, 71(3):75–83, 1993.
24. A. Nordmann, N. Hochgeschwender, and S. Wrede. A Survey on Domain-Specific Languages in Robotics. In D. Brugali, J. F. Broenink, T. Kroeger, and B. A. MacDonald, editors, *Simulation, Modeling, and Programming for Autonomous Robots*, volume 8810 of *Lecture Notes in Computer Science*, pages 195–206. Springer Cham Heidelberg, Germany, Springer International Publishing, 2014.
25. M. Radestock and S. Eisenbach. Coordination in Evolving Systems. In *Trends in Distributed Systems CORBA and Beyond*, volume 1161 of *Lecture Notes in Computer Science*. O. Spaniol, C. Linnhoff-Popien, B. Meyer (eds), Springer, Berlin, Heidelberg, 1996.

26. C. Schlegel. *Navigation and Execution for Mobile Robots in Dynamic Environments: An Integrated Approach*. PhD thesis, Ulm, Germany, University of Ulm, 2004.
27. C. Schlegel, T. Hassler, A. Lotz, and A. Steck. Robotic Software Systems: From Code-Driven to Model-Driven Designs. In *Proceedings of the International Conference on Advanced Robotics*, Munich, Germany, pages 1–8, June 2009.
28. C. Schlegel, A. Lotz, M. Lutz, and D. Stampfer. Supporting Separation of Roles in the SmartMDSD-Toolchain: Three Examples of Integrated DSLs. In *(invited talk) 5th International Workshop on Domain-specific Languages and Models for Robotic Systems (DSLRob) in Conjunction with the International Conference on Simulation, Modeling, and Programming for Autonomous Robots (SIMPAR)*, Bergamo, Italy, October 2014.
29. C. Schlegel, A. Lotz, M. Lutz, D. Stampfer, J. F. Inglés-Romero, and C. Vicente-Chicote. Model-Driven Software Systems Engineering in Robotics: Covering the Complete Life-Cycle of a Robot. In *Informatik 2013, Workshop Roboter-Kontrollarchitekturen*. Koblenz, Germany, Springer LNI der GI, September 2013.
30. C. Schlegel, A. Lotz, and A. Steck. SmartSoft: The State Management of a Component. Technical report, University of Applied Sciences Ulm, January 2011.
31. C. Schlegel and D. Stampfer. The SmartMDSD Toolchain: Supporting Dynamic Reconfiguration by Managing Variability in Robotics Software Development. Tutorial on Managing Software Variability in Robot Control Systems. In *Robotics: Science and Systems Conference (RSS 2014)*, Berkeley, CA, July 2014.
32. C. Schlegel, A. Steck, and A. Lotz. Model-Driven Software Development in Robotics: Communication Patterns as Key for a Robotics Component Model. In Daisuke Chugo and Sho Yokota, editors, *Introduction to Modern Robotics*. Hong Kong, iConcept Press, 2012. ISBN:978-0980733068.
33. C. Schlegel, A. Steck, and A. Lotz. Robotic Software Systems: From Code-driven to Model-driven Software Development. In Ashish Dutta, editor, *Robotic Systems - Applications, Control and Programming*. Rijeka, Croatia, InTech, 2012. ISBN:978-953-307-941-7.
34. C. Schlegel and R. Wörz. Interfacing Different Layers of a Multilayer Architecture for Sensorimotor Systems Using the Object-Oriented Framework SMARTSOFT. In *Third European Workshop on Advanced Mobile Robots (Eurobot '99)*, pages 195–202, Zurich, 1999. IEEE.
35. C. Schlegel and R. Wörz. The Software Framework SMARTSOFT for Implementing Sensorimotor Systems. In *Proceedings IEEE/RSJ International Conference on Intelligent Robots and Systems (IROS)*, Kyongju, Korea, volume 3, pages 1610–1616, 1999.
36. E. Scioni, N. Huebel, S. Blumenthal, A. Shakhimardanov, M. Klotzbuecher, H. Garcia, and H. Bruyninckx. Hierarchical Hypergraphs for Knowledge-centric Robot Systems: A Composable Structural Meta Model and its Domain Specific Language NPC4. *JOSER - Special Issue on Domain-Specific Languages and Models for Robotic Systems*, 7(1):55–74, 2016.
37. SPARC - The Partnership for Robotics in Europe. Strategic Research Agenda (SRA) for Robotics in Europe 2014-2020, 2013 & 2014. euRobotics aisbl.
38. D. E. Sprott and L. Wilkes. Understanding Service-Oriented Architecture. *The Architecture Journal*, 1(1):10–17, 2004.
39. T. Stahl, M. Völter, and K. Czarnecki. *Model-Driven Software Development: Technology, Engineering, Management*. John Wiley & Sons, Inc., Hoboken, NJ, 2006.
40. D. Stampfer. *Contributions to System Composition using a System Design Process driven by Service Definitions for Service Robotics*. Dissertation, Technische Universität München, München, 2018.

41. D. Stampfer, A. Lotz, M. Lutz, and C. Schlegel. The SmartMDSD Toolchain: An Integrated MDSD Workflow and Integrated Development Environment (IDE) for Robotics Software. *JOSER - Special Issue on Domain-Specific Languages and Models for Robotic Systems*, 7(1):3–19, 2016.
42. D. Stampfer and C. Schlegel. Dynamic State Charts: Composition and Coordination of Complex Robot Behavior and Reuse of Action Plots. *Journal of Intelligent Service Robotics*, 7(2):53–65, 2014.
43. A. Steck and C. Schlegel. Managing Execution Variants in Task Coordination By Exploiting Design-Time Models at Run-Time. In *IEEE International Conference on Intelligent Robots and Systems (IROS)*, San Francisco, CA, USA, 2011.
44. C. A. Szyperski, D. Gruntz, and S. Murer. *Component Software - Beyond Object-oriented Programming, 2nd Edition.* Addison-Wesley component software series. Addison-Wesley, Boston, 2002.
45. M. Völter. Model-Driven Software Development - Introduction and Overview, 2006. www.voelter.de/data/presentations/InfWest-IntroToMDSD.ppt.

9 Network Robotics

Lorenzo Sabattini, Cristian Secchi, and Claudio Melchiorri

CONTENTS

9.1 INTRODUCTION

Robotic systems have been extensively used, since the middle of the 20th century, in industrial environments for automated production lines. Typically, robotic systems are utilized for high speed-production lines for the production of complex items such as automobiles, electronic goods, etc., where high productivity is mandatory.

While robotic lines are very effective in obtaining high productivity, their deployment requires huge economic investments. Hence, they are not a practical solution for small-size factories, where typical production consists in small batches of rapidly changing products, such as in an artisan workshop. In those situations, it is necessary to maximize the flexibility of the production equipment with general-purpose tools, whose use can be combined based on the particular *production cycle*.

Along these lines, we envision an automated robotic solution that is suitable for this situation. In particular, consider a team of mobile robots, each of which is equipped with a (or multiple) general-purpose tool(s) (e.g., a drill, a hammer, a solder, a screwdriver, ...). The sequence of operations to be performed depends on the particular object to be produced. Considering mobile robots the sequence of operations can be constantly changed, as tools are not constrained on a fixed infrastructure. Hence, the production cycle of the object defines the *trajectories* to be followed by the mobile robots.

In typical production cycles, operations cannot be executed in any order: their execution needs to be coordinated. Therefore, it is necessary to introduce a coordination strategy among the mobile robots, in such a way that the desired sequence is obtained. Coordination in multi-robot systems has been extensively investigated in the literature: typically, the objective is to implement decentralized control strategies for regulating the overall state of the multi-robot system to some desired configuration, thus obtaining coordinated behaviors such as aggregation, swarming, formation control, coverage, and synchronization [2,6,8,14,21,36].

Conversely, in this application, we are interested in solving a different problem; that is, controlling each robot to follow a predefined trajectory. A few strategies can be found in the literature that consider the problem of coordinating multiple robots along complex trajectories. For instance, Tsiotras and Castro [49] extends the standard consensus protocol to obtain periodic geometric patterns, while Pimenta et al. [24] and Sabattini et al. [37] present decentralized strategies for the coordination of groups of mobile robots moving along arbitrary non-trivial paths. Hence, the objective can be fulfilled by programming all the trajectories related to every different production cycle on each mobile robot.

However, this solution is not scalable and becomes rapidly impractical as the number of robots increases, in particular if production cycles frequently change. In fact, robots need to agree on the currently implemented production cycle, and each robot needs to have knowledge of the trajectories that characterize each single production cycle. This can be realized by equipping each robot with advanced sensing, communication, and elaboration devices that lead to the possibility of implementing decentralized high-level decision-making processes.

Conversely, we propose to adopt a heterogeneous solution, in which the multi-robot system is partitioned into two groups. In particular, a small group of *independent robots* act as supervisors: they are in charge of defining the particular production cycle to be executed. Their motion is then defined in order to provide an input for the remaining robots, namely, the *dependent robots*, which are those robots that are actually equipped with tools and act as workers. It is worth noting that in this scenario, dependent robots do not need to perform high-level decision-making operations and hence, do not need high-performance sensing and communication devices.

9.2 STATE OF THE ART

9.2.1 GRAPH THEORY AND CONSENSUS

In this section, we summarize some of the main notions on graph theory used hereafter. Further details can be found, for instance, in Godsil and Royle [9].

Let $\mathcal{G}$ indicate a generic undirected graph: we will always refer to undirected graphs unless otherwise specified. Let $\mathsf{V}(\mathcal{G})$ and $\mathsf{E}(\mathcal{G})$ be the vertex set and the edge set of the graph $\mathscr{G}$, respectively. Moreover, let N be the cardinality of $\mathsf{V}(\mathscr{G})$ (i.e., the number of vertices, or nodes, of the graph), and let M be the cardinality of $\mathsf{E}(\mathscr{G})$ (i.e., the number of edges, or links, of the graph). Clearly, $\mathsf{E} \subseteq \mathsf{V} \times \mathsf{V}$.

Let $v_i, v_j \in \mathsf{V}(\mathscr{G})$ be the ith and the jth vertices of the graph, respectively. Then, v_i and v_j are neighbors if $(v_i, v_j) \in \mathsf{E}(\mathscr{G})$. Given an undirected graph, $(v_i, v_j) \in \mathsf{E}(\mathscr{G})$ if and only if $(v_j, v_i) \in \mathsf{E}(\mathscr{G})$.

Define then an indexing of the edges of the graph, namely:

$$\mathsf{E} = \{e_1, \ldots, e_M\} \tag{9.1}$$

Defining then an arbitrary orientation of each edge, the incidence matrix $\mathscr{I}(\mathscr{G}) \in \mathbb{R}^{N \times M}$ can be defined as a matrix whose (i,k)th element ι_{ik} is [10]

$$\iota_{ik} = \begin{cases} -1 \text{ if } v_i \text{ is the head of } e_k \\ 1 \text{ if } v_i \text{ is the tail of } e_k \\ 0 \text{ otherwise} \end{cases} \tag{9.2}$$

In an edge-weighted graph, a positive number (the weight) is associated with each edge of the graph. Let $w_k > 0$ be the weight associated with the kth edge, and let $w = [w_1, \ldots, w_M] \in \mathbb{R}^M$. Then, the weight matrix $\mathscr{W}(\mathscr{G}) \in \mathbb{R}^{M \times M}$ is defined as $\mathscr{W}(\mathscr{G}) = \operatorname{diag}(w)$.

The (weighted) Laplacian matrix $\mathscr{L}(\mathscr{G}) \in \mathbb{R}^{N \times N}$ of the graph $\mathscr{G}$, associated with the weight matrix $\mathscr{W}(\mathscr{G})$, can then be defined as follows:

$$\mathscr{L}(\mathscr{G}) = \mathscr{I}(\mathscr{G}) \mathscr{W}(\mathscr{G}) \mathscr{I}^T(\mathscr{G}) \tag{9.3}$$

Graphs are often exploited for modeling the communication between robots in multi-robot systems, and for this reason, they are also called *communication graphs*. A vertex i represents a robot, and the presence of an edge between vertices i and j means that robots i and j can exchange information.

The *consensus problem* [21] is a well-known and widely studied problem in the field of decentralized control of multi-robot systems. Consider a group of N robots that are modeled as holonomic kinematic systems:

$$\dot{\chi}_i = u_i \tag{9.4}$$

where $\chi_i \in \mathbb{R}^\nu$ is the state of the ith agent. The consensus problem for a multi-robot system, whose goal is to drive the whole system to a final common state, can be solved, in a completely decentralized way, with the following Laplacian-based feedback interconnection law:

$$\dot{\chi}_i = -\sum_{j \in \mathscr{N}_i} w_{ij} (\chi_i - \chi_j) \tag{9.5}$$

where $w_{ij} > 0$ is the edge weight of the edge that connects the ith and the jth robots. Consider now the scalar case, namely, $\chi_i \in \mathbb{R}$, and let $\chi = [\chi_1, \ldots, \chi_N]^T$ be the state of the multi-robot system. Then, as shown in Ji and Egersted [10], the control law in Equation 9.5 can be rewritten in the following matrix form:

$$\dot{\chi} = -\mathscr{L}(\mathscr{G})\chi \tag{9.6}$$

where $\mathscr{L}(\mathscr{G})$ is the weighted Laplacian matrix defined as in Equation 9.3. For further details on the consensus problem, the reader is referred to Olfati-Saber et al. [21].

Several control strategies can be found in the literature that exploit consensus for addressing several problems, in the field of multi-robot systems, in a decentralized manner [29]. For instance, rendezvous is the problem of controlling the robots in such a way that based on locally available quantities, their positions converge to a common value [16,20]. Consensus-based techniques can be also used for formation control, that is, the problem of driving the robots to create a desired geometric shape [5,6,17,30], and for flocking, that is, the problem of synchronizing their velocities [13,23,47,48].

9.2.2 COOPERATIVE MULTI-ROBOT BEHAVIORS

The idea of implementing complex cooperative behaviors has been recently gaining attention in the research community interested in multi-robot systems.

The work in Pimenta et al. [24] models the motion of a swarm of multiple robotic systems exploiting models from fluid dynamics. In particular, the objective is to define a feedback control law to make a group of mobile robots create a desired geometric formation and move in a complex environment while avoiding collisions. A solution for this problem is provided by considering each robot as a particle of the fluid, and defining artificial potential fields that define the overall motion of the swarm, based on local interaction rules.

Artificial potential fields are exploited also in Sabattini et al. [37] for coordinating mobile robots along arbitrarily defined closed curve trajectories. Without any global synchronization, a dynamically changing number of robots is able to spread along a curve and track it while guaranteeing collision avoidance.

Control strategies based on the consensus protocol [21] are often exploited for decentralized control of multi-robot systems, typically for synchronization [22,28] or for formation control [5,6]. The standard consensus protocol is extended in Tsiotras and Castro [49], defining a local feedback control law that includes also gyroscopic effects, thus leading to the definition of complex periodic geometric orbits.

As shown in Egerstedt et al. [4], the interconnection by means of consensus protocol can be exploited to model a multi-robot system as a classical linear time-invariant (LTI) system, where the characteristic matrices are related to the interconnection topology. Subsequently, it is possible to extend classical results related to modeling and control of LTI systems to the case of multi-robot systems. In particular, it is possible to apply the classical notions of controllability and observability. Specifically, it turns out that in the case of multi-robot systems, these properties are heavily influenced by the topology of the underlying communication graph. Several works can be found in the literature that develop strategies to infer the controllability property of multi-robot systems. For instance, Ji and Egersted [10] and Rahmani et al. [26] provide sufficient conditions on the graph topology to verify the controllability property. On the same lines, the work presented in Franceschelli et al. [7] provides necessary and sufficient conditions on the spectrum of the Laplacian matrix to verify the controllability of a multi-robot system.

When weighted graphs are considered, the concept of structural controllability can be exploited. Considering a linear system defined by a generic structure, and a set of variable parameters, the concept of *structural controllability* [15,41] refers to the possibility of ensuring the controllability property with a particular choice of the parameters. This concept has been recently applied to multi-robot systems [18,43,45,46,51] interconnected by means of an edge-weighted graph. Specifically, structural controllability of a weighted graph identifies the possibility of making a multi-robot system controllable with an opportune choice of the edge-weights. As shown in Zamani and Lin [51], it is possible to demonstrate that a multi-robot system is structurally controllable if and only if the underlying graph is connected. The control strategy introduced in Sabattini et al. [38] guarantees, in a completely decentralized fashion, the controllability of the multi-robot system almost totally. This strategy is based on a connectivity maintenance algorithm first introduced in Sabattini et al. [32,33] and on the choice of random edge-weights.

Duality principle can be invoked to show that a multi-robot system is controllable if and only if it is observable. Once observability and controllability have been guaranteed, then classical feedback control law design strategies can be extended to multi-robot systems in order to obtain the desired cooperative behavior.

Starting from a controllable and observable multi-robot system, we propose a decentralized methodology to solve a tracking problem for multi-robot systems in a decentralized manner. Specifically, the proposed methodology exploits the well-known Francis' *regulator equations* to design a decentralized control law to make a multi-robot system follow a predefined setpoint. In particular, periodic setpoints are defined for each dependent robot by means of an exosystem, as the linear combination of a given number of harmonics. Following the methodology first introduced in Cocetti et al. [3] and Sabattini et al. [35], it is possible to define the set of admissible setpoint functions that can be followed by the multi-robot system once the topology of the graph has been defined.

A methodology for defining the input for the system by means of the solution of the regulator equations was introduced in Cocetti et al. [3] and Sabattini et al. [35] considering the independent robots as virtual agents, whose position could be directly controlled. This method is extended here, modeling the independent robots as single integrator agents and thus controlling their velocity.

The proposed methodology is implemented in a decentralized manner, exploiting an estimation scheme developed according to the methodology first introduced in Sabattini et al. [39]. In particular, this estimation scheme exploits the constrained consensus-based estimation procedure proposed in Ren [27] to let each independent robot estimate the complete (time-varying) output vector and the complete (time-varying) state of the independent robots. The estimation procedure is defined assuming that a connected communication graph is available among the independent robots. This can be ensured in a decentralized manner by exploiting a connectivity maintenance control algorithm [1,31–33,52]). A Luenberger state observer is subsequently implemented to let the independent robots estimate the overall state of the dependent robots based only on local measurements.

9.2.3 NOTATION AND MATHEMATICAL OPERATORS

In this section, we define some symbols that will be used hereafter.

The symbols $\mathbf{1}_\rho$ and $\mathbf{0}_\rho$ will be used to indicate a vector of all ones and a vector of all zeros, respectively, in $\mathbb{R}^\rho$. Moreover, the symbol $\mathbb{I}_\rho$ will be used to indicate the identity matrix in $\mathbb{R}^{\rho\times\rho}$, while the symbols $\mathbb{O}_\rho$ and $\mathbb{O}_{\rho,\sigma}$ will be used to indicate a square and a rectangular zero matrix in $\mathbb{R}^{\rho\times\rho}$ and in $\mathbb{R}^{\rho\times\sigma}$, respectively.

Let $\Omega \in \mathbb{R}^{\rho\times\sigma}$ be a generic matrix. We define $\Omega[i,:] \in \mathbb{R}^\sigma$, $\Omega[:,j] \in \mathbb{R}^\rho$ and $\Omega[i,j] \in \mathbb{R}$ as the ith row, the jth column, and the element (i,j) of Ω, respectively. On the same lines, given a vector $\omega \in \mathbb{R}^\rho$, we define $\omega[i] \in \mathbb{R}$ as the ith entry of ω.

The symbol $\text{vec}(\cdot)$ will be used to indicate the vectorization operations: namely, given a matrix $\Omega \in \mathbb{R}^{\rho\times\sigma}$, then $\text{vec}(\Omega) \in \mathbb{R}^{\rho\sigma}$ is a vector obtained by stacking the columns of Ω, namely,

$$\text{vec}(\Omega) = \left[\Omega[:,1]^T \ldots \Omega[:,\sigma]^T\right]^T$$

The symbol $\cdot \otimes \cdot$ will be used to indicate the Kronecker product. This operator exhibits the following *mixed product property* [11]:

Property 1. *Let $\mathscr{A}$, $\mathscr{B}$, $\mathscr{C}$, $\mathscr{D}$ be matrices of opportune dimension, defined in such a way that $\mathscr{A}\mathscr{C}$ and $\mathscr{B}\mathscr{D}$ exist. Then,*

$$(\mathscr{A} \otimes \mathscr{B})(\mathscr{C} \otimes \mathscr{D}) = \mathscr{A}\mathscr{C} \otimes \mathscr{B}\mathscr{D} \tag{9.7}$$

9.3 DECENTRALIZED TRAJECTORY TRACKING FOR MULTI-ROBOT SYSTEMS

9.3.1 MODEL OF THE SYSTEM

Consider a group of N interconnected robots, whose communication structure can be represented as a undirected constant graph $\mathscr{G}$, whose topology is supposed to be known. Let $\chi_i \in \mathbb{R}^m$ be the state of the ith robot: we will hereafter consider the case where the state corresponds to each robot's position and where the velocity of each robot can be directly controlled. Namely, we consider a group of N single integrator agents: it is worth remarking that by endowing a robot with a sufficiently good cartesian trajectory tracking controller, it is possible to use this simple model to represent the kinematic behavior of several types of mobile robots, like wheeled mobile robots [44] and UAVs [12]. We consider then the following kinematic model:

$$\dot{\chi}_i = \mu_i \qquad i = 1,\ldots,N \tag{9.8}$$

where $\mu_i \in \mathbb{R}^\nu$ is the ith robot's input.

Without loss of generality, and for ease of notation, we will hereafter refer to the scalar case, namely, $\chi_i, \mu_i \in \mathbb{R}$. It is, however, possible to extend all the results to the multi-dimensional case, considering each component independently.

Considering the communication graph $\mathscr{G}$, let $\mathsf{N}_i \subseteq \mathsf{V}(\mathscr{G})$ be the neighborhood of the ith robot, defined as the set of the robots that are interconnected to the ith one, namely:

$$\mathsf{N}_i = \{j \in \mathsf{V}(\mathscr{G}) \text{ such that } (v_i, v_j) \in \mathsf{E}(\mathscr{G})\} \tag{9.9}$$

Assume now that the goal is to control the state of the multi-robot system: for this purpose, define a few independent robots, whose motion can be explicitly controlled, and that can then be exploited for injecting a control action into the system. The state of the other robots, referred to as the dependent robots, evolves according to the consensus protocol.

More specifically, let $\mathsf{V}_I(\mathscr{G}) \subset \mathsf{V}(\mathscr{G})$ be the set of the independent robots, and let $\mathsf{V}_D(\mathscr{G}) = \mathsf{V}(\mathscr{G}) - \mathsf{V}_I(\mathscr{G})$ be the set of the dependent robots. Subsequently, consider the following interconnection:

$$\begin{cases} \dot{\chi}_i &= -\sum_{j \in \mathsf{N}_i} w_{ij}(\chi_i - \chi_j) & \text{if } v_i \in \mathsf{V}_D(\mathscr{G}) \\ \dot{\chi}_h &= -\sum_{j \in \mathsf{N}_h} w_{hj}(\chi_h - \chi_j) + u_h & \text{if } v_h \in \mathsf{V}_I(\mathscr{G}) \end{cases} \tag{9.10}$$

where $w_{ij}, w_{hj} > 0$ is the edge weight, and $u_h \in \mathbb{R}$ is a control input. It is worth noting that if the control input $u_h = 0$, $\forall v_h \in \mathsf{V}_I(\mathscr{G})$, then the interconnection reported in Equation 9.10 represents the standard well-known (weighted) consensus protocol [21], that drives the interconnected robots to converge to a common configuration.

Define now the overall state of the multi-robot system $\chi = [\chi_1, \ldots, \chi_N] \in \mathbb{R}^N$, and the overall input vector $u = [u_1, \ldots, u_I] \in \mathbb{R}^I$. It is always possible to index the robots such that the first D are the dependent robots, and the last I are the independent robots. Then, as shown in Egerstedt et al. [4], it is possible to decompose the Laplacian matrix $\mathscr{L}(\mathscr{G})$ as follows:

$$\mathscr{L}(\mathscr{G}) = -\left[\begin{array}{c|c} \mathscr{A} & \mathscr{B} \\ \hline \mathscr{B}^T & \mathscr{E} \end{array}\right] \tag{9.11}$$

where $\mathscr{A} = \mathscr{A}^T \in \mathbb{R}^{D \times D}$ represents the interconnection among the dependent robots, $\mathscr{B} \in \mathbb{R}^{D \times I}$ represents the interconnection among independent and dependent robots, and $\mathscr{E} = \mathscr{E}^T \in \mathbb{R}^{I \times I}$ represents the interconnections among the independent robots.

The dynamics of the multi-robot system Equation 9.10 can then be rewritten as follows:

$$\dot{\chi} = \left[\begin{array}{c|c} \mathscr{A} & \mathscr{B} \\ \hline \mathscr{B}^T & \mathscr{E} \end{array}\right] \chi + \begin{bmatrix} \mathbb{O}_{D,I} \\ \mathbb{I}_I \end{bmatrix} u \tag{9.12}$$

The independent robots are exploited as an input for the multi-robot system. Considering the interconnection introduced in Equations 9.10 and 9.12, the independent robots influence the dynamics of the dependent robots based on their position. In particular, let $x = [\chi_1, \dots, \chi_D] \in \mathbb{R}^D$ be the collective state of the dependent robots, and let $\eta = [\chi_{D+1}, \dots, \chi_N] \in \mathbb{R}^I$ be the collective state of the independent robots. Then, the dynamics of the multi-robot system can be rewritten as follows:

$$\begin{cases} \dot{x} &= \mathscr{A}x + \mathscr{B}\eta \\ y &= \mathscr{B}^T x \end{cases} \tag{9.13}$$

which represents a standard LTI system, where the state vector x represents the state of the dependent robots, the input vector η is the state of the independent robots, and $y \in \mathbb{R}^I$ is the output vector. It is worth noting that y is defined as a linear combination of the states of the dependent robots that are neighbors of the independent ones: namely, the elements of y are quantities that are measurable by the independent robots. Therefore, the output vector is well posed.

We will hereafter make the following assumption:

Assumption 1. *A* ***connected*** *undirected communication graph exists among the independent robots.*

The overall multi-robot system can then be represented as in Figure 9.1.

We now introduce the following quantities related to the multi-robot system, which will be used in the following sections:

$$\begin{aligned} \underline{x} &= \mathbf{1}_I \otimes x & \underline{\eta} &= \mathbf{1}_I \otimes \eta \\ \underline{y} &= \mathbf{1}_I \otimes y & & \\ \underline{\mathscr{A}} &= \mathbb{I}_I \otimes \mathscr{A} & \underline{\mathscr{B}} &= \mathbb{I}_I \otimes \mathscr{B} \end{aligned} \tag{9.14}$$

From Equation 9.13 it is then possible to obtain the following:

$$\dot{\underline{x}} = \underline{\mathscr{A}}\,\underline{x} + \underline{\mathscr{B}}\,\underline{\eta} \tag{9.15}$$

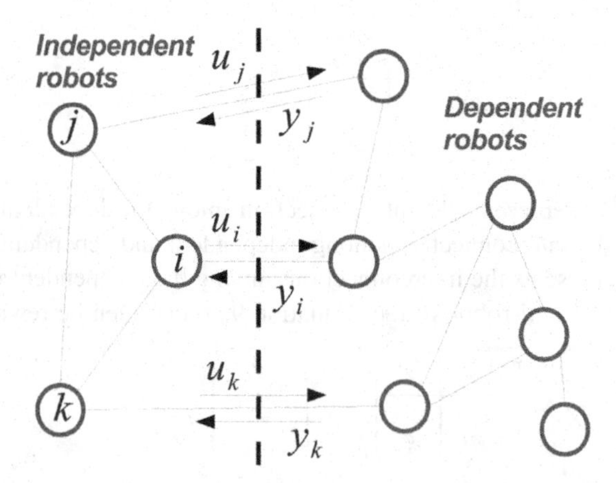

Figure 9.1 Scheme of the overall multi-robot system.

9.3.2 PERIODIC SETPOINT DEFINITION

In this section we will define an exosystem [40], whose state is exploited to generate a desired periodic setpoint for the multi-robot system. More specifically, we will show how to generate a setpoint that approximates a generic periodic function as the linear combination of $n \in \mathbb{N}$ harmonics.

For this purpose, consider a generic periodic function $f(t) \in \mathbb{R}$, with period $T > 0$, that is

$$f(t) = f(t+T) \ \forall t \in \mathbb{R} \tag{9.16}$$

Define then the exosystem as an autonomous system whose state vector $\xi \in \mathbb{R}^{\bar{n}}$, where $\bar{n} = 2n+1$, evolves according to the following dynamics:

$$\dot{\xi} = \mathscr{G}\xi \tag{9.17}$$

where the matrix $\mathscr{G} \in \mathbb{R}^{\bar{n}\times\bar{n}}$ is a block diagonal matrix, namely:

$$\mathscr{G} = \begin{bmatrix} \tilde{\mathscr{G}}_0 & 0 & \dots & 0 \\ 0 & \tilde{\mathscr{G}}_1 & \dots & 0 \\ \vdots & \vdots & \ddots & \vdots \\ 0 & 0 & \dots & \tilde{\mathscr{G}}_n \end{bmatrix} \tag{9.18}$$

Let each block $\tilde{\mathscr{G}}_p$ be defined as follows:

$$\tilde{\mathscr{G}}_p = \begin{cases} 0 & \text{if } p = 0 \\ \begin{bmatrix} 0 & p\frac{2\pi}{T} \\ -p\frac{2\pi}{T} & 0 \end{bmatrix} & \text{if } p = 1,\dots,n \end{cases} \tag{9.19}$$

According to the definition given in Equations 9.17 through 9.19, it is possible to show that the exosystem exhibits the following property:

Property 2. *Any periodic function with period T may be approximated with a linear combination of the elements of the vector* ξ.

In fact, consider a generic periodic function $f(t)$, with period $T > 0$, defined as in Equation (9.16). As is well known, any periodic function can be decomposed in its Fourier series and can then be rewritten as an infinite sum of trigonometric functions, namely:

$$f(t) = \frac{\alpha_0}{2} + \sum_{p=1}^{\infty}\left[\alpha_p \cos\left(\frac{2\pi}{T}pt\right) + \beta_p \sin\left(\frac{2\pi}{T}pt\right)\right] \tag{9.20}$$

where the terms α_p and β_p are denoted as the Fourier coefficients and are defined as follows, $\forall p \geq 0$:

$$\begin{aligned} \alpha_p &= \frac{2}{T}\int_0^T f(t)\cos\left(\frac{2\pi}{T}pt\right)\mathrm{d}t \\ \beta_p &= \frac{2}{T}\int_0^T f(t)\sin\left(\frac{2\pi}{T}pt\right)\mathrm{d}t \end{aligned} \tag{9.21}$$

Considering a finite number n of harmonics, a periodic function $f(t)$ can be approximated with the following finite sum:

$$f(t) \approx \frac{\alpha_0}{2} + \sum_{p=1}^{n} \left[\alpha_p \cos\left(\frac{2\pi}{T} pt\right) + \beta_p \sin\left(\frac{2\pi}{T} pt\right) \right] \tag{9.22}$$

Considering the definition given in Equations 9.18 and 9.19, the matrix $\mathcal{G}$ is expressed in Jordan normal form.

We define the initial state of the exosystem $\xi_0 \in \mathbb{R}^{\bar{n}}$ as follows:

$$\xi_0 = [1, 0, 1, 0, \dots, 1]^T \tag{9.23}$$

Hence, the solution of the differential equation in Equation 9.17 is the following:

$$\xi(t) = e^{\mathcal{G}t} \xi_0 = \left[1 \ \sin\left(\frac{2\pi}{T} t\right) \ \cos\left(\frac{2\pi}{T} t\right) \ \dots \ \sin\left(n\frac{2\pi}{T} t\right) \ \cos\left(n\frac{2\pi}{T} t\right) \right] \tag{9.24}$$

Then, according to Equation 9.22, a periodic function $f(t)$ can be approximated with a linear combination of the components of the state vector of the exosystem ξ.

Consider then a generic periodic setpoint $x_s(t) \in \mathbb{R}^D$, which has to be tracked by the state $x(t) \in \mathbb{R}^D$ of the dependent robots. According to Property 2, it is possible to define a matrix $\mathcal{J} \in \mathbb{R}^{D \times \bar{n}}$ such that

$$x_s(t) = \mathcal{J} \xi(t) \tag{9.25}$$

Therefore, once the period T and the number of harmonics n have been defined, the choice of the matrix $\mathcal{J}$ leads to the definition of a particular periodic setpoint x_s.

In the next section, we will define a methodology to solve the *tracking problem for the multi-robot system* (Equation 9.13): specifically, assuming the state of the exosystem to be available to all the independent robots, the problem is that of defining the input $u(t)$ in such a way that the state $x(t)$ of the system (i.e., the position of the dependent robots) converges to the desired setpoint $x_s(t)$.

9.3.3 CENTRALIZED CONTROL LAW FOR SETPOINT TRACKING

In this section, we introduce a methodology to define a control law that makes the dependent robots track a periodic setpoint defined according to Equation 9.25. For this purpose, let the tracking error $e(t) \in \mathbb{R}^D$ be defined as follows:

$$e(t) = x(t) - \mathcal{J} \xi(t) \tag{9.26}$$

Then, considering Equations 9.13, 9.17, and 9.26, the dynamics of the multi-robot system can be rewritten as follows:

$$\begin{cases} \dot{x} &= \mathcal{A} x + \mathcal{B} \eta \\ y &= \mathcal{B}^T x \\ e &= x - \mathcal{J} \xi \\ \dot{\xi} &= \mathcal{G} \xi \end{cases} \tag{9.27}$$

Define then $\bar{\eta} \in \mathbb{R}^I$ as the desired input for Equation 9.27, that is, the input that we would like to obtain if the position of the independent robots could be directly controlled. As is well known from standard linear control theory [40], in order to make the dependent robots track the desired setpoint trajectories defined according to Equation 9.25, the desired control law $\bar{\eta}$ can be defined as

$$\bar{\eta} = \mathscr{F}x + (\Gamma - \mathscr{F}\Pi)\,\xi \tag{9.28}$$

where $\mathscr{F}$ is an arbitrary matrix, chosen such that $(\mathscr{A} + \mathscr{B}\mathscr{F})$ is Hurwitz stable, and $\Gamma \in \mathbb{R}^{I \times \bar{n}}$ and $\Pi \in \mathbb{R}^{D \times \bar{n}}$ are the solution of the *regulator equations* that, considering the dynamical system described in Equation 9.27, can be written as follows:

$$\begin{cases} \mathscr{A}\Pi + \mathscr{B}\Gamma = \Pi\mathscr{G} & (9.29\text{a}) \\ \quad\;\; \Pi - \mathscr{J} = 0 & (9.29\text{b}) \end{cases}$$

Matrix $\mathscr{F} \in \mathbb{R}^{I \times D}$ has to be chosen such that $(\mathscr{A} + \mathscr{B}\mathscr{F})$ is Hurwitz stable. As is well known from basic linear control theory, such a matrix can always be found if the pair $(\mathscr{A}, \mathscr{B})$ is controllable. It is worth noting that the concept of *structural controllability* [15] can be invoked to demonstrate that under a connected communication topology, controllability can be guaranteed almost completely with a random choice of the edge weights [7,18,19,38].

Moreover, it is worth noting that Equation 9.29a defines a generalized Sylvester equation [50], while Equation 9.29b can be solved by satisfying the following equality:

$$\mathscr{J} = \Pi \tag{9.30}$$

The equality in Equation 9.30 represents a constraint on matrix $\mathscr{J}$, which according to Equation 9.25, implies a constraint on the choice of the setpoint.

Considering then the definition of the setpoint trajectories $x_s(t)$ given in Equation 9.25, and considering the regulator equations in Equation 9.29, we now introduce the following definition of *admissible setpoint functions*.

Definition 9.1. *The* set of admissible setpoint trajectories $\mathsf{S}_a \in \mathbb{R}^D$ *is defined as follows:*

$$\mathsf{S}_a = \{x_s(t) = \mathscr{J}\xi(t) \text{ such that } \mathscr{J} = \Pi\} \tag{9.31}$$

in such a way that a solution (Π, Γ) *exists to the regulator equations Equation 9.29.*

The set of admissible setpoint trajectories can be characterized [3,35] once the interconnection topology has been defined. In particular, the methodology derived in Wu et al. [50, theorem 1] can be exploited for finding (Π, Γ) that solve the regulator equations (Equation 9.29).

According to Equation 9.12, the dynamics of the independent robots can be written as follows:

$$\dot{\eta} = \mathscr{B}^T x + \mathscr{E}\eta + u \tag{9.32}$$

In order to implement the desired input $\bar{\eta}$ introduced in Equation 9.28, the input u can then be defined as follows:

$$u = -\mathscr{B}^T x - \mathscr{E}\eta + \dot{\bar{\eta}} - \mathscr{H}(\eta - \bar{\eta}) \tag{9.33}$$

with $\mathscr{H} \in \mathbb{R}^{I \times I}$. Considering Equation 9.32, the dynamics of the independent robots can then be rewritten as follows:

$$\dot{\eta} = \dot{\bar{\eta}} - \mathscr{H}(\eta - \bar{\eta}) \tag{9.34}$$

Define now $\delta \in \mathbb{R}^I$ as the difference between the desired input and the actual state of the independent robots, namely,

$$\delta = \eta - \bar{\eta} \tag{9.35}$$

The subsequent results immediately follow from the dynamics of the independent robots (Equation 9.34).

Lemma 9.1

Consider the dynamics of the independent robots in Equation 9.34. Then, the difference between the desired input and the actual state of the independent robots, namely δ defined in Equation 9.35, asymptotically vanishes for any $\mathscr{H} > 0$. ■

This result ensures that for any choice of $\mathscr{H} > 0$, Equation 9.34 evolves in such a way that the independent robots implement the desired input $\bar{\eta}$.

Considering the definition of $\bar{\eta}$ in Equation 9.28, and considering the dynamics of the exosystem (Equation 9.17), then

$$\dot{\bar{\eta}} = \mathscr{F}\mathscr{A}x + \mathscr{F}\mathscr{B}\eta + (\Gamma - \mathscr{F}\Pi)\mathscr{G}\xi \tag{9.36}$$

Thus, the dynamics of the independent robots (Equation 9.34) can be rewritten as follows:

$$\dot{\eta} = (\mathscr{F}\mathscr{A} + \mathscr{H}\mathscr{F})x + (\mathscr{F}\mathscr{B} - \mathscr{H})\eta + [(\Gamma - \mathscr{F}\Pi)\mathscr{G} + \mathscr{H}(\Gamma - \mathscr{F}\Pi)]\xi \tag{9.37}$$

9.3.4 DECENTRALIZED IMPLEMENTATION

In this section, we will show how to implement the control strategy introduced in Section 9.3.3 in a decentralized manner. In fact, it is worth noting that Equation 9.37 is the sum of three terms: a feedback of the state of the dependent robots x, a feedback of the state of the independent robots η, and a feedforward term. Therefore, the implementation of the control law (Equation 9.33) requires the complete knowledge of x and η. However, in a decentralized implementation, each independent robot has access only to information regarding its neighboring robots.

Nevertheless, it is worth remarking that the system in Equation 9.13 represents a standard LTI system. Hence, the classical notions of controllability and observability can be applied to the multi-robot system itself. In particular, the following property can be derived:

Property 3. *A multi-robot system whose dynamics are written according to Equation 9.13 is observable if and only if it is controllable.*

This property can be demonstrated by observing that the input matrix (i.e., $\mathscr{B}$) is the transpose of the output matrix (i.e., $\mathscr{B}^T$).

Since matrices $\mathscr{A}$ and $\mathscr{B}$ are extracted from the Laplacian matrix of the communication graph, the interconnection topology heavily influences the controllability and observability properties of the multi-robot system. In particular, as shown in Sabattini et al. [38], a random choice of the edge weights over a connected communication graph guarantees controllability and observability of the multi-robot system almost completely.

Thus, we will hereafter assume that the topology is defined in such a way that the multi-robot system is observable. It is then possible to design a standard Luenberger state observer [40] for obtaining an estimate of x, which will be hereafter referred to as $\hat{x} \in \mathbb{R}^D$. Specifically, opportunely defining the gain matrix $\mathscr{K}_l \in \mathbb{R}^{D\times I}$, the following update law may be defined for the state observer:

$$\dot{\hat{x}} = \mathscr{A}\hat{x} + \mathscr{B}\eta - \mathscr{K}_l\left(y - \mathscr{B}^T\hat{x}\right) \tag{9.38}$$

The estimation error $\hat{e} \in \mathbb{R}^D$ can then be defined as follows:

$$\hat{e} = x - \hat{x} \tag{9.39}$$

The estimation error converges to zero if and only if the eigenvalues of the matrix $\left(\mathscr{A} + \mathscr{K}_l\mathscr{B}^T\right)$ are all negative. As is well known, if the system is observable, then it is always possible to find a matrix $\mathscr{K}_l$ that satisfies this condition.

However, it is worth remarking that the observer defined in Equation 9.38 cannot be implemented in a decentralized manner: in fact, according to the model of the system described in Section 9.3.1, each independent robot acquires only one component of the output vector y (i.e., a linear combination of the states of its neighboring dependent robots) and has just a local knowledge of η.

In the next subsections, we will introduce a decentralized estimation methodology, which will be exploited by each independent robot to compute an estimate of the complete output vector y, and of the independent robot state η. This estimate will then be used for implementing a decentralized state observer. The overall estimation and control scheme is represented in Figure 9.2.

9.3.4.1 Decentralized Output Estimation

Under Assumption 1, we will now introduce a decentralized estimation scheme to let all the independent robots estimate the full output vector y. For this purpose,

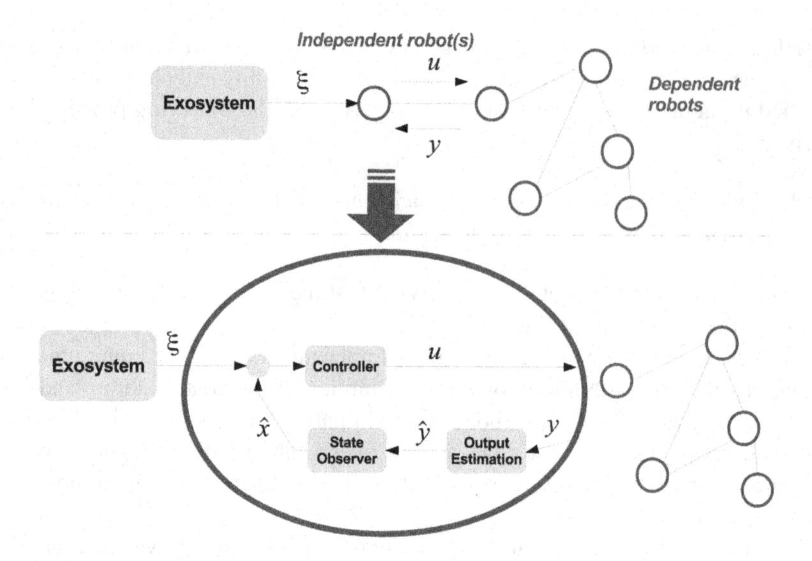

Figure 9.2 Decentralized estimation and control scheme.

define $\hat{y}_h \in \mathbb{R}^I$ as the hth independent robot's estimate of the output y, and let $\hat{y} = \left[\hat{y}_1^T \ldots \hat{y}_I^T\right]^T$ be the stacked vector of the output estimates. Moreover, let $\tilde{\varepsilon}_h = y - \hat{y}_h$ be the error in the estimation of the output obtained by the hth independent robot, and let $\tilde{\varepsilon} = \left[\tilde{\varepsilon}_1^T \ldots \tilde{\varepsilon}_I^T\right]^T$. Hence:

$$\tilde{\varepsilon} = \underline{y} - \hat{y} \tag{9.40}$$

The output vector y is composed of I elements: it is possible to implement a decentralized estimator for each element of y in such a way that every independent robot can obtain an estimate of the entire vector y.

In particular, the hth estimator is in charge of providing $\hat{y}_k[h]$ as the kth independent robot's estimate of the hth element of y, $\forall k = 1, \ldots, I$. The hth estimator is designed in such a way that

$$\hat{y}_k[h] \longrightarrow y[h] \qquad \forall k = 1, \ldots, I \tag{9.41}$$

This is ensured by the fact that the hth independent robot has access to the hth element of y and can then inject the time-varying reference $y[h]$ into the estimation process.

Inspired by Ren [27], the following constrained consensus update law is now introduced:

$$\begin{cases} \dot{\hat{y}}_k[h] & = \dfrac{1}{\left|\mathsf{N}_k^*\right|} \sum_{j \in \mathsf{N}_k^*} \left[\dot{\hat{y}}_j[h] - \alpha\left(\hat{y}_k[h] - \hat{y}_j[h]\right)\right] \qquad \forall k \neq h \\ \dot{\hat{y}}_h[h] & = \dot{y}[h] - \alpha\left(\hat{y}_h[h] - y[h]\right) \sum_{j \in \mathsf{N}_h^*} \alpha\left(\hat{y}_h[h] - \hat{y}_j[h]\right) \end{cases} \tag{9.42}$$

$\forall h = 1,\ldots,I$, where $\alpha > 0$ is a design parameter. The set N_k^* is defined as the set of independent robots that are neighbors of the kth one, and $\left|\mathsf{N}_k^*\right| = \sum_{j \in \mathsf{N}_k^*} w_{kj}$.

Lemma 9.2

Consider the output estimation algorithm defined in Equation 9.42. Then, the output estimation error $\tilde{\varepsilon}$ defined in Equation 9.40 asymptotically vanishes. ■

The proof is based on the results in Ren [27, theorem 3.3], where a decentralized estimation scheme was defined, in a multi-agent system, for guaranteeing convergence to a time-varying reference, which is supposed to be measurable by one of the agents. It is worth noting that the convergence speed can be made arbitrarily fast by increasing the value of the gain $\alpha > 0$.

9.3.4.2 Decentralized Independent Robot State Estimation

On the same lines, under Assumption 1, it is possible to introduce a similar decentralized estimation scheme to let all the independent robots estimate the full vector η. For this purpose, define $\hat{\eta}_h \in \mathbb{R}^I$ as the hth independent robot's estimate of the output η, and let $\hat{\eta} = \left[\hat{\eta}_1^T \ldots \hat{\eta}_I^T\right]^T$ be the stacked vector of the output estimates. Moreover, let $\psi_h = \eta - \hat{\eta}_h$ be the error in the estimation of the independent robot state, obtained by the hth independent robot, and let $\psi = \left[\psi_1^T \ldots \psi_I^T\right]^T$. Hence:

$$\psi = \underline{\eta} - \hat{\eta} \tag{9.43}$$

We can then introduce a decentralized estimation scheme similar to Equation 9.42, as follows:

$$\begin{cases} \dot{\hat{\eta}}_k[h] &= \dfrac{1}{\left|\mathsf{N}_k^*\right|} \sum_{j \in \mathsf{N}_k^*} \left[\dot{\hat{\eta}}_j[h] - \gamma(\hat{\eta}_k[h] - \hat{\eta}_j[h])\right] \qquad \forall k \neq h \\ \dot{\hat{\eta}}_h[h] &= \dot{\eta}[h] - \gamma(\hat{\eta}_h[h] - \eta[h]) \sum_{j \in \mathsf{N}_h^*} \gamma(\hat{\eta}_h[h] - \hat{\eta}_j[h]) \end{cases} \tag{9.44}$$

$\forall h = 1,\ldots,I$, where $\gamma > 0$ is a design parameter. The set N_k^* is defined as the set of independent robots that are neighbors of the kth one, and $\left|\mathsf{N}_k^*\right| = \sum_{j \in \mathsf{N}_k^*} w_{kj}$.

Lemma 9.3

Consider the output estimation algorithm defined in Equation 9.44. Then, the independent robot state estimation error ψ defined in Equation 9.43 asymptotically vanishes. ■

The proof is based on the results in Ren [27, theorem 3.3], where a decentralized estimation scheme was defined, in a multi-agent system, for guaranteeing convergence to a time-varying reference, which is supposed to be measurable by one of the agents. It is worth noting that the convergence speed can be made arbitrarily fast by increasing the value of the gain $\gamma > 0$.

9.3.4.3 Decentralized State Estimation and Control Strategy

This section describes how to exploit the output estimation strategy introduced so far for the decentralized implementation of the Luenberger state observer (Equation 9.38).

In particular, assume that each independent robot i can compute its own estimate of y of η, namely $\hat{y}_i$ and $\hat{\eta}_i$, respectively. We will hereafter derive a methodology for computing $\hat{x}_i$, namely the ith independent robot's estimate of x, with $i = 1, \ldots, I$.

Exploiting the state's estimate $\hat{x}$, the ith independent robot can define the decentralized realization of the ith component of $\bar{\eta}$, computed replacing x with $\hat{x}_i$, namely:

$$\bar{\eta}\,[i] = \mathscr{F}\,[i,:]\,\hat{x}_i + (\Gamma\,[i,:] - (\mathscr{F}\Pi)\,[i,:])\,\xi \tag{9.45}$$

Let now $\hat{x}$ be the stacked vector of the state estimates, namely $\hat{x} = \left[\hat{x}_1^T \ldots \hat{x}_I^T\right]^T$. Moreover, let $\tilde{\mathscr{F}} \in \mathbb{R}^{I\times(DI)}$ be the following matrix:

$$\tilde{\mathscr{F}} = \begin{bmatrix} \mathscr{F}\,[1,:] & \mathbb{O} & \mathbb{O} & \ldots & \mathbb{O} \\ \mathbb{O} & \mathscr{F}\,[2,:] & \mathbb{O} & \ldots & \mathbb{O} \\ \vdots & \vdots & \vdots & \vdots & \vdots \\ \mathbb{O} & \mathbb{O} & \ldots & \mathbb{O} & \mathscr{F}\,[I,:] \end{bmatrix} \tag{9.46}$$

Therefore, according to Equation 9.45, the desired input $\bar{\eta}$ defined in Equation 9.28 can be computed in a decentralized manner by each independent robot as follows:

$$\bar{\eta} = \tilde{\mathscr{F}}\hat{x} + (\Gamma - \mathscr{F}\Pi)\,\xi \tag{9.47}$$

It is worth noting that each independent robot i has access only to its own estimate of the state, namely $\hat{x}_i$. Subsequently, the ith independent robot's estimate of $\bar{\eta}$, namely $\tilde{\eta}_i \in \mathbb{R}^I$, can be computed replacing $\hat{x}$ with $\hat{x}_i$ (i.e., as if all the independent robots had the same estimate of the state), namely

$$\tilde{\eta}_i = \mathscr{F}\hat{x}_i + (\Gamma - \mathscr{F}\Pi)\,\xi \tag{9.48}$$

The state observer introduced in Equation 9.38 is then implemented by each independent robot in a decentralized manner, as follows:

$$\dot{\hat{x}}_i = \mathscr{A}\hat{x}_i + \mathscr{B}\tilde{\eta}_i - \mathscr{K}_I\left(\hat{y}_i - \mathscr{B}^T\hat{x}_i\right) \tag{9.49}$$

Define now $\hat{x} = \left[\hat{x}_1^T \ldots \hat{x}_I^T\right]^T$. Then, from Equation 9.49 it is possible to obtain the following:

$$\begin{aligned} \dot{\hat{x}} = {} & (\mathbb{I}_I \otimes \mathscr{A})\,\hat{x} + (\mathbb{I}_I \otimes \mathscr{B}\mathscr{F})\,\hat{x} + (\mathbb{I}_I \otimes (\mathscr{B}(\Gamma - \mathscr{F}\Pi)))\,\underline{\xi} - (\mathbb{I}_I \otimes \mathscr{K}_I)\,\hat{y} \\ & + \left(\mathbb{I}_I \otimes \left(\mathscr{K}_I\mathscr{B}^T\right)\right)\hat{x} \end{aligned} \tag{9.50}$$

Let $\tilde{e}_i = x - \hat{x}_i$ be the state estimation error obtained by the ith independent robot. Moreover, let $\tilde{e} = \left[\tilde{e}_1^T \dots \tilde{e}_I^T\right]^T$ be the stacked vector of the estimation errors. Hence:

$$\tilde{e} = \underline{x} - \hat{x} \tag{9.51}$$

We now introduce the *selection matrix* $\mathscr{S}_i \in \mathbb{R}^{(DI)\times(DI)}$, defined as a block diagonal matrix, whose blocks are all zeros except for the ith block of each row, which is the identity matrix. Namely:

$$\mathscr{S}_i = \left[\mathbb{O}_{DI,D(i-1)} \quad \mathbf{1}_I \otimes \mathbb{I}_D \quad \mathbb{O}_{DI,D(I-i)}\right] \tag{9.52}$$

It is then possible to define the matrix $\Omega \in \mathbb{R}^{(DI)\times(DI)}$ as follows:

$$\Omega = \begin{bmatrix} \mathscr{B}\tilde{\mathscr{F}}\left(\mathscr{S}_1 - \mathbb{I}_{DI}\right) \\ \vdots \\ \mathscr{B}\tilde{\mathscr{F}}\left(\mathscr{S}_I - \mathbb{I}_{DI}\right) \end{bmatrix} \tag{9.53}$$

The following proposition shows that the closed loop dynamics of the multi-robot system can be written as an LTI system with a persistent exogenous input given by the setpoint and a vanishing exogenous input given by the estimation error.

Proposition 1. *Consider the dynamics of the multi-robot system described in Equation 9.15, the dynamics of the state estimate in Equation 9.50, the state estimation error $\tilde{e}$ defined in Equation 9.51, the input decentralized realization of the desired input $\bar{\eta}$ defined in Equation 9.47, and the difference δ between the desired input $\bar{\eta}$ and the actual state of the independent robots η defined in Equation 9.35. Then, the overall dynamics of the multi-robot system can be rewritten as follows:*

$$\begin{bmatrix} \dot{x} \\ \dot{\tilde{e}} \end{bmatrix} = \Phi \begin{bmatrix} x \\ \tilde{e} \end{bmatrix} + g(\xi) + f(\tilde{\varepsilon}, \delta) \tag{9.54}$$

where $\Phi \in \mathbb{R}^{(2DI)\times(2DI)}$ is an opportunely defined matrix, $g(\xi)$ is a function of the setpoint, and $f(\tilde{\varepsilon}, \delta)$ is a vanishing term.

Proof. In order to prove the statement, consider the desired input $\bar{\eta}$, which was defined in Equation 9.47 as follows:

$$\bar{\eta} = \tilde{\mathscr{F}}\hat{x} + (\Gamma - \mathscr{F}\Pi)\xi$$

It will now be rewritten in terms of the estimation error $\tilde{e}$. It is worth noting that, $\forall i = 1, \dots, I$, there exists a vector $\Delta_i \in \mathbb{R}^I$ such that

$$\tilde{\mathscr{F}}\hat{x} = \mathscr{F}\hat{x}_i + \Delta_i \quad \forall i = 1, \dots, I \tag{9.55}$$

Considering the definition of $\tilde{\mathscr{F}}$ given in Equation 9.46, the term Δ_i can be defined as follows:

$$\Delta_i = \tilde{\mathscr{F}}\hat{x} - \mathscr{F}\hat{x}_i = \begin{bmatrix} \mathscr{F}[1,:][\hat{x}_1 - \hat{x}_i] \\ \vdots \\ \mathscr{F}[I,:][\hat{x}_I - \hat{x}_i] \end{bmatrix} \tag{9.56}$$

According to the definition of $\tilde{\mathscr{F}}$ given in Equation 9.46, it follows that

$$\Delta_i = \tilde{\mathscr{F}} \begin{bmatrix} \hat{x}_1 - \hat{x}_i \\ \vdots \\ \hat{x}_I - \hat{x}_i \end{bmatrix} = \tilde{\mathscr{F}} \left(\hat{x} - \mathbf{1} \otimes \hat{x}_i \right) \tag{9.57}$$

It is now possible to rewrite Δ_i in terms of the estimation error $\tilde{e}$:

$$\begin{aligned} \Delta_i &= \tilde{\mathscr{F}} \left(\hat{x} - \underline{x} + \underline{x} - \mathbf{1}_I \otimes \hat{x}_i \right) = \\ &= \tilde{\mathscr{F}} \left(-\tilde{e} + \mathbf{1}_I \otimes x - \mathbf{1}_I \otimes \hat{x}_i \right) = \\ &= \tilde{\mathscr{F}} \left(\mathbf{1}_I \otimes \tilde{e}_i - \tilde{e} \right) = \tilde{\mathscr{F}} \left(\mathscr{S}_i - \mathbb{I}_{DI} \right) \tilde{e} \end{aligned} \tag{9.58}$$

where $\mathscr{S}_i \in \mathbb{R}^{(DI) \times (DI)}$ is the *selection matrix* defined as in Equation 9.52.

The desired input $\bar{\eta}$ in Equation 9.47 can then be rewritten as follows:

$$\begin{aligned} \bar{\eta} &= \mathscr{F} \hat{x}_i + \tilde{\mathscr{F}} \left(\mathscr{S}_i - \mathbb{I}_{DI} \right) \tilde{e} + \left(\Gamma - \mathscr{F} \Pi \right) \xi \\ &= \mathscr{F} x - \mathscr{F} \tilde{e}_i + \tilde{\mathscr{F}} \left(\mathscr{S}_i - \mathbb{I}_{DI} \right) \tilde{e} + \left(\Gamma - \mathscr{F} \Pi \right) \xi \end{aligned} \tag{9.59}$$

Consider now the dynamics of the state x defined in Equation 9.13 as follows:

$$\dot{x} = \mathscr{A} x + \mathscr{B} \eta = \mathscr{A} x + \mathscr{B} \bar{\eta} + \mathscr{B} \delta$$

Considering the desired input as defined in Equation 9.59, and considering $\delta = \eta - \bar{\eta}$, then we obtain

$$\begin{aligned} \dot{x} &= \mathscr{A} x + \mathscr{B} \eta = \mathscr{A} x + \mathscr{B} \bar{\eta} + \mathscr{B} \delta \\ &= \left(\mathscr{A} + \mathscr{B} \mathscr{F} \right) x - \mathscr{B} \mathscr{F} \tilde{e}_i + \mathscr{B} \tilde{\mathscr{F}} \left(\mathscr{S}_i - \mathbb{I}_{DI} \right) \tilde{e} + \mathscr{B} \left(\Gamma - \mathscr{F} \Pi \right) \xi + \mathscr{B} \delta \\ \forall i &= 1, \ldots, I \end{aligned} \tag{9.60}$$

Subsequently, considering the dynamics of the stacked vector of the state $\underline{x}$ given in Equation 9.15, we obtain

$$\begin{aligned} \dot{\underline{x}} &= \mathbb{I}_I \otimes \left(\mathscr{A} + \mathscr{B} \mathscr{F} \right) \underline{x} - \begin{bmatrix} \mathscr{B} \mathscr{F} \tilde{e}_1 \\ \vdots \\ \mathscr{B} \mathscr{F} \tilde{e}_I \end{bmatrix} + \begin{bmatrix} \mathscr{B} \tilde{\mathscr{F}} \left(\mathscr{S}_1 - \mathbb{I}_{DI} \right) \tilde{e} \\ \vdots \\ \mathscr{B} \tilde{\mathscr{F}} \left(\mathscr{S}_I - \mathbb{I}_{DI} \right) \tilde{e} \end{bmatrix} \\ &\quad + \left(\mathbb{I}_I \otimes \left(\mathscr{B} \left(\Gamma - \mathscr{F} \Pi \right) \right) \xi \right) + \left(\mathbb{I}_I \otimes \mathscr{B} \right) \delta = \\ &= \mathbb{I}_I \otimes \left(\mathscr{A} + \mathscr{B} \mathscr{F} \right) \underline{x} - \left(\mathbb{I}_I \otimes \left(\mathscr{B} \mathscr{F} \right) \right) \tilde{e} + \Omega \tilde{e} + \left(\mathbb{I}_I \otimes \left(\mathscr{B} \left(\Gamma - \mathscr{F} \Pi \right) \right) \xi \right) + \left(\mathbb{I}_I \otimes \mathscr{B} \right) \delta \end{aligned} \tag{9.61}$$

where the matrix $\Omega \in \mathbb{R}^{(DI) \times (DI)}$ is defined as in Equation 9.53.

Consider the estimation error $\tilde{e}$ defined in Equation 9.51. Then, according to Equations 9.50 and 9.61, the following dynamics can be computed:

$$\begin{aligned}
\dot{\tilde{e}} &= \dot{\underline{x}} - \dot{\hat{x}} \\
&= (\mathbb{I}_I \otimes (\mathscr{A} + \mathscr{B}\mathscr{F}))\underline{x} - (\mathbb{I}_I \otimes (\mathscr{B}\mathscr{F}))\tilde{e} + \Omega\tilde{e} - (\mathbb{I}_I \otimes (\mathscr{A} + \mathscr{B}\mathscr{F}))\hat{x} \\
&\quad - \left(\mathbb{I}_I \otimes \left(\mathscr{K}_I\mathscr{B}^T\right)\right)\hat{x} + (\mathbb{I}_I \otimes \mathscr{K}_I)\hat{y} + (\mathbb{I}_I \otimes \mathscr{B})\,\delta \\
&= (\mathbb{I}_I \otimes (\mathscr{A} + \mathscr{B}\mathscr{F}))\tilde{e} - (\mathbb{I}_I \otimes (\mathscr{B}\mathscr{F}))\tilde{e} + \Omega\tilde{e} + (\mathbb{I}_I \otimes \mathscr{K}_I)\hat{y} - \left(\mathbb{I}_I \otimes \left(\mathscr{K}_I\mathscr{B}^T\right)\right)\hat{x} \\
&\quad + (\mathbb{I}_I \otimes \mathscr{B})\,\delta \\
&= (\mathbb{I}_I \otimes \mathscr{A} + \Omega)\,\tilde{e} + (\mathbb{I}_I \otimes \mathscr{K}_I)\hat{y} - \left(\mathbb{I}_I \otimes \left(\mathscr{K}_I\mathscr{B}^T\right)\right)\hat{x} + (\mathbb{I}_I \otimes \mathscr{B})\,\delta
\end{aligned} \tag{9.62}$$

We will now rewrite the dynamics of the state estimation error $\tilde{e}$ in terms of the output estimation error $\tilde{\varepsilon}$. For this purpose, from Equation 9.40, considering that

$$\underline{y} = \left(\mathbb{I}_I \otimes \mathscr{B}^T\right)\underline{x} \tag{9.63}$$

then it is possible to rewrite Equation 9.62 as follows:

$$\begin{aligned}
\dot{\tilde{e}} &= (\mathbb{I}_I \otimes \mathscr{A} + \Omega)\,\tilde{e} + (\mathbb{I}_I \otimes \mathscr{K}_I)\left(\mathbb{I}_I \otimes \mathscr{B}^T\right)\underline{x} - (\mathbb{I}_I \otimes \mathscr{K}_I)\,\tilde{\varepsilon} - \left(\mathbb{I}_I \otimes \left(\mathscr{K}_I\mathscr{B}^T\right)\right)\hat{x} \\
&\quad + (\mathbb{I}_I \otimes \mathscr{B})\,\delta
\end{aligned} \tag{9.64}$$

Subsequently, according to Equation 9.7 (Property 1), we obtain the following:

$$\begin{aligned}
\dot{\tilde{e}} &= (\mathbb{I}_I \otimes \mathscr{A} + \Omega)\,\tilde{e} + \left(\mathbb{I}_I \otimes \left(\mathscr{K}_I\mathscr{B}^T\right)\right)\underline{x} - \left(\mathbb{I}_I \otimes \left(\mathscr{K}_I\mathscr{B}^T\right)\right)\hat{x} - (\mathbb{I}_I \otimes \mathscr{K}_I)\,\tilde{\varepsilon} \\
&\quad + (\mathbb{I}_I \otimes \mathscr{B})\,\delta \\
&= (\mathbb{I}_I \otimes \mathscr{A} + \Omega)\,\tilde{e} + \left(\mathbb{I}_I \otimes \left(\mathscr{K}_I\mathscr{B}^T\right)\right)\tilde{e} - (\mathbb{I}_I \otimes \mathscr{K}_I)\,\tilde{\varepsilon} + (\mathbb{I}_I \otimes \mathscr{B})\,\delta \\
&= \left(\mathbb{I}_I \otimes \left(\mathscr{A} + \mathscr{K}_I\mathscr{B}^T\right)\right)\tilde{e} - (\mathbb{I}_I \otimes \mathscr{K}_I)\,\tilde{\varepsilon} + (\mathbb{I}_I \otimes \mathscr{B})\,\delta
\end{aligned} \tag{9.65}$$

The overall dynamics of the system can then be written as in Equation 9.66.

$$\begin{aligned}
\begin{bmatrix} \dot{\underline{x}} \\ \dot{\tilde{e}} \end{bmatrix} &= \begin{bmatrix} \mathbb{I}_I \otimes (\mathscr{A} + \mathscr{B}\mathscr{F}) & -\mathbb{I}_I \otimes (\mathscr{B}\mathscr{F}) + \Omega \\ \mathbb{O}_{ID} & \mathbb{I}_I \otimes \left(\mathscr{A} + \mathscr{K}_I\mathscr{B}^T\right) + \Omega \end{bmatrix} \begin{bmatrix} \underline{x} \\ \tilde{e} \end{bmatrix} \\
&\quad + \begin{bmatrix} \mathbb{I}_I \otimes [(\mathscr{B}(\Gamma - \mathscr{F}\Pi))\,\xi] \\ \mathbf{0}_{ID} \end{bmatrix} + \begin{bmatrix} \mathbf{0}_{ID} \\ -(\mathbb{I}_I \otimes \mathscr{K}_I)\,\tilde{\varepsilon} \end{bmatrix} + \begin{bmatrix} \mathbf{0}_{ID} \\ -(\mathbb{I}_I \otimes \mathscr{B})\,\delta \end{bmatrix}
\end{aligned} \tag{9.66}$$

The statement is then proven with

$$\Phi = \begin{bmatrix} \mathbb{I}_I \otimes (\mathscr{A} + \mathscr{B}\mathscr{F}) & -\mathbb{I}_I \otimes (\mathscr{B}\mathscr{F}) + \Omega \\ \mathbb{O}_{ID} & \mathbb{I}_I \otimes \left(\mathscr{A} + \mathscr{K}_I\mathscr{B}^T\right) + \Omega \end{bmatrix} \tag{9.67}$$

$$g(\xi) = \begin{bmatrix} \mathbb{I}_I \otimes [(\mathscr{B}(\Gamma - \mathscr{F}\Pi))\,\xi] \\ \mathbf{0}_{ID} \end{bmatrix} \tag{9.68}$$

$$f(\tilde{\varepsilon}, \delta) = \begin{bmatrix} \mathbf{0}_{ID} \\ -(\mathbb{I}_I \otimes \mathscr{K}_I)\,\tilde{\varepsilon} \end{bmatrix} + \begin{bmatrix} \mathbf{0}_{ID} \\ -(\mathbb{I}_I \otimes \mathscr{B})\,\delta \end{bmatrix} \tag{9.69}$$

In fact, according to the results given in Lemma 9.2, the output estimation error $\tilde{\varepsilon}$ is vanishing. Moreover, according to Lemma 9.1, the difference δ between the desired input $\bar{\eta}$ and the actual state of the independent robots η is vanishing as well. Subsequently, it is possible to conclude that function $f(\tilde{\varepsilon}, \delta)$ is vanishing as well. □

Under the assumption of controllability of the LTI system defined in Equation 9.13, the gain matrices $\mathscr{F}$ and $\mathscr{K}_I$ can be chosen in such a way that the state matrix in Equation 9.66 is Hurwitz stable, and that, consequently, the desired setpoint is asymptotically tracked. Further details on the use of LMI techniques for tuning these gain matrices can be found in Sabattini et al. [39].

9.4 SIMULATIONS AND EXPERIMENTS

The proposed control strategy was validated by means of repeated simulations and experiments, with a variable number of dependent and independent robots, and under varying initial conditions.

In particular, simulations were carried out utilizing Matlab for simulating a group of robots (modeled as kinematic agents) moving in a three-dimensional environment. The results of a representative example are summarized in Figure 9.3.

The interconnection topology among the robots is depicted in Figure 9.3a. In the picture, Di indicates the ith dependent robot, and Ij indicates the jth independent robot. The lines are used for representing the interconnection among dependent robots and independent robots.

The three-dimensional setpoint trajectories $x_s(t)$ are represented in Figure 9.3b: each colored line represents the setpoint for one of the dependent robots. The objective of the control system is then to make each dependent robot track one of the setpoint trajectories. This is obtained by means of the independent robots, which act as the control input for the system: in particular, the desired input for the system, namely $\bar{\eta}$ computed according to Equation 9.28, is depicted in Figure 9.3c. As expected, the dependent robots correctly track the desired setpoint trajectories: as

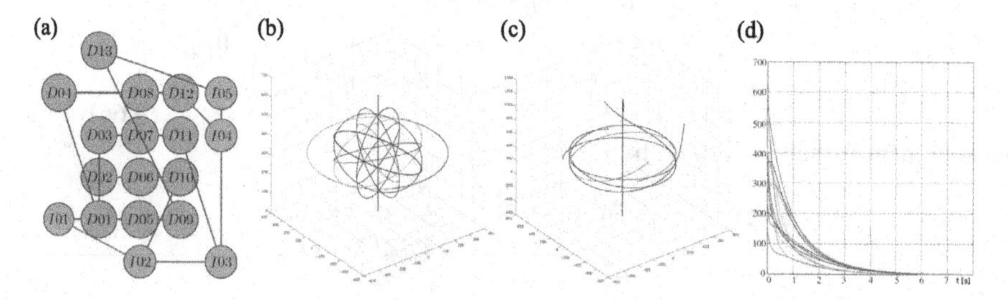

Figure 9.3 Simulation performed with $D = 13$ dependent robots, and $I = 5$ independent robots. (a) Interconnection topology. (b) Setpoint trajectories $x_s(t)$, defined as in Equation 9.25. (c) Desired input $\bar{\eta}(t)$ for the system, computed as in Equation 9.28. (d) Tracking error $e(t)$, defined as in Equation 9.26.

shown in Figure 9.3d, the tracking error $e(t)$, defined as in Equation 9.26, asymptotically vanishes.

A simplified experimental setup was implemented for validating the proposed control strategies on real robots. In particular, we exploited the Robot Operating System (ROS) [25] to control a group of differential drive mobile robots moving in a planar environment. It is worth remarking that the proposed control strategy was developed for single integrator agents: therefore, feedback linearization [42] was exploited for implementing the proposed control strategy on nonholonomic robots.

The experimental setup is composed of two or three differential drive robots, as shown in Figure 9.4: those robots were used as dependent robots to be controlled. The motion of the dependent robots was controlled by means of *virtual* independent robots, whose kinematic behavior was simulated on an external computer. Dependent robots and virtual independent robots are represented in Figure 9.5, where virtual independent robots are depicted as spheres, and the lines are used to represent dependent–dependent robot and dependent–independent robot connections, respectively.

(a) $D = 2$ (b) $D = 3$

Figure 9.4 Experimental setup composed of two or three differential drive ground robots.

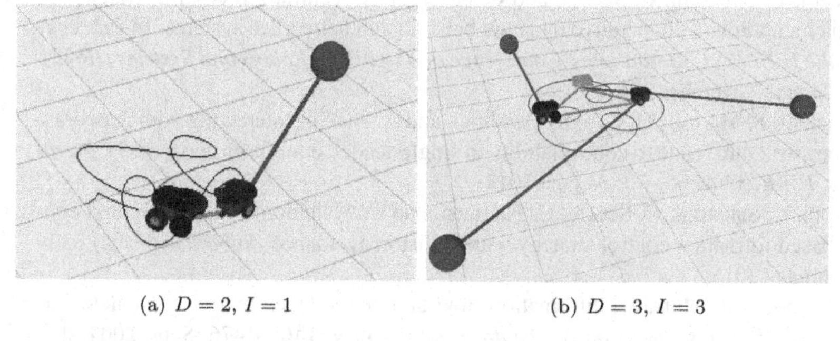

(a) $D = 2, I = 1$ (b) $D = 3, I = 3$

Figure 9.5 Real dependent robots and virtual independent robots (spheres): the lines between robots represent dependent–dependent robot connection, and the lines ending in spheres represent dependent–independent robot connections.

9.5 CONCLUSIONS

The proposed methodology aims at implementing, in a decentralized manner, complex dynamic behaviors in multi-robot systems. In particular, the multi-robot system is partitioned into two groups: independent and dependent robots. Exploiting local interaction rules, the input velocity for the independent robots is then defined in order to control the dependent robots to cooperatively track desired periodic trajectories.

A decentralized estimation procedure is defined to let each independent robot compute an estimate of the overall state of the multi-robot systems: this estimate is then exploited for defining a feedback control action obtained by means of the solution of the regulator equations.

The existence of a solution is guaranteed only if the desired setpoint trajectories are included within an admissibility set, which can be computed once the interconnection topology has been defined. Preliminary results were introduced in Sabattini et al. [34] for solving the inverse problem, namely defining the most suitable topology, given the desired setpoint trajectory to be tracked.

The proposed methodology relies on the assumption of having a time-invariant interconnection topology among the robots. This assumption is clearly restrictive in real-world applications, where the possibility of exchanging information is generally dependent on the relative positions. Therefore, current effort aims at extending the scope of proposed estimation and control strategy for considering time-varying communication topologies, which imply a time-varying model of the system.

REFERENCES

1. A. Ajorlou, A. Momeni, and A.G. Aghdam. A class of bounded distributed control strategies for connectivity preservation in multi–agent systems. *IEEE Transactions on Automatic Control*, 55:2828–2833, 2010.
2. G. Antonelli, F. Arrichiello, F. Caccavale, and A. Marino. Decentralized time-varying formation control for multi-robot systems. *The International Journal of Robotics Research*, 33(7), 2014.
3. M. Cocetti, L. Sabattini, C. Secchi, and C. Fantuzzi. Decentralized control strategy for the implementation of cooperative dynamic behaviors in networked systems. In *Proceedings of the IEEE/RSJ International Conference on Intelligent Robots and Systems (IROS)*, Tokyo, Japan, November 2013.
4. M. Egerstedt, S. Martini, M. Cao, K. Camlibel, and A. Bicchi. Interacting with networks: How does structure relate to controllability in single-leader, consensus networks? *Control Systems, IEEE*, 32(4):66 –73, August 2012.
5. R. Falconi, L. Sabattini, C. Secchi, C. Fantuzzi, and C. Melchiorri. Edge-weighted consensus based formation control strategy with collision avoidance. *Robotica*, 33(02):332–347, February 2015.
6. J.A. Fax and R.M. Murray. Information flow and cooperative control of vehicle formations. *IEEE Transactions on Automatic Control*, 49(9):1465–1476, Sept. 2004, doi: 10.1109/TAC.2004.834433.
7. M. Franceschelli, S. Martini, M. Egerstedt, A. Bicchi, and A. Giua. Observability and controllability verification in multi-agent systems through decentralized laplacian

spectrum estimation. In *49th IEEE Conference on Decision and Control (CDC)*, Atlanta, GA, 2010, pages 5775 –5780, doi: 10.1109/CDC.2010.5717400.
8. A. Ganguli, J. Cortes, and F. Bullo. Multirobot rendezvous with visibility sensors in nonconvex environments. *Robotics, IEEE Transactions on*, 25(2):340 –352, April 2009.
9. C. Godsil and G. Royle. *Algebraic Graph Theory*. Springer-Verlag, New York, 2001.
10. M. Ji and M. Egersted. A graph-theoretic characterization of controllability for multi-agent systems. In *American Control Conference, 2007. ACC '07*, New York, NY, pages 4588–4593, July 2007.
11. A.J. Laub. *Matrix Analysis for Scientists and Engineers*. Siam, Philadelphia, PA, 2005.
12. D. Lee, A. Franchi, H.I. Son, C. Ha, H.H. Bulthoff, and P. Robuffo Giordano. Semiautonomous haptic teleoperation control architecture of multiple unmanned aerial vehicles. *IEEE/ASME Transactions on Mechatronics*, 18(4):1334–1345, August 2013.
13. D. Lee and M.W. Spong. Stable flocking of multiple inertial agents on balanced graphs. *IEEE Transactions on Automatic Control*, 52(8):1469–1475, 2007.
14. Naomi Ehrich Leonard and Edward Fiorelli. Virtual leaders, artificial potentials and coordinated control of groups. In *Proceedings of the 40th IEEE Conference on Decision and Control (CDC)*, Orlando, FL, volume 3, pages 2968–2973. IEEE, 2001.
15. C.T. Lin. Structural controllability. *Automatic Control, IEEE Transactions on*, 19(3):201 – 208, June 1974.
16. J. Lin, A.S. Morse, and B.D.O. Anderson. The multi-agent rendezvous problem. In *Proceedings of the 42nd IEEE Conference on Decision and Control*, Maui, HI, volume 2, pages 1508–1513, December 2003.
17. Z. Lin, B. Francis, and M. Maggiore. Necessary and sufficient graphical conditions for formation control of unicycles. *IEEE Transactions on Automatic Control*, 50(1):121–127, 2005.
18. X. Liu, H. Lin, and B.M. Chen. A graph-theoretic characterization of structural controllability for multi-agent system with switching topology. In *Decision and Control, 2009 Held Jointly with the 2009 28th Chinese Control Conference. CDC/CCC 2009. Proceedings of the 48th IEEE Conference on*, Shanghai, China, pages 7012–7017, December 2009.
19. Y.Y. Liu, J.J. Slotine, and A.L. Barabasi. Controllability of complex networks. *Nature*, 473:167–173, 2011.
20. Sonia Martınez, Jorge Cortes, and Francesco Bullo. On robust rendezvous for mobile autonomous agents. In *16th IFAC World Congress*, Praha, CZ, 2005.
21. R. Olfati–Saber, J.A. Fax, and R.M. Murray. Consensus and cooperation in networked multi–agent systems. *Proceedings of the IEEE*, 95(1):215–233, 2007.
22. R. Olfati–Saber and R.M. Murray. Consensus problems in networks of agents with switching topology and time–delays. *IEEE Transactions on Automatic Control*, 9:1520–1533, 2004.
23. R. Olfati-Saber. Flocking for multi-agent dynamic systems: Algorithms and theory. *IEEE Transactions on Automatic Control*, 51(3):401–420, 2006.
24. L.C.A. Pimenta, G.A.S. Pereira, N. Michael, R.C. Mesquita, M.M. Bosque, L. Chaimowicz, and V. Kumar. Swarm coordination based on smoothed particle hydrodynamics technique. *IEEE Transactions on Robotics*, 29(2):383–399, April 2013.
25. Morgan Quigley, Ken Conley, Brian Gerkey, Josh Faust, Tully Foote, Jeremy Leibs, Rob Wheeler, and Andrew Y Ng. ROS: An open-source robot operating system. In *ICRA Workshop on Open Source Software*, Kobe, Japan, volume 3, page 5, 2009.
26. A. Rahmani, M. Ji, M. Mesbahi, and M. Egerstedt. Controllability of multi-agent systems from a graph-theoretic perspective. *SIAM Journal on Control and Optimization*, 48(1):162–186, February 2009.

27. W. Ren. Multi-vehicle consensus with a time-varying reference state. *Systems and Control Letters*, 56(7):474–483, 2007.
28. W. Ren and R.W. Beard. Consensus seeking in multiagent systems under dynamically changing interaction topologies. *IEEE Transactions on Automatic Control*, 50(5):655–661, 2005.
29. W. Ren, R.W. Beard, and E.M. Atkins. Information consensus in multivehicle cooperative control. *IEEE Control Systems*, 27(2):71–82, April 2007.
30. Wei Ren. Consensus based formation control strategies for multi-vehicle systems. In *American Control Conference*, Minneapolis, MN, pages 6. IEEE, 2006.
31. P. Robuffo Giordano, A. Franchi, C. Secchi, and H.H. Bülthoff. A passivity-based decentralized strategy for generalized connectivity maintenance. *The International Journal of Robotics Research*, 32(3):299–323, 2013.
32. L. Sabattini, N. Chopra, and C. Secchi. Decentralized connectivity maintenance for cooperative control of mobile robotic systems. *The International Journal of Robotics Research (SAGE)*, (12):1411–1423, October 2013.
33. L. Sabattini, C. Secchi, N. Chopra, and A. Gasparri. Distributed control of multi-robot systems with global connectivity maintenance. *IEEE Transactions on Robotics*, (5):1326–1332, October 2013.
34. L. Sabattini, C. Secchi, M. Cocetti, and C. Fantuzzi. Implementation of arbitrary periodic dynamic behaviors in networked systems. In *Proceedings of the IEEE International Conference on Robotics and Automation (ICRA)*, Hong Kong, China, June 2014.
35. L. Sabattini, C. Secchi, M. Cocetti, A. Levratti, and C. Fantuzzi. Implementation of coordinated complex dynamic behaviors in multi-robot systems. *IEEE Transactions on Robotics*, 31(4):1018–1032, August 2015.
36. L. Sabattini, C. Secchi, and C. Fantuzzi. Arbitrarily shaped formations of mobile robots: Artificial potential fields and coordinate transformation. *Autonomous Robots (Springer)*, 30(4):385–397, May 2011.
37. L. Sabattini, C. Secchi, and C. Fantuzzi. Closed–curve path tracking for decentralized systems of multiple mobile robots. *Journal of Intelligent and Robotic Systems*, 71(1):109–123, 2013.
38. L. Sabattini, C. Secchi, and C. Fantuzzi. Controllability and observability preservation for networked systems with time varying topologies. In *Proceedings of the IFAC World Congress*, Cape Town, South Africa, August 2014.
39. L. Sabattini, C. Secchi, and C. Fantuzzi. Cooperative dynamic behaviors in networked systems with decentralized state estimation. In *Proceedings of the IEEE/RSJ International Conference on Intelligent Robots and Systems (IROS)*, pages 3782–3787, Chicago, IL, USA, September 2014.
40. A. Saberi, A.A. Stoorvogel, and P. Sannuti. *Control of Linear Systems with Regulation and Input Constraints*. Springer-Verlag, New York, LLC, 2000.
41. R. Shields and J. Pearson. Structural controllability of multiinput linear systems. *Automatic Control, IEEE Transactions on*, 21(2):203–212, April 1976.
42. B. Siciliano, L. Sciavicco, L. Villani, and G. Oriolo. *Robotics: Modelling, Planning and Control*. Springer, London, UK, 2009.
43. F. Sorrentino. Effects of the network structural properties on its controllability. *Chaos*, 17(3):033102, 2007.
44. R. Soukieh, I. Shames, and B. Fidan. Obstacle avoidance of non-holonomic unicycle robots based on fluid mechanical modeling. In *Proceedings of the European Control Conference*, Budapest, Hungary, 2009.

45. S. Sundaram and C.N. Hadjicostis. Structural controllability and observability of linear systems over finite fields with applications to multi-agent systems. *Automatic Control, IEEE Transactions on*, 58(1):60–73, January 2013.
46. N. Tan, H. Lin, and Z. Ji. New results on controllability of multi-agent systems. In *Intelligent Control and Automation (WCICA), 2010 8th World Congress on*, Jinan, China, pages 574–579, July 2010.
47. H.G. Tanner, A. Jadbabaie, and G.J. Pappas. Stable flocking of mobile agents part i: Dynamic topology. In *Proceedings of the 42nd IEEE Conference on Decision and Control*, Maui, HI, volume 2, pages 2016–2021. IEEE, 2003.
48. H.G. Tanner, A. Jadbabaie, and G.J. Pappas. Stable flocking of mobile agents, part i: Fixed topology. In *Proceedings of the 42nd IEEE Conference on Decision and Control*, Maui, HI, volume 2, pages 2010–2015. IEEE, 2003.
49. P. Tsiotras and L.I.R. Castro. Extended multi-agent consensus protocols for the generation of geometric patterns in the plane. In *American Control Conference (ACC), 2011*, San Francisco, CA, pages 3850–3855, 29 2011-July 1.
50. A.G. Wu, G.R. Duan, and B. Zhou. Solution to generalized sylvester matrix equations. *Automatic Control, IEEE Transactions on*, 53(3):811–815, April 2008.
51. M. Zamani and H. Lin. Structural controllability of multi-agent systems. In *American Control Conference, 2009. ACC '09.*, St. Louis, MO, pages 5743–5748, June 2009.
52. M.M. Zavlanos, M.B. Egerstedt, and G.J. Pappas. Graph–theoretic connectivity control of mobile robot networks. *Proceedings of the IEEE*, 99:1525–1540, 2011.

10 Intelligent, Adaptive Humanoids for Human Assistance

Darwin G Caldwell, Arash Ajoudani, Jinoh Lee, and Nikos Tsagarakis

CONTENTS

10.1 HUMANOIDS – AN INTRODUCTION

The notion of machines able to help their human masters goes back to the earliest recorded history, and for more than 200 years, advanced automatons have been developed to exhibit a variety of human-like abilities [1], but it is only in the past 40 years, with developments in computing, control, actuation, mechanical design, materials science, sensors, software, etc., that the prospect of humanoid robots has become a genuine possibility. The early history of humanoids was dominated by the work of Kato at Waseda University with his pioneering work on the WABOT-1 [2] and in a more indirect way, Vukobratović with his work on exoskeletons [3]. Their ground-breaking work laid down many of the key principles of humanoid design and ambulation, which have been the basis of many robots up to the current time.

Inspired by the initial success and great potential of Kato's work, Honda began a long research program that included the development of robots such as E0 (1986), E1-E2-E3 (1987–1991), E4-E5-E6 (1991–1993), and P1-P2-P3 (1993–1997), through to the original ASIMO (2000) and the new ASIMO (2011) [4,5].

The P3, unveiled worldwide in 1998 [4], was a pivotal development in humanoid design, proving for the first time the viability of free-moving humanoid platforms and inspiring and initiating research on a number of other key platforms.

In Japan, work on the Humanoid Robot Platform (HRP), adapted from the Honda P3, started in the late 1990s and subsequently evolved into a series of models (HRP-2L/2P/2/3/4) [6,7], with each generation showing new and exciting advances. The HRP series of robots formed part of a joint French–Japanese collaboration and has become one of the most extensively researched and published humanoid platforms with an impressive range of capabilities.

During this time, the researchers at Waseda University had not been inactive. The developers of the original WABOTs continued their research and development through several further generations, leading to their most recent robot, WABIAN-2R [8].

Outside Japan, interest in the development of humanoids was also increasing. In Korea, researchers at Korea Advanced Institute of Science and Technology (KAIST) designed and built KHR-1/2/3, which ultimately became Hubo [9]. Hubo has now been turned into one of the first commercial humanoid products and is available internationally, with several models used around the world.

Europe also has had its share of humanoid projects. Among the earliest (and most impressive) was the "Johnnie" Robot from the Technical University of Munich. This robot achieved remarkable walking and fast walking capabilities [10]. In the late 2000s, this robot evolved into LOLA, which (unlike Johnnie) had a powered upper body.

While Johnnie was produced as part of a national (German) initiative, the iCub humanoid was developed in the EU project RobotCub, the first joint European effort to develop a humanoid. At the start of the RobotCub project, the primary aim was to design a "child-like" and child-sized humanoid platform that could be used to study cognitive systems [11]. As the iCub was designed to perform the functions of an (18-month-old) child, at the project outset, crawling rather than walking was seen as the primary ambulation; however, as the project progressed, the iCub "aged," and eventually, the robot was sized for a 4-year-old, with a capacity for standing and simple walking.

Kato's early humanoids used hydraulics as the primary actuation (due to the high power to weight performance needed in walking), but from the early Honda robots, the trend, particularly in Asia, was toward electric systems due to the improvements in electric motor power and control and perceived weaknesses in the control and operation of hydraulics. However, this trend, although common in mainstream industrial robots, was not universal, and some key robotic developments still made use of hydraulics. In the United States, researchers at Sarcos striving for ever higher power to weight performance returned to the hydraulic drive option for their CB robot [12]. This robot did achieve much higher power potential and for the first time in humanoids, started to introduce the concept of joint force as opposed to joint position control; however, the overall ambulation performance fell well short of the electric designs available at that time.

Nonetheless, the force control concept for humanoids was now recognized as an important upgrade, particularly with the expectation that humanoid robots would be required to physically interact with people in order to serve as intelligent assistants. To produce the necessary adaptability, robustness, and resilience to be deployed and effectively used in an assistive human-centered scenario, position-controlled humanoid robots were being replaced by controllably compliant (torque-controlled) systems. Two approaches arose, one based solely on sensing and software and a second based on sensing/software and structural elements–compliant actuators.

The first approach, typical in hydraulic robots such as CB, Atlas, Big Dog, and HyQ [13] but also found in the Toro Humanoid from DLR [14] (built using expertise gained from the lightweight arm [15]), uses sensing and software to regulate the robot's very high mechanical impedance and replicate compliant behaviors. In these systems, there is no inherent compliance. When the actuation is hydraulic, this typically creates robots that are very powerful, and this means that safety is a very real concern, making them unsuitable when operating in human-centered environments. Although the Toro robot is powerful, the output from the electric systems is not quite as concerning from a safety perspective.

The second common actuation approach uses physically compliant actuation systems. Here, elasticity is introduced between the load and the actuator to effectively decouple the high inertia of the actuator from the link. The series elastic actuator (SEA) [16], which has a fixed compliance element between a high-impedance actuator and the load, was one of the earliest of these designs. Later work introduced the concept of variable impedance into robots targeted at even closer human interaction. Variable impedance has the potential to improve the ability to withstand large force peaks and contact uncertainties (absorbing impacts) and increase energy efficiency [17]. Some of these concepts were incorporated into bipeds such as the M2V2 bipedal robot [18], the FLAME humanoid from Delft [19], and the MABEL biped [20]. At the Italian Institute of Technology (IIT), the cCub biped (an electric robot) was developed with active compliance regulation in all its joints and passive/active compliance in the ankles and knees [21]. This ultimately evolved into the COMAN (COMpliant huMANoid), which had active compliance in all joints of the lower and upper body and passive/active compliance in the ankle, knee, hip, waist, and all the joints of the upper body [22].

Back in Japan, researchers at Shaft, a company set up by engineers and scientists who had gained their knowledge and expertise working at the University of Tokyo's JSK lab, developed a high-performance humanoid that used liquid cooling to achieve greatly enhanced performance from electric motors. The Shaft robot went on to win the first round of the Defense Advanced Research Projects Agency (DARPA) Robotics Challenge (DRC) before being bought by Google.

For over 2 years, the DRC [34] formed a focus for many of the leading robotics groups working on humanoid technology and included the Atlas robot (a hydraulic humanoid designed and developed by Boston Dynamics), several variants of the HuBO, and WalkMan, a further evolution of the iCub/cCub/COMAN. Some of these robots are the current manifestation of a long developmental trend and will point to the future of intelligent assistive humanoid technology.

Following the completion of the DRC (won by the KAIST Hubo), work on humanoids did not slow, with new generations of torque-controlled electric systems such as WalkMan v2 [23]], Valkyrie, and COMAN+ [24]. However, although electrics has been the prime focus of most researchers, there is no doubt that the most impressive examples of humanoid performance in recent times have been with the hydraulic Atlas robot, which has demonstrated walking, running, jumping, and even somersaults. Clearly, the future of humanoid robots has a bright, highly dynamic outlook that will hold great potential and challenges for researchers, industry, and society.

10.2 COMAN HUMANOID – DESIGN AND CONSTRUCTION

The COMAN compliant humanoid is a fully torque-controlled robot, developed at the IIT within the EU project AMARSi (www.amarsi-project.eu) and evolved from previous work on the iCub humanoid (EU project RobotCub—www.robotcub.org) and the compliant biped cCub [20–22]. To the center of its neck, COMAN is 945 mm high, although with a head, this increases to 1.2 m tall (approx. the size of a 4-year-old child). The width and depth at the hips are 147 and 110 mm, respectively (giving the COMAN a very narrow waist compared with the iCub and cCub). The distance between the centers of the shoulders is 312 mm. The total weight of the robot is 34 kg, with the legs/waist weighing 18.5 kg, and the torso and the arms weighing 15.5 kg. COMAN has 33 degrees of freedom (D.O.F.) distributed across the body, as described in Table 10.1.

10.2.1 COMPLIANT ACTUATION UNIT

One of the key defining features of the highly integrated technology within COMAN is the novel series of elastic elements formed into a compliant actuation module (CompACT^{TM}) [25]. Each ComPACT^{TM} SEA weighs 0.52 kg, with a maximum output torque of 55 Nm, a peak output speed of 10.7 rad/s, and a nominal power of 190 W. This ComPACT^{TM} SEA module (developed in the EU project VIACTORS—www.viactors.org) uses high-performance, low-mass frameless brushless motors with a patented rotational series elastic element (CompACT^{TM}) [26]. This element can within milliseconds provide joint compliance variation over a range of over 60:1 (10 to 600 Nm/rad). This is greater than the range of human muscle (typically 20:1). To optimize compactness while maintaining rotary stiffness, the compliant module is formed as a spoke structure; see Figure 10.1. The output link (three-spoke element) rotates with respect to the input pulley and is coupled to it by six linear springs.

10.2.2 TORSO AND ARM DESIGN

Aluminum alloy 7075 (Ergal) is used for the main mechanical components (actuator housings, limbs, and chassis). Highly stressed elements, e.g., joint shafts and torque sensors, are made from stainless steel 17-4PH, which has excellent oxidation and

Table 10.1
Specification of the COMAN (COmpliant huMANoid) Robot

Property	Value	Robot
Dimensions	0.95 m (without head), 120 cm tall (with head)	
Weight	34 kg	
Degrees of Freedom	33: Leg 2 × 6 o Waist 3 o Arms 2 × 7 o Neck 2 o Hand 2 × 1 (SoftHand)	
Compliance	Fully torque controlledActive compliance in all jointsPassive compliance: Legs (ankle, knee and hip) Waist (roll, pitch and yaw) Arm (all joints)	
Actuators and Gearing	DC brushless motors (Kollmorgen) 55 Nm peak torque Harmonic gear (100:1 ratio)	
Battery	Lithium polymer ion (29 V 10 Ah—80 Wh/kgCapable of over 2 hrs continuous operation	
Construction	• Body (aluminum 7075) • Shell (ABS) • High-stress sections (Steel 17-4PH) • Torso core (titanium)	
Body Housing	All internal electrical wiringFully covered—no exposed components/wires	
Onboard Sensing and Perception	Position (relative and absolute encoder)Custom-designed individual joint torque sensors 2 × 6 axis force/torque sensor in anklePelvis mounted inertial measurement Unit (IMU)	
Onboard Computer	Dual core 2 Pentium PC104 running real-time Linux (Xenomai)	

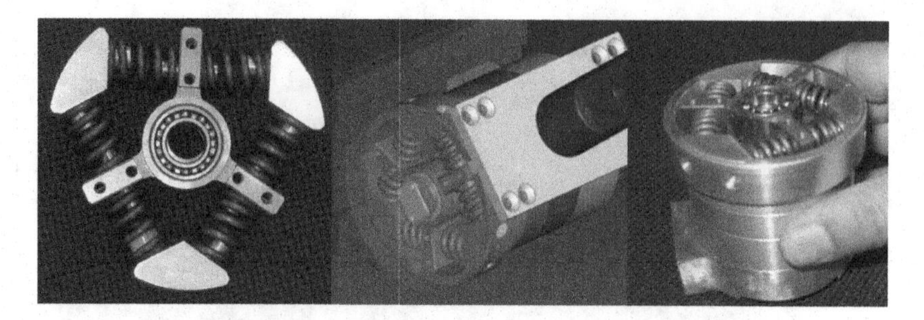

Figure 10.1 CompACT^{TM} Actuator Modules.

corrosion resistance coupled with high strength. The outer shell of the robot is made of ABS plastic (which provides good protection from dust and wipe down but is not designed to prevent the ingress of water). The central torso, which is made from titanium for strength and low weight, acts as the chassis for the onboard processing unit and battery/power management system. This central chassis also forms the mounting support for the shoulder flexion D.O.F. in each arm and the neck module and acts as the interface attachment between the upper body and the waist. The upper arms/shoulder have 4 D.O.F. in a pitch-roll-yaw kinematic arrangement and an elbow flexion/extension joint. The shoulder flexion/extension unit is entirely housed within the torso and is actuated by a ComPACT^{TM} SEA unit (peak torque of 55 Nm). The elbow is directly driven by a compliant ComPACT^{TM} module mounted at the center of the elbow joint. The forearm has 3 D.O.F., corresponding to the natural motions of the human arm (forearm roll and wrist flexion/extension and ad/abduction). The hand uses a newly developed high-robustness "softhand" developed within the EU project SoftHands (www.softhands.eu), which gives excellent dexterity in a compact, robust, low-mass configuration [27]. Previously, high-dexterity (21 D.O.F.) units had been tested successfully [28], but for most applications, this complexity is not needed due to the utility of the SoftHands concept.

10.2.3 LEG, HIP, AND WAIST DESIGN

COMAN's legs have an anthropomorphic kinematic structure with hips, a thigh with integrated knee joints, a calf with integral ankle joints, and a foot. The design of the leg is based on the "cCub" prototype developed by Hurst [20] but has an additional passively/actively compliant joint in the hip (flexion-extension). This gives the COMAN a greater range of motion than the cCub (Table 10.2).

The hip pitch motion is driven by the ComPACT^{TM} module (actively and passively compliant), while the roll and yaw motion actuators are of the purely active torque design. The hip roll motor is placed below the hip center, transmitting its torque to the hip (around the hip center) using a four-bar mechanism. This design permits integration of the SEA module at the hip pitch actuator without increasing the distance between the two hip centers. The hip yaw motion is powered by an ac-

Table 10.2
Specifications of COMAN Robot Joints

Joint	Motion Range (Degrees)	Max. Torque (Nm)	Max. Speed @36 V (rad/s)	Max. @48 V (rad/s) Speed
Hip Flexion/Extension	+45, −110	55	6.2	9.0
Hip Abduction/Adduction	+60, −20	55	6.2	9.0
Hip Rotation	+50, −50	55	6.2	9.0
Knee Flexion/Extension	+110, −10	55	6.2	9.0
Ankle Flexion/Extension	+70, −50	55	6.2	9.0
Ankle Abduction/Adduction	+35, −35	55	6.2	9.0
Waist Pitch	+50, −20	55	6.2	9.0
Waist Roll	+30, −30	55	6.2	9.0
Waist Yaw	+80, −80	55	6.2	9.0
Shoulder Flexion/Extension	+95, −195	55	6.2	9.0
Shoulder Abduction/Adduction	+120, −18	55	6.2	9.0
Shoulder Rotation	+90, −90	55	6.2	9.0
Elbow Flexion/Extension	+135, 0	55	6.2	9.0

tuator encased within the thigh. The knee is directly driven by a SEA mounted at the center of the knee joint, with the ankle pitch motion similarly using a SEA actuation unit but mounted in the calf. This calf-mounted motor transfers torque to the ankle through a four-bar link transmission. The ankle inversion/eversion uses a stiff (actively compliant) actuator located on the foot plate and directly coupled to the ankle roll joint. All motors in the legs/hips/waist have a peak torque of 55 Nm.

10.2.4 ONBOARD PROCESSING

Onboard processing is based on an embedded dual core Pentium PC104 embedded in the torso and running at 2.5 GHz. The computational environment provides onboard computation (multi onboard PC modules combined with GPUs) connected via an Ethercat network to distributed digital signal processors (DSPs) located at each joint. The DSPs run the servo control firmware at 1 kHz in real time and provide torque-level controllers.

10.2.5 SENSING

COMAN has traditional encoder-based joint position sensing on every joint in the legs, body, and arms, giving it the capacity to move and position all its limbs accurately; however, in addition, it has custom-designed high-performance torque sensors integrated into every motor of every joint, giving full active torque (compliance) regulation. This means that the robot can (under software control) respond precisely to unmodeled contacts and collisions. This gives COMAN unequaled tolerance to single and multiple, sequential and simultaneous impacts and disturbances over all of the body [29]. Moreover, in-house-developed 6 D.O.F. force/torque sensors are mounted at both ankles and wrists to measure the interaction forces between the foot and the ground or between the hand and the environment. In addition to its joint sensing, COMAN has an IMU (inertial measurement unit) mounted in the waist to give a global sense of balance, and vision sensors, including stereo cameras (RGB-D [Red, Green, Blue plus Depth]–Asus Xtion Pro), are mounted above the torso to provide point cloud information, with Lidar to provide feedback to a teleoperator or to build models of the environment for autonomous problem solving.

10.3 ROBOT INTERFACES

In a human-centered world, the range of tasks performed on a daily basis is immense and includes many apparently mundane activities such as walking and climbing, opening, closing, reaching, picking up, placing, retrieving, pushing and pulling objects, and manipulating tools. For humans, consideration of how we achieve these tasks seldom goes beyond the "difficulty" of initially learning the task sequence, but for humanoid robots, this remains one of the most daunting of undertakings, requiring multiple co-ordinations of multiple robots (e.g., 2 arms, 2 legs, 10 fingers, etc.). Given the complexity of the tasks, the traditional approach has been to decompose the activity into a series of decoupled actions that are typically separated into locomotion and manipulation subtasks. These can then be executed by controlling the robot legs, arms, or hands separately; however, if the robot operates in a real-world environment that contains large and/or heavy objects, then the picture changes completely. Indeed, humans naturally and efficiently perform heavy tasks by using their legs, arms, hands, and trunk in a coordinated whole-body movement, producing locomotion and manipulation (loco-manipulation) while maintaining equilibrium. Contacts on the more proximal limb parts are often intentionally sought, as their reduced mobility can be turned into an advantage in terms of stronger and more robust actions (grasping, pushing, pulling, etc.). To build a robot that can effectively assist humans, it is necessary to address such challenges, but although clearly advantageous, the use of the whole-body for loco-manipulation introduces a host of completely new problems on the modeling and control side.

10.3.1 CONTROL ARCHITECTURE

The control architecture used for the COMAN is shown in Figure 10.2. This architecture is capable of controlling the actions of the robot in both autonomous and mixed autonomy (shared tele-operation) modes to accomplish the highly coordinated tasks described earlier, e.g., valve turning, door opening, tool grasping and manipulation, debris clearance, or undertaking physically interactive and coordinated exchanges with a human "co-worker". Within the COMAN architecture, high-level interfacing with the robot uses the middleware language YARP [32], while sensory perception data communication is handled within ROS [33]. As previously mentioned in Section 10.2.4, COMAN has a central computation core that runs and regulates all the higher-level (cognitive) processes, but each joint/actuator is also controlled by a dedicated DSP, which controls all local operations in a 1 kHz loop. This local control enables real-time decentralized joint position and impedance regulation.

Within the overall control architecture, the *COMANInterface* module (Figure 10.2) acts as the link between the low-level library *Robolli* and high-level (YARP) functionalities and incorporates the low-level library *Robolli*. This module communicates directly with the DSP boards and exports the low-level DSP-generated data to the higher YARP-based control modules.

The task (manipulation) module, controlling the actual activity to be completed, e.g., (ValveTurnModule, openDoorModule, DrillHole, RemoveDebris, etc.), is written in YARP, while the kinematic and dynamic model of the robot is specified using the Unified Robot Description Format (URDF2) parsed by the *IDynTree* library to generate forward kinematics, Jacobians, and dynamics data. This module follows a simple communication protocol to control state transitions using an internal state machine.

The manipulation action is regulated by the target module (e.g., OpenDoorModule and ValveTurnModule in Figure 10.2) using data on the characteristics of the object to be "manipulated" (size, shape, center point, rotation angle, etc.). The actions in each module are decomposed to a series of primitives to execute the task. For instance, in the "Reach" primitive, the end goal of the robot can be tailored to the task demands (valve turning, door opening, debris removal, tool grasping, object manipulation, etc.) by simply changing the parameters of this module. This does not impact on the remainder of the control architecture and its interconnectivity and ensures the robustness of the overall set-up. Clearly, the accuracy of this motion can be

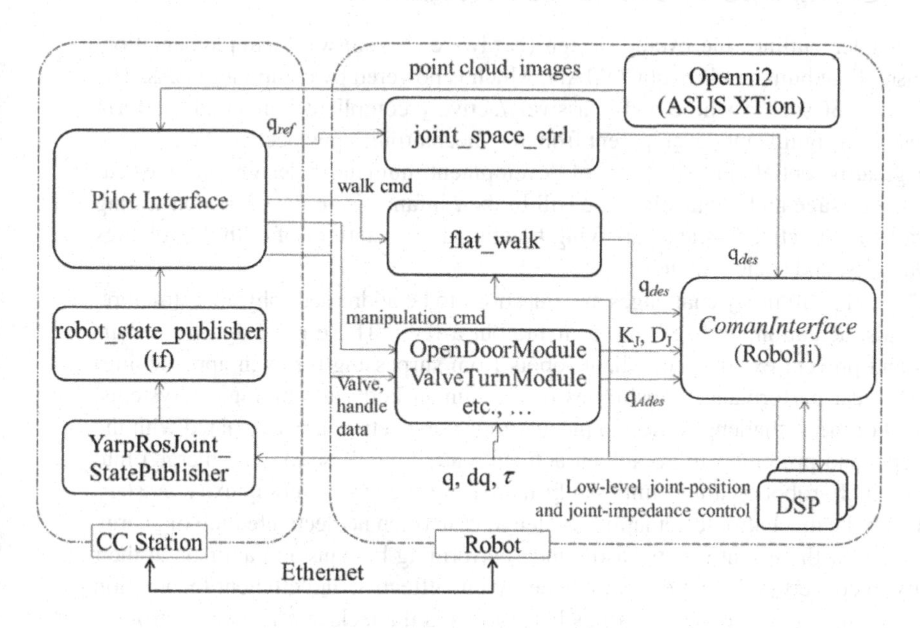

Figure 10.2 COMAN control architecture.

critical, yet the imprecise nature of the positioning of the robot and the target object (valve in this example), coupled with position changes that can result from the need for a humanoid to maintain and adjust its balance, is very problematic for a traditional position-controlled humanoid (or robot in general). However, the compliance regulation (active and passive) and torque control that are critical features of the COMAN robot make this humanoid extremely well suited to imprecise (or unmodeled) interactions.

The visual perception (Openni2) module provides the position and orientation of the target object or tool, e.g., valve or door handle, with respect to the base reference frame, and this is sent to the YARP port to be used for trajectory generation. Finally, the command port accepts manipulation primitives to reach for the object (e.g., door handle, valve, etc.), grasp, manipulate, rotate, release, pull, or disengage as required by the demands of the task. The motion primitives commands are sent to the robot using the TCP protocol composed of strings and doubles [35,36].

The target (valve, door handle, work tool, etc.) is physically reached when the robot moves (walks) to that target point, and this is controlled using the walking (flatWalk) module, which accepts forward, lateral (side step), and rotate in place commands. The *Pilot Interface* shows the actual configuration of the robot and the point cloud of the scene given by the three-dimensional (3-D) camera, which is used to recognize the object/tool with respect to a robot local frame of reference (see Figure 10.3).

10.4 CONCLUSIONS AND FUTURE CHALLENGES

This chapter outlines the overall design (hardware and software), implementation, and use of the humanoid robot COMAN, which is powered by compliant SEAs. The realization of such a whole-body passively/actively compliant humanoid platform represents a significant development from the mechatronics perspective. Yet, despite their great potential, and the years of development, humanoid design, and particularly their usage and application, are still in their infancy in terms of demonstrating operational effectiveness and satisfying functional expectations of both the robotics community, and society in general.

There are still many challenges and questions to be addressed (physical structure, materials, actuation, and sensing, to name but a few). These pose significant barriers and prevent existing compliant robots from surpassing (or even approaching) the physical performance capabilities of the human body and biological systems. Several of these challenges, from a mechatronic perspective, are associated with the concepts and principles of compliant actuation and how these may be merged into multi-D.O.F. robots without impossibly increasing the system complexity. Associated with the mechatronic/actuation challenge is an often neglected feature of energy efficiency, with current robots vastly underperforming humans and animals in their ability to convert (still limited) stored energy into effective and efficient locomotion and motion. One of the main reasons behind this is the lack of effective controllers that can explore the physical elasticity of actuators and robots.

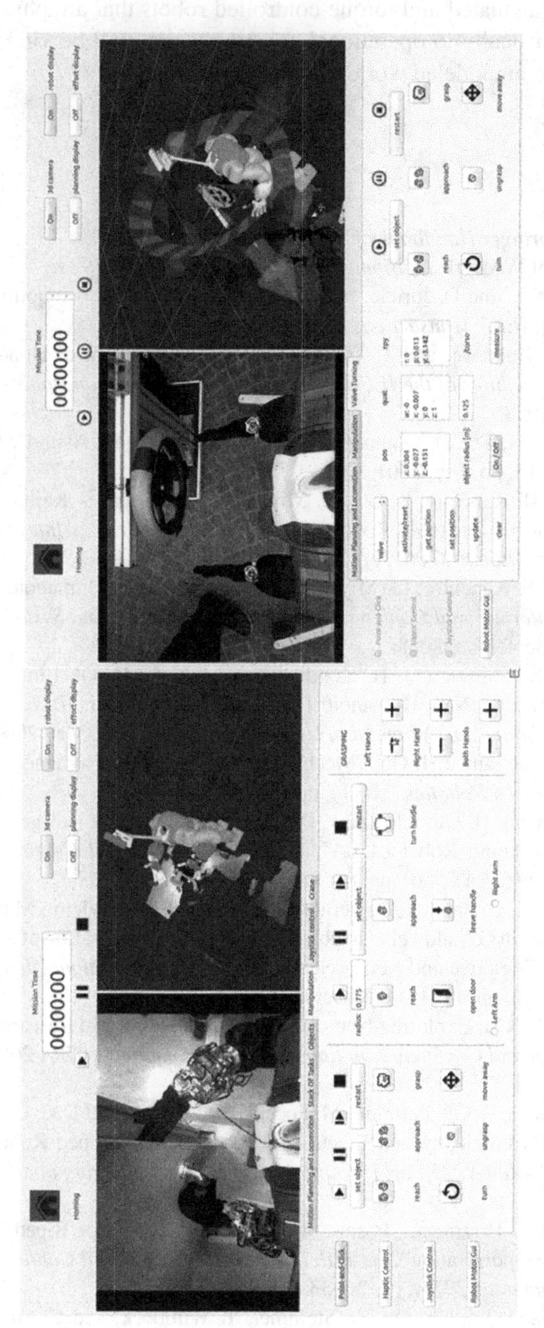

Figure 10.3 The pilot interface: a 3-D object (e.g., door handle, valve, drill, etc.) model is overlaid on the 3-D scene, and commands are sent from a GUI.

These form only some of the future research directions needed to create a new generation of compliantly actuated and torque-controlled robots that are physically robust and can adapt intrinsically to operational constraints that will be vital when interacting in, and with, an unmodeled world or assisting humans.

REFERENCES

1. Shimon Y. Nof (Ed.), *Springer Handbook of Automation*, Springer, 2009.
2. I. Kato, "Development of WABOT-1", *Biomechanism*, 2, pp. 173–214, 1983.
3. M. Vukobratovic, A. Frank, and D. Juricic, "On the Stability of Biped Locomotion", In: *Biomedical Engineering, IEEE Transactions On*, pp. 25–36, 1970.
4. K. Hirai, M. Hirose, Y. Haikawa, and T. Takaneka, "The Development of Honda Humanoid Robot", In: *Proceedings of the IEEE International Conference on Robot. & Automat*, pp. 1321–1326, 1998.
5. R. Hirose, and T. Takenaka, T., "Development of the Humanoid Robot Asimo", *Honda R&D Technical Review*, 13, pp. 1–6, 2001.
6. K. Akachi, K. Kaneko, H. Kanehira, S. Ota, G. Miyamori, M. Hirata, S. Kajita, and F. Kanehiro, "Development of Humanoid Robot Hrp-3p", In: *5th IEEE-Ras International Conference On Humanoid Robots, 2005*, pp. 50–55, 2005.
7. K. Kaneko, K. Harada, F. Kanehiro, G. Miyamori, and K. Akachi, "Humanoid Robot Hrp-3", In: *IEEE/RSJ International Conference On Intelligent Robots And Systems (Iros 2008)*, pp. 2471–2478, September 2008.
8. Y. Ogura, H. Aikawa, K. Shimomura, H. Kondo, A. Morishima, H. Ok Lim, and A. Takanishi, "Development of a New Humanoid Robot WABIAN-2", In: *IEEE International Conference on Robotics and Automation (ICRA 2006)*, pp. 76–81, May 2006.
9. I.W. Park, J.Y. Kim, J. Lee, and J.H. Oh, "Mechanical Design of the Humanoid Robot Platform, HUBO", *Advanced Robotics*, 21(11), pp. 1305–1322, 2007.
10. S. Lohmeier, T. Buschmann, H. Ulbrich, and F. Pfeiffer, "Modular Joint Design for Performance Enhanced Humanoid Robot LOLA", In: *IEEE International Conference on Robotics and Automation (ICRA 2006)*, pp. 88–93, May 2006.
11. N.G. Tsagarakis, G. Metta, G. Sandini, D. Vernon, R. Beira, J. Santos-Victor, M.C. Carrazzo, F. Becchi, and Darwin G. Caldwell, "iCub – The Design and Realisation of an Open Humanoid Platform for Cognitive and Neuroscience Research", *International Journal of Advanced Robotics*, 21(10), pp. 1151–1175, October 2007.
12. G. Cheng, S. Hyon, et al. "CB: Exploring Neuroscience with a Humanoid Research Platform", In: *IEEE International Conference on Robotics and Automation, (ICRA 2008)*, pp. 1772–1773, 2008.
13. C. Semini, N.G. Tsagarakis, E. Guglielmino, M. Focchi, F. Cannella, and D.G. Caldwell, "Design of HyQ – A Hydraulically and Electrically Actuated Quadruped Robot", In: *Proceedings of IMechE Volume 225 Part I: Journal of Systems Engineering and Control*, 2011.
14. C. Ott, M.A. Roa, and G. Hirzinger, "Posture and Balance Control for Biped Robots Based on Contact Force Optimisation", In: *11th IEEE-RAS International Conference on Humanoid Robots (Humanoids 2011)*, pp. 26–33, 2011.
15. A. Albu-Schäffer, S. Haddadin, Ch. Ott, A. Stemmer, T. Wimböck, and G. Hirzinger, "The DLR Lightweight Robot – Design and Control Concepts for Robots in Human Environments", *Industrial Robot: An International Journal*, 34(5), 2007.

16. G.A. Pratt, and M.M. Williamson, "Series Elastic Actuators", In: *IEEE/RSJ International Conference on Intelligent Robots and Systems–Workshop on 'Human Robot Interaction and Cooperative Robots'*, Pittsburg, USA, 1995.
17. B. Vanderborght, A. Albu-Schaeffer, A. Bicchi, E. Burdet, D.G. Caldwell, R. Carloni, M. Catalano, O. Eiberger, W. Friedl, G. Ganesh, M. Garabini, M. Grebenstein, G. Grioli, S. Haddadin, H. Hoppner, A. Jafari, M. Laffranchi, D. Lefeber, F. Petit, S. Stramigioli, N. Tsagarakis, M. Van Damme, R. Van Ham, L.C. Visser, and S. Wolf, "Variable Impedance Actuators: A Review", *Robotics and Autonomous Systems*, 61(12), pp. 1601–1614, 2013.
18. J. Pratt, T. Koolen, T. De Boer, J. Rebula, S. Cotton, J. Carff, M. Johnson, and P. Neuhaus, "Capturability-Based Analysis and Control of Legged Locomotion, Part 2: Application to M2V2, a Lower-Body Humanoid", *International Journal of Robotics Research*, 31(10), pp. 1117–1133, 2012.
19. D. Hobbelen, T. De Boer, and M. Wisse, "System Overview of Bipedal Robots Flame and Tulip: Tailor-Made for Limit Cycle Walking", In: *IEEE/RSJ Interantional Conferenec on Intelligent Robots and Systems IROS 2008*, pp. 2486–2491, 2008.
20. J. Hurst, The Electric Cable Differential Leg: A Novel Design Approach for Walking and Running, *International Journal of Humanoid Robotics*, 8(02), pp. 301–321, 2011.
21. N.G. Tsagarakis, Zhibin Li, J.A. Saglia, and D.G. Caldwell, "The Design of the Lower Body of the Compliant Humanoid Robot 'cCub'", In: *IEEE International Conference on Robotics and Automation, ICRA 2011*, Shanghai, China, pp. 2035–2040, May 2011.
22. N. Tsagarakis, S. Morfey, G.A. Medrano-Cerda, Z. Li, and D.G. Caldwell, "Development of Compliant Humanoid Robot COMAN: Body Design and Stiffness Tuning", In: *IEEE International Conference on Robotics and Automation 2013 (ICRA 2013)*, Karlsruhe, Germany, pp. 665–670, May 2013.
23. N.G. Tsagarakis, D.G. Caldwell, et al. "WALK-MAN: A High-Performance Humanoid Platform for Realistic Environments", *Journal of Field Robotics*, June 2017. doi:10.1002/rob.21702.
24. E. Rolley-Parnell, D. Kanoulas, et al. "Bi-Manual Articulated Robot Teleoperation Using an External RGB-D Range Sensor", In: *15th International Conference on Control, Automation, Robotics and Vision (ICARCV)*, Singapore, pp. 298–304, November 2018.
25. N.G. Tsagarakis, G. Metta, G. Sandini, D. Vernon, R. Beira, J. Santos-Victor, M.C. Carrazzo, F. Becchi, and Darwin G. Caldwell, "iCub – The Design and Realisation of an Open Humanoid Platform for Cognitive and Neuroscience Research", *International Journal of Advanced Robotics*, 21(10), pp. 1151–1175, October 2007.
26. N.G. Tsagarakis, M. Laffranchi, B. Vanderborght, and D.G. Caldwell, "A Compact Soft Actuator Unit for Small Scale Human Friendly Robots", In: *ICRA 2009*, Kobe, Japan, pp. 4356–4362, May 2009.
27. M.G. Catalano, G. Grioli, A. Serio, E.C. Farnioli, A. Piazza, and Bicchi, "Adaptive Synergies for a Humanoid Robot Hand", In: *IEEE-RAS International Conference on Humanoid Robots, Humanoids 2012*, Osaka, Japan, pp. 7–14, December 2012.
28. S. Davis, N.G. Tsagarakis, and D.G. Caldwell, "The Initial Design and Manufacturing Process of a Low Cost Hand for the Robot iCub", In: *IEEE Humanoids 2008*, Daejeon, Korea, pp. 40–45, December 2008.
29. Z. Li, N. Tsagarakis, and D.G. Caldwell, "A Passivity Based Cartesian Admittance Control for Stabilizing the Compliant Humanoid COMAN", In: *2012 IEEE-RAS International Conference on Humanoid Robots, IEEE Humanoids'12*, Osaka, Japan, pp. 44–49, November 2012.
30. G. Metta, P. Fitzpatrick, and L. Natale, "YARP: Yet Another Robot Platform", *International Journal of Advanced Robotic Systems*, 3(1), pp. 43–48, 2006.

31. Morgan Quigley, P. Conley, et al. "ROS: An Open-Source Robot Operating System", In: *ICRA Workshop on Open Source Software*, Vol. 3, No. 3, p. 2, 2009.
32. Jean-Claude Samin, *Symbolic Modeling of Multibody Systems*, Vol. 112. Springer Science & Business Media, 2003.
33. http://gazebosim.org.
34. http://www.theroboticschallenge.org/.
35. J. Lee, A. Ajoudani, E. Mingo, A. Rocchi, A. Settimi, M. Ferrati, A. Bicchi, N.G. Tsagarakis, and D.G. Caldwell, "Upper-Body Impedance Control with Variable Stiffness for a Door Opening Task", In: *2014 IEEE-RAS International Conference on Humanoid Robots (Humanoids 2014)*, Madrid, Spain, pp. 713–719, November 18–20, 2014.
36. A. Ajoudani, J. Lee, A. Rocchi, M. Ferrati, E. Mingo, A. Settimi, D.G. Caldwell, A. Bicchi, and N.G. Tsagarakis, "A Manipulation Framework for Compliant Humanoid COMAN: Application to a Valve Turning Task", In: *IEEE-RAS International Conference on Humanoid Robots*, Spain, 2014.

11 Advanced Sensors and Vision Systems

Van-Dung Hoang and Kang-Hyun Jo

CONTENTS

11.1 OVERVIEW

Research on machine vision and some advanced sensors is important for improving the existing applications in product inspections, automation, intelligent transportation systems (ITS), human life assistance, and other intelligent systems [1–4]. The technological applications of vision sensors are to permit adaptive capability in a variety of automatic machine designs, such as robots, mobile devices, unmanned ground/aerial robots (UGV/UAV) in ITS, and so on. In the field of robotic systems, visual odometry (VO) is the process of analyzing the associated camera images to discover the location and orientation of a mobile robot [5,6].

An autonomous vehicle is a competent self-driving car fully implementing the intelligent transportation capabilities of the traditional vehicle. As an autonomous vehicle, it has the capacity for sensing the outdoor environment and auto-navigating without human decisions. Safe driving is one of the most important problems in intelligent transportation in modern life. There are many researchers focusing on developing and constructing an assistant driving system for vehicle navigation. Vehicle producing companies as well as users are interested in creating a process to assist in safer and more comfortable driving.4

Autonomous robot/vehicle navigation has been an important objective of research, with study of applications ranging from autonomous driver assistance in the civil engineering sector to military supply convoys and to space exploration. A critical requirement of high-level navigation applications is that the vehicle must have some reasonable knowledge of its current position with respect to certain benchmarks. For example, in autonomous navigation applications that plan a global path for an entailed motion in order to reach a target position, the system needs to estimate accurately the localization of its trajectory for tracking the path to the destination and for confirming the success of travel [7–12]. In addition, the system should have the capability to recognize obstacles in order to avoid crashing into them [13–16]. The process of position estimation with respect to a fixed reference frame is defined as localization of the vehicle with global coordination. Developing an intelligent navigation system for vehicles is necessary because this allows drivers to have a better knowledge of the route when controlling their vehicles in outdoor environments, especially in urban regions. The developed navigation system has many potential applications, ranging from the construction of realistic object models for civil engineering and other industries to the quantitative recovery of metric information for scientific and engineering data analysis. Path planning and motion estimation are important for both human and robot/vehicle navigation. As a result, autonomous driving maps can guide vehicles to their destinations within a given route. Similarly, topological maps of a road network inform an autonomous vehicle about where it can move to and what actions should be performed. By providing the necessary information about the driving environment, mapping the path and trajectory optimizes both manual and autonomous navigation. In the autonomous vehicle field, a key ability for intelligent vehicles is autonomous navigation. Motivated by these reasons, we resolved the general problem of autonomous vehicle navigation design based on the combination of multiple sensors using different image processing techniques.

There are some discussions on applications of multiple advanced sensors to the field of automotive navigation in ITS. The motion and localization estimation of an outdoor robot are addressed using a combination of cameras and laser range finder (LRF) sensors based on an incremental method combined with global positioning system (GPS) information. The perspective camera and omnidirectional camera as well as LRF sensors are investigated to utilize the strengths of each device. Motion estimation based on sequential images for autonomous navigation is an incremental approach because of the processing time when the robot moves over a long distance. One solution to speed up the computing system, the vision-based compass approach, is also proposed and applied. The cumulative error of visual odometry is included using GPS correction under maximum likelihood estimation in a filter framework, such as the extended Kalman filter (EKF) technique. Practical experiments in a variety of conditions demonstrate the accuracy of robot localization using the method based on integration of multiple sensors. Furthermore, a robust global motion method is presented using corresponding sequential images based on a bundle adjustment technique combined with GPS information.

In recent years, many methods have been developed for robots, autonomous robots, navigation, and motion planning [10,17–19], which can be divided into several categories. The first group of methods uses only vision systems, e.g., monocular camera, stereo cameras, and catadioptric cameras. The second one uses other electromagnetic devices, e.g., stocktickerGPS, inertial measurement units (IMU), wheel odometers, and LRF. Other researchers have combined vision sensors and other electromagnetic devices. Additionally, the integration of cameras and LRF devices has been applied to much research on UGV, such as autonomous navigation of outdoor robots.

11.2 VISUAL ODOMETRY

Visual odometry is the process of recovering the related position and orientation of a robot by analyzing the associated vision devices. Alternatively, visual odometry is defined as the estimation of the egomotion of the robot using only cameras mounted on the robot [20,21]. It aims to recover the parameters of the equivalent odometry data using sequential camera images to estimate the motion of travel by the robot. The visual odometry technique allows enhancement of the navigation precision of robots in a variety of indoor or outdoor environments and terrain. An example of visual odometry estimates a robot's motion using camera images to extract relative features between two sequential images. Vision-based motion estimation is a very important task in applications of autonomous navigation of robots. The general flowchart of motion estimation based on vision sensors is illustrated in Figure 11.1.

In structure from motion (SfM) technology, the general idea is that a camera captures the sequential images of the world scene in different positions. The core idea is finding the relations of images in order to extract the camera position and thus, the robot location. Therefore, the position of the sensor is recovered based on the geometrical constraints in sequential images. The principles for three-dimensional (3-D) views and multiple views are more complicated. The sequential images can

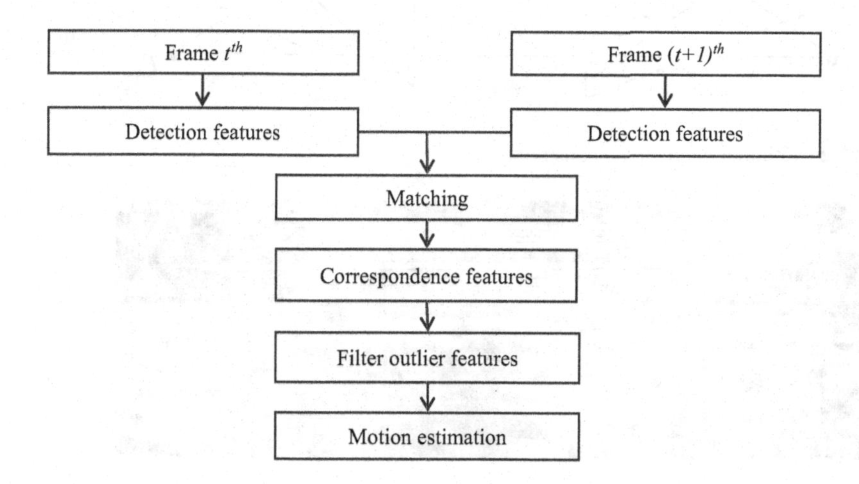

Figure 11.1 General flowchart for motion estimation based on vision sensors.

be achieved using a perspective camera or a catadioptric camera such as an omnidirectional camera. This presentation considers both the perspective camera and the omnidirectional camera. The general idea is also to use correspondence constraints in the sequential images, which are achieved using an omnidirectional camera [22–24]. In the multiple view geometry, there are two kinds of approach for solving the problem: the non-optimization and the optimization approach. There is a tradeoff in the solution of the problem; the optimization approach achieves higher accuracy but has a higher computational cost than non-optimization. Therefore, the non-optimization solution can be considered to apply in real-time application systems.

Let us consider this canonical problem in a mathematical equation. Given the correspondence features r and r', the problem is how to find the 3-D point P and camera positions with the smallest error. Therefore, the correspondence points are essential to extract the extrinsic parameters of visual odometry–based movement estimation (Figures 11.2 and 11.3).

For presenting geometrical constraints, the visual odometry system is composed of sequential image pair constraints. These constraints are analyzed based on the epipolar constraint using the essential matrix E. Figure 11.4 shows two corresponding rays of r and r' from the focal point to the world point P. The rays of r and r' are observed from the camera at two camera positions, and their relative constraint:

$$r'^{T} E r = 0 \tag{11.1}$$

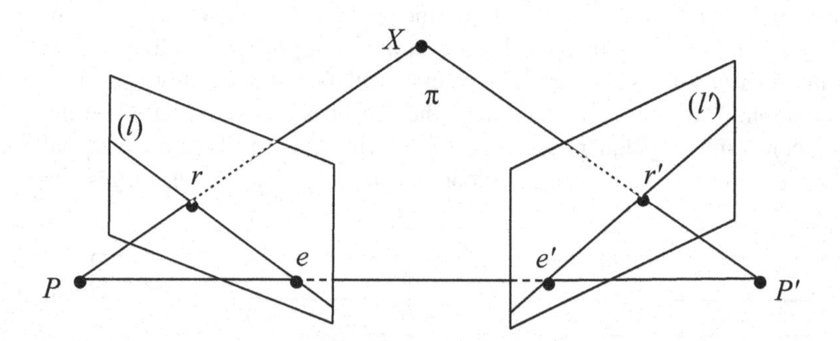

Figure 11.2 Epipolar geometrical constraint.

Figure 11.3 The correspondence points in sequential images.

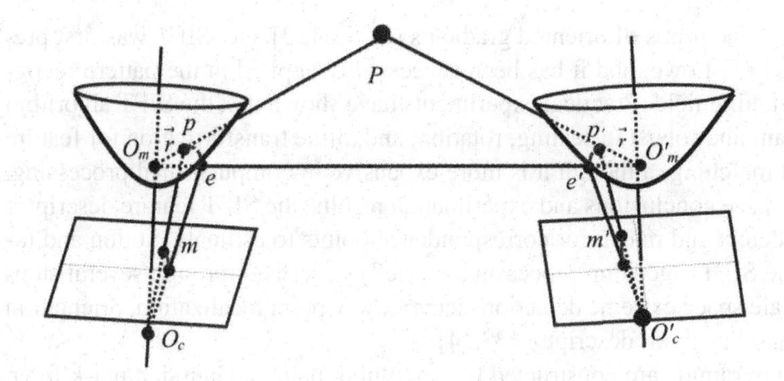

Figure 11.4 Epipolar geometry principle in the catadioptric camera model.

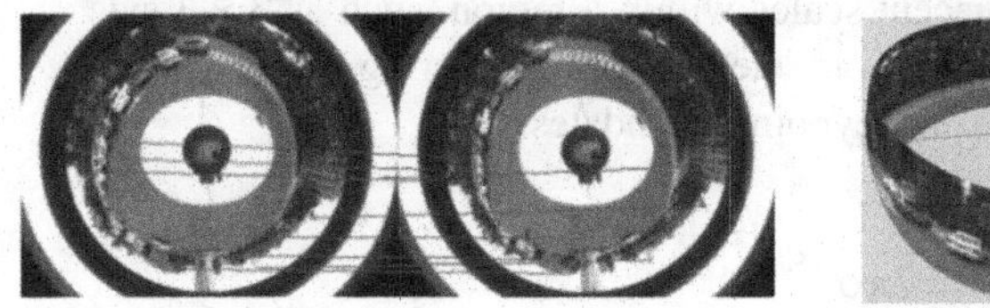

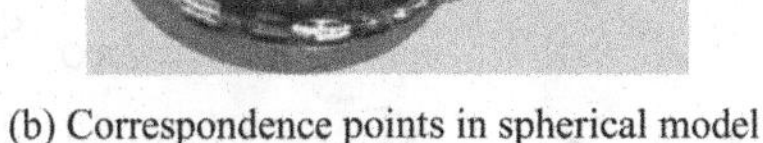

(a) Correspondence points in omni-images (b) Correspondence points in spherical model

Figure 11.5 Correspondence points in sequential omnidirectional images.

where the essential matrix E is defined based on a rotation matrix and a translation vector, $E = [T] \times R$. The translation vector $T = [T_X, T_Y, T_Z]^T$ and the rotation matrix $R = R_Y R_Z R_X$, where R_Y, R_Z, R_X are pitch, yaw, and roll rotation components, respectively.

$$R = \begin{bmatrix} \cos\alpha\cos\beta & \sin\beta\sin\gamma - \sin\alpha\cos\beta\cos\gamma & \sin\beta\cos\gamma + \sin\alpha\cos\beta\sin\gamma \\ \sin\alpha & \cos\alpha\cos\gamma & -\cos\alpha\sin\gamma \\ -\cos\alpha\sin\beta & \cos\beta\sin\gamma + \sin\alpha\sin\beta\cos\gamma & \cos\beta\cos\gamma - \sin\alpha\sin\beta\sin\gamma \end{bmatrix} \tag{11.2}$$

Six components of camera transformation at sequential camera positions are estimated by solving Equation 11.1. There are several methods to solve this problem, among them the eight-point [25] and five-point algorithms [26,27] (Figure 11.5).

In this section, three techniques are presented to compensate for motion estimation: the scale-invariant interest points (SIFT) keypoint feature descriptor, the vision-like compass, and vanishing points detection using omnidirectional vision for rotation estimation.

11.2.1 FEATURE POINT DESCRIPTOR

Many kinds of methods for extraction and matching of feature descriptors have been proposed: Harris corner [28], distinctive image features from SIFT [29], speeded up robust features (SURF) [30], gradient location and orientation histogram

(GLOH) [31], histograms of oriented gradients (HOG) [32], etc. SIFT was first presented by David G Lowe, and it has been successfully applied in the pattern recognition and matching field. Practical experiments have shown that the SIFT algorithm is very invariant and robust in scaling, rotation, and affine transformation for feature detection and matching, although it is more expensive in computational processing. According to these conclusions and experimental results, the SIFT feature descriptor is utilized to detect and match for correspondence points to estimate motion and localization. The SIFT algorithm processing is briefly described through several steps as follows: scale space extreme detection, accurate key point localization, orientation assignment, and key point descriptor [33,34].

The image pyramids are constructed by smoothing using a Gaussian mask filter. The image pyramids of difference of Gaussian (DoG) are obtained by subtraction of adjacent Gaussian smoothed images. Each cell element on each scale level is collated to the upper and lower adjacent scales within a region, such as 3×3 size of pixels. The extreme value among them is selected for discovering the keypoints. These elements are also regarded to the keypoint candidates.

$$G(x,y,\sigma) = \frac{1}{2\pi\sigma^2} e^{-(x^2+y^2)/2\sigma^2} \tag{11.3}$$

The convolution of the original image I with the Gaussian smooth $G(x,y,k\sigma)$ at scale k is

$$L(x,y,k\sigma) = G(x,y,k\sigma) \otimes I(x,y) \tag{11.4}$$

where $\otimes$ is the convolution operation in x and y.

The DoG processing is described by the following formulation:

$$D(x,y,\sigma) = L(x,y,k_1\sigma) - L(x,y,k_2\sigma) \tag{11.5}$$

The set of candidate keypoints located at the central sample points are interpolated with adjacent elements to determine their positions precisely. The quadratic function is applied to fit the local sample points to determine the true location of the keypoints; the Taylor expansion can be used as follows:

$$D(x) = D + \frac{\partial D^T}{\partial x} x + \frac{1}{2} x^T \frac{\partial^2 D}{\partial x^2} x \tag{11.6}$$

where D and its derivatives are evaluated at the sample points. $(x,y,\sigma)^T$ is the offset of the candidate keypoint, and by taking the derivative of this function with respect to x and setting it to zero, the location of the extreme $\hat{x}$ can be determined, following the formulation in Figure 11.6:

$$\hat{x} = -\frac{\partial^2 D^{-1}}{\partial x^2} \frac{\partial D}{\partial x} \tag{11.7}$$

The aim of the next step is to eliminate unstable points from the set of candidate keypoints. Finding low contrast point is evaluated $D(\hat{x})$ value with threshold. The

Figure 11.6 Keypoint detection. (a) Pyramid of Gaussian images. (b) Difference of Gaussian images. (c) Candidate keypoints.

values of candidate keypoints in the DoG pyramid at the extreme are achieved by substituting Equations 11.6 and 11.7 into the following formulation:

$$D(\hat{x}) = D + \frac{1}{2}\frac{\partial D^{-1}}{\partial x}\hat{x} \tag{11.8}$$

The candidate keypoint is discarded if the value of $D(\hat{x})$ is lower than a filter threshold.

The next step aims to discard poorly localized extremes; the principal curvature appearing in the perpendicular direction of the DoG function is greater than the large principal curvature along the edge. The second order of Hessian matrix $H_{2\times 2}$ is used to find the principal curvature, as depicted in the following form:

$$H = \begin{bmatrix} D_{xx} & D_{xy} \\ D_{xy} & D_{yy} \end{bmatrix} \tag{11.9}$$

The ratio of the trace of H and its determinant is at a minimum when the eigenvalues are equal to each other. The candidate keypoints are recognized as poor localizations and can be discarded from the candidate keypoints in the case of meeting the criterion

$$\frac{(D_{xx} + D_{yy})^2}{D_{xx}D_{yy} - (D_{xy})^2} > \frac{(r_{th} + 1)^2}{r_{th}} \tag{11.10}$$

In order to assign a consistent orientation for each key point using the local image properties, the histogram of its orientation is computed based on the orientations of sample gradient within a region around the keypoint locations. The gradient magnitude and the gradient orientation of the keypoint locations are computed as follows:

$$m(x,y) = \sqrt{(L(x+1,y)-L(x-1,y))^2 + (L(x,y+1)-L(x,y-1))^2} \quad (11.11)$$

$$\theta(x,y) = \tan^{-1}\left(\frac{(L(x,y+1)-L(x,y-1))}{(L(x+1,y)-L(x-1,y))}\right) \quad (11.12)$$

In the original method, the authors proposed using 36 bins of orientation histogram to cover the 360 degrees of gradient, so that each bin covers 10 degrees of direction. In total, there are 36 directional bins of feature descriptor. The gradient magnitude of local keypoints and a Gaussian-weighted circular window with σ equal to 1.5 times the scale of the keypoint are weighted for histogram binning. The highest detected peaks in these histograms represent dominant orientations. All local peaks that reach more than 80% of the highest peak value are considered keypoint features.

Finally, the descriptor of keypoint features is computed by using region 16×16 pixels around a keypoint. Each histogram consists of a 4×4 sub-region, known as a cell, of the original neighborhood regions. The eight-orientated bins of the histogram are used for computing each cell, which means that there are eight elements representing each cell. Therefore, there are $4 \times 4 = 16$ histograms with each eight bins, so the feature vector of the keypoint consists of 128 elements. Finally, the feature vector is normalized for invariance with illumination intensity (Figure 11.7).

11.2.2 OMNIDIRECTIONAL VISION–BASED ROTATION ESTIMATION

Taking advantage of omnidirectional vision for robust motion estimation, a vision-like compass is proposed to estimate the rotation of the robot. In an omnidirectional image, there is a special phenomenon, as depicted in Figure 11.8. Scenes on two sides of the camera mirror change in opposite directions when the camera moves straight. The landmarks in the front and rear of the robot mirror change quite slowly and separately in the two sides of the mirror. Landmarks are uniformly changed under camera rotation. The relation between robot rotation and the appearance of scenes at the front/rear of the robot would be utilized to estimate robot rotation. An omnidirectional camera is used as a visual compass for rotation estimation [35,36]. The correspondence edges of landmarks in the front and rear scenes of a robot are matched using the Chamfer matching method [37]. The method is briefly presented as follows. The Chamfer edge matching can be directly implemented on the omnidirectional image or the panoramic image. In this chapter, we present a method for processing on the panoramic image. In order to do this task, the omnidirectional image is unwrapped to the panoramic image based on previously proposed methodology [6].

Figure 11.7 Keypoint detection and matching. (a) Extracted keypoints of two sequential images. (b) Matching correspondence keypoints of two sequential images.

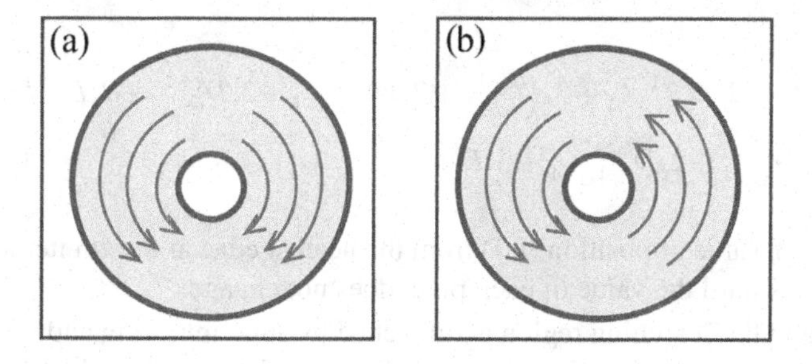

Figure 11.8 Image landmark transformation flowing motion. (a) Translation. (b) Rotation.

The template region in the front scenes of the robot in the current frame and the searching region in the next frame are extracted (Figure 11.9a,b. To reduce processing time, the searching regions are restricted to the field of view of about 30° around the front scene of the robot, which is shown in the red color window in Figure 11.9(b). The edges of both the template and the searching region are extracted using any kind of edge detection method.

The distance transformation of both template and searching regions is computed. Initially, all edge pixels are set to zero, and the remaining non-edge pixels are set to infinity. The integer approximation of Euclidian distance was used to compute the distance transform (DT) by the following expression:

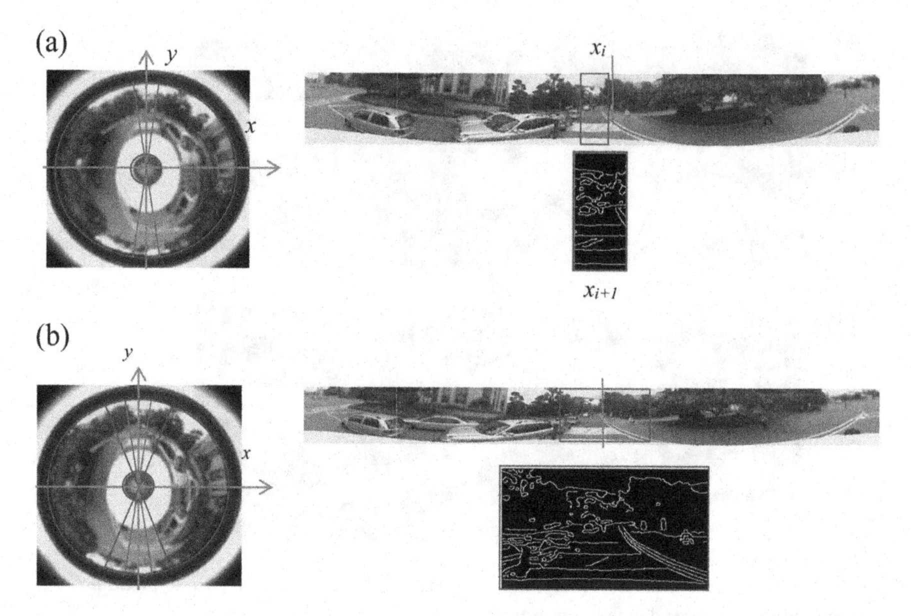

Figure 11.9 Extracting candidate regions for matching. (a) The template window extract from current frame. (b) The searching region extracts from the next frame.

$$D^k_{i,j} = \min(D^{k-1}_{i-1,j} + 2, D^{k-1}_{i-1,j-1} + 3, D^{k-1}_{i,j-1} + 2, D^{k-1}_{i+1,j-1} + 3, D^{k-1}_{i+1,j} + 2, D^{k-1}_{i+1,j+1} + 3, D^{k-1}_{i,j+1} + 2, D^{k-1}_{i-1,j+1} + 3, D^{k-1}_{i,j}) \quad (11.13)$$

where $D^k_{i,j}$ is the DT value of position (i, j) from the nearest edge at the kth iteration. This process iterates until the value of each pixel does not change.

The template and the searching region are matched by superimposing and sliding the template onto the searching region. The mean square of subtraction pixel values between the distance transformation of the template and that of the searching region is computed. It is called the edge distance. This value is zero in the case of a perfect match. Finally, the rotation angle between successive camera positions is computed using the following equation:

$$\alpha = \frac{180(x_{i+1} - x_i)}{(\pi R)} \quad (11.14)$$

where

R is the outer radius of the bounding circle with respect to the width of the panoramic image.

x_i and $x_{i+1}$are the center matching locations on the horizontal axis in the current image and the next image.

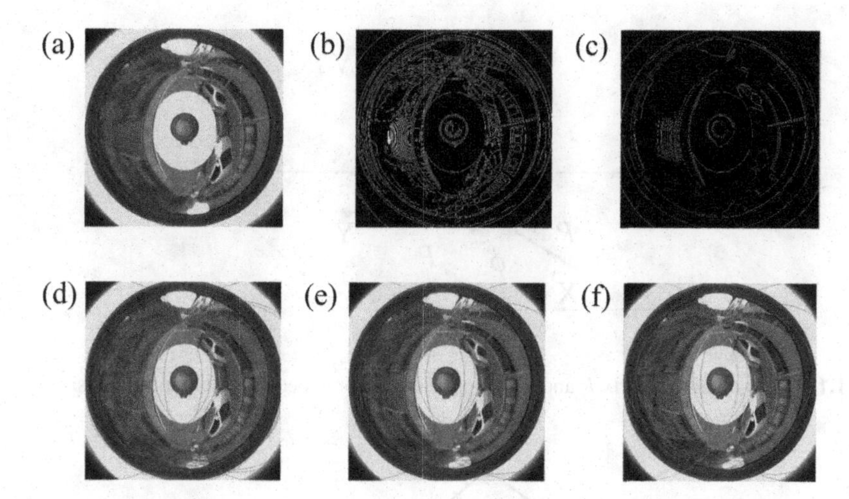

Figure 11.10 Extraction of parallel lines and vanishing point. (a) Original image. (b) Ege detector images. (c) Extracted chains. (d) Catadioptric line detection. (e) catadioptric line results. (f) vanishing point.

11.2.3 VANISHING POINTS FOR ROTATION ESTIMATION

In order to estimate vanishing points for rotation estimation using an omnidirectional image, it is necessary to extract parallel lines by using its properties in the image plane [33]. Assume that $P_1, P_2, \ldots P_n$ belong to a line in the omnidirectional image. Without loss of generality, the terminal points are denoted as P_1 and P_n. They are reprojected on the sphere model. Two points and the sphere center constitute a plane. A point is on a line if its distance to a plane is less than the threshold. The endpoints of the chain are presented as $P_1 = (x_1, y_1, z_1)$ and $P_n = x_n, y_n, z_n)$, and the normal vector of the plane is $\vec{n} = \overrightarrow{O_1P_1} \times \overrightarrow{O_1P_N} = (n_x, n_y, n_z)^T$. The intersected circle of the sphere and the plane is defined as follows:

$$\begin{cases} n_xX + n_yY + n_zZ = 0 \\ (X,Y,Z) \notin S^2 \end{cases} \tag{11.15}$$

A comparison threshold of the distance of a point $P_i = (x_i, y_i, z_i)$ to the plane is presented via a normal vector as follows (Figure 11.10):

$$\left| n_xX_i + n_yY_i + n_zZ_i \right| \leq \lambda \tag{11.16}$$

To find the corresponding vanishing points in the consecutive frames, the correspondence of the bundle of lines needs to be extracted. As analyzed earlier, each curve belongs to one circle in the sphere model. Therefore, a line coincides with the plane normal vector containing the circle. Let u_i be the ith direction detected in the first image with $i = 1, 2 \ldots . n$. The continuity constraint simply matches the pair $(u_i, u_{i'})$ having the lowest angle (Figure 11.11).

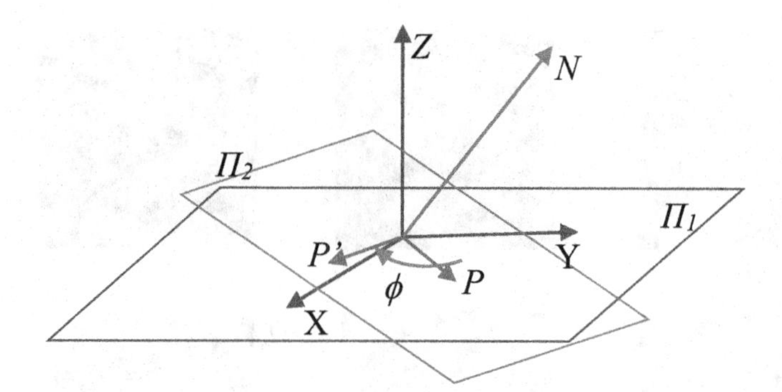

Figure 11.11 The rotation matrix R and the translation vector T between catadioptric positions.

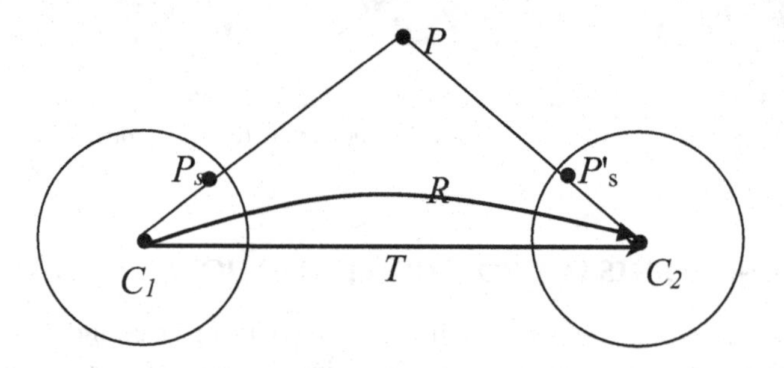

Figure 11.12 Rotation defined by Rodrigues's form.

The correspondence parallel lines set in the first and second images are represented by $s_l = \{l_1, l_2, \ldots l_n\}$ and $s_l{}' = \{l_1{}', l_2{}', \ldots l_n{}'\}$, corresponding to normal vectors $n = \{n_1, n_2, \ldots, n_m\}$ and $n' = \{n_1{}', n_2{}', \ldots, n_m{}'\}$. The geometrical constraint is represented as $s_l{}' = Rs_l$, where R is a rotation matrix. This rotation matrix is formed by the Rodrigues formula with a rotation axis N and a rotation angle ϕ (Figure 11.12).

$$u' = \cos\phi \cdot u + (1 - \cos\phi)(u \cdot N)N + \sin\phi[N \times u] \tag{11.17}$$

$$N = \frac{(u_1 - u_1') \times (u_2 - u_2')}{\left\|(u_1 - u_1') \times (u_2 - u_2')\right\|} \tag{11.18}$$

$$\cos\phi = \frac{u_2' \cdot u_2 - (u_2 \cdot N)^2}{1 - (u_2 \cdot N)^2} \tag{11.19}$$

$$\sin\phi = \frac{u_2' \cdot (N \times u_2)}{\|N \times u_2\|^2} \tag{11.20}$$

11.3 ADVANCED SENSORS FOR AUTONOMOUS NAVIGATION

In recent years, in the localization and mapping field applied in autonomous navigation, the problem of the combination of multiple sensors has been considered as a solution for improving precise capacity and speeding up processing time. When a robot travels a long distance, the estimated trajectory will diverge compared with the real path of movement, which is caused by accumulative error of motion estimation. This error is a traditional problem in motion estimation based on relative sensors, such as camera, LiDAR, IMU, etc. The cumulative errors of visual odometry and other relative sensors are solved by a method based on the combination of local motion estimation cameras/LRFs and GPS. In this section, some kinds of advanced sensors are presented with their applications to robot motion estimation and navigation.

11.3.1 LASER RANGE FINDER FOR MOTION ESTIMATION

The application of an LRF to estimating robot motion is briefly summarized. An LRF is a rangefinder using laser beams to determine the distance from the device center to an object, as depicted in Figure 11.13. The LRF estimate is based on the time-of-flight principle by sending laser beams toward objects and measuring the time for the reflection to return to the sensor. A light pulse of a defined time length is reflected off an object and is received via the same path as it was sent. The precision of LRF is estimated by the rise or fall time of the laser beam and the speed of the receiver.

LRF sensors are used in autonomous robot systems for motion estimation, obstacle detection, and obstacle avoidance. In robotic systems, the LRF sensor is also employed to extract point clouds of 3-D scenes for map reconstruction and to estimate potential obstacles to the robot during navigation. Here, we briefly present the Iterative closest point (ICP) method for extracting transformation information, such as rotation and translation parameters. An example experimental result for robot motion is illustrated as follows:

Input: Lines scan P, L

1. Initial transform R, t, and error E=INF.
2. Repeat
3. $E \longrightarrow E'$;
4. For all $x_i \in P_k$, find closest point $y \in L$;
5. Transform F(R,T): $P_k \longrightarrow P_{k+1}$ to minimize distances between each x_{i+1} and y;
6. Computing the error E;
7. Until $|E\text{-}E'| \leq$ threshold
8. Output: R, t

11.3.2 GPS FOR RECTIFYING VISUAL ODOMETRY

GPS information allows the receiver position to be estimated based on an unobstructed signal from least four GPS satellites. GPS-based localization provides important capabilities, which have been applied in many fields, both military and civilian. In the field of research on outdoor autonomous navigation applications, GPS

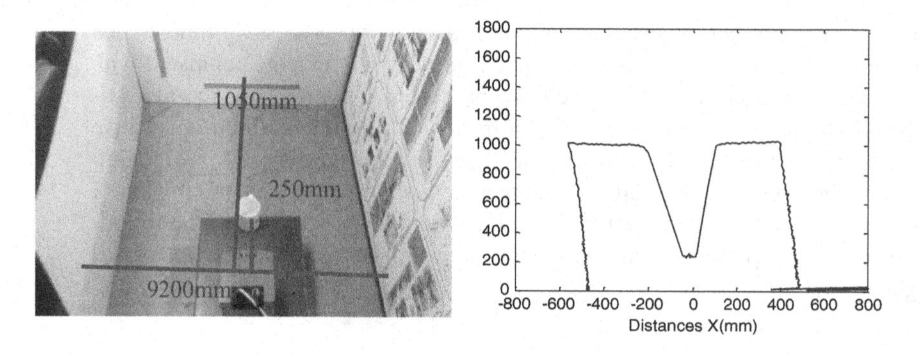

Figure 11.13 LRF measures distance from objects to device.

is considered as information for rectifying the global position and orientation. However, a high-precision GPS device is more expensive, while the standard GPS receiver is usually uncertain of local position in detail. Especially if the robot is moving in eclipsed areas such as high buildings or tunnels, the GPS positioning will be affected by significant error. To improve GPS positioning, it can be combined with other sensors for fixing global and local position estimation [38]. The observation equation of GPS positioning is formed as follows (Figure 11.14):

$$P_{gps}(t) = (x_{gps}(t), y_{gps}(t), \theta_{gps}(t))^T \tag{11.21}$$

where the GPS observation $(x_{gps}(t), y_{gps}(t), \theta_{gps}(t))$ is GPS position measurement. The GPS measurement may be affected by noise, so the GPS position error can be

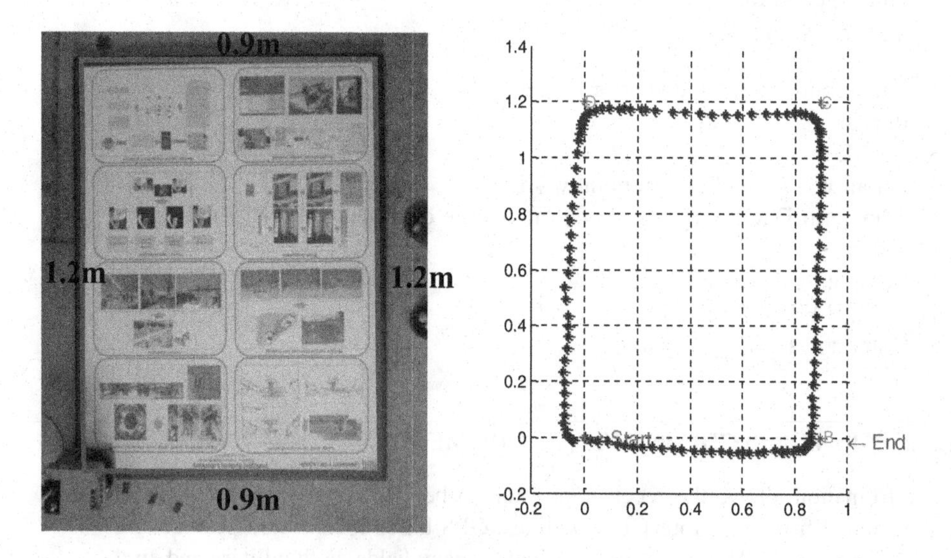

Figure 11.14 Trajectory estimation of movement using LRF sensor.

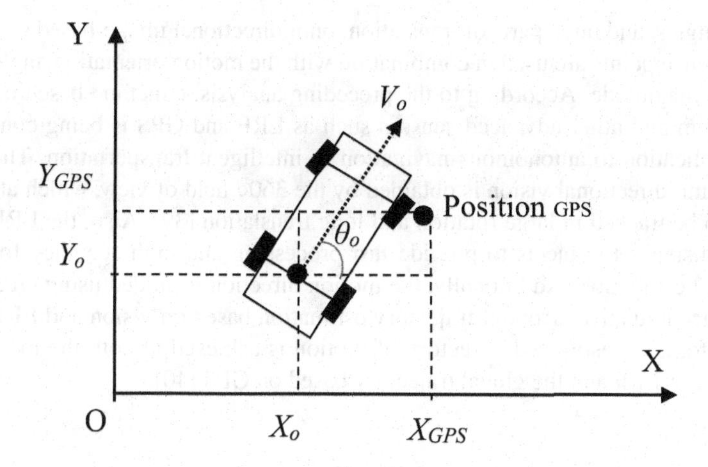

Figure 11.15 Combining the GPS and other sensors for localization estimation.

described as follows:

$$W_{gps}^{-1} = \begin{pmatrix} \sigma_{xgps}^{2}{}^{-1} & 0 & 0 \\ 0 & \sigma_{ygps}^{2}{}^{-1} & 0 \\ 0 & 0 & \sigma_{\theta gps}^{2}{}^{-1} \end{pmatrix} \tag{11.22}$$

The Mahalanobis distance is used to evaluate the disparity of the GPS and visual odometer position as follows (Figure 11.15):

$$d_{xy}(t) = \sqrt{\frac{(X_{gps}(t) - X_o(t))^T(\Sigma_o(t) + \Sigma_{gps}(t))^{-1}(X_{gps}(t) - X_o(t))}{2}} \tag{11.23}$$

Similarly, the direction distance is computed as follows:

$$d_{\theta}(t) = \sqrt{\frac{(\theta_{gps}(t) - \theta_o(t))^T(\sigma_{\theta o}^2(t) + \sigma_{\theta gps}^2)^{-1}(\theta_{gps}(t) - \theta_o(t))}{2}} \tag{11.24}$$

The correlation and covariance matrices Σ_{gps} and $\sigma_{\theta gps}$ are determined from the measurement error of GPS observation. In the case of a distance lower than the threshold value, GPS location is a reasonable response and is used for visual trajectory correction.

11.4 INTELLIGENT TRANSPORTATION SYSTEMS BASED ON A COMBINATION OF MULTIPLE SENSORS

The system consists of an omnidirectional camera, a two-dimensional (2-D) LRF, and a GPS, all mounted on the robot. The 2-D LRF is used to estimate the orientation of motion and translation magnitude of the robot. For estimating rotation components of yaw, pitch angles, and three parts of translation, omnidirectional image-based single correspondence points are used in combination with the motion orientation angle and translation magnitude. According to the preceding analysis, a method based on the vision system and other advanced sensors such as LRF and GPS is being considered for application to autonomous navigation in intelligent transportation. The advantage of omnidirectional vision is obtained by the 360o field of view, which allows objects to be tracked in large rotation and long translation [39]. Also, the LRF measures the distance to objects to provide fast processing and high accuracy for absolute motion estimation. Additionally, the motion direction is guided using GPS for rectifying accumulative errors of trajectory estimation based on vision and LRF sensors. Therefore, the estimated trajectory of motion is achieved in both the local accuracy of vision-LRF and the global trajectory based on GPS [40].

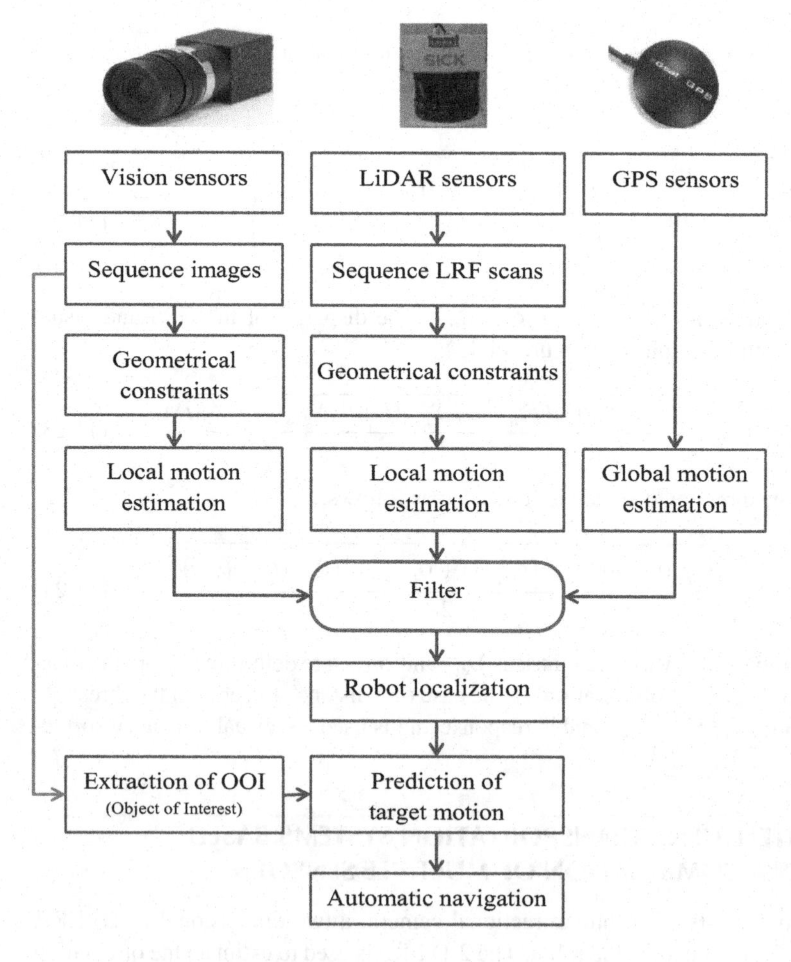

Figure 11.16 General flowchart of combination of multiple sensors.

The combination of visual odometry with LRF achieves significant accuracy and processing time in the estimation of robot movement. The vision-based method provides advantages for correcting estimation error in local positions and short travel distances, but it encounters an accumulating error problem.

On the contrary, the GPS yields acceptable global positioning, but it has insufficient precision in local detail to assist in autonomous navigation. The precision of the GPS signal is affected by multiple paths and diffraction; therefore, position prediction error increases significantly when the robot is moving near high buildings or in tunnels. Taking into account the advantages of each kind of sensor, the positional prediction is estimated based on a combination method using the EKF filter. This multiple-sensor system is also used for mutual compensation in the case that one sensor malfunctions. The overview method for the combination of multiple sensors for localization estimation is depicted in Figure 11.16.

The result of GPS positioning is used to correct the result of visual and LRF estimation. The equation of maximum likelihood based on the EKF framework is applied as follows:

$$\hat{P}_f(t) = P(t) + \Sigma_f(t) W_{gps}^{-1}(P_{gps}(t) - P(t)) \tag{11.25}$$

$$\Sigma_f(t) = (\Sigma_p(t)^{-1} + W_{gps}^{-1})^{-1} \tag{11.26}$$

where $\hat{P}_f(t)$ and $\Sigma_f(t)$ represent the corrected robot position and the error covariance matrix. For correcting the robot position estimated by odometry, each $P(t)$ and $\Sigma_p(t)$ is updated with $\hat{P}_f(t)$ and $\Sigma_f(t)$.

11.5 DISCUSSION

Nowadays, scene understanding and localization based on the combination of multiple sensors has become a topic of great importance in outdoor mobile robot research and applications. Robot localization and mapping using the combined multiple-sensor system is of significant importance. A system based on the combination of vision and advanced sensors is proposed to estimate the motion of a robot over a long distance. Some advantages were pointed out through our arguments and experiments. The vision-like compass works accurately in the complicated scene structure and varying conditions of outdoor environments. The robustness and rapidity of this approach can be promoted for real-time application. An accumulative error in visual odometry leading to a diverging trajectory is solved by compensating with GPS-based location. The combination of multiple sensors for local and global location systems not only reduces cumulative error but also corrects the GPS-based position.

This chapter has presented a set of methods for motion and localization estimation. We expect that others will build on these ideas and other existing work in the community of intelligent transportation systems and autonomous navigation to advance the study of related fields. Several ideas for future work could spring from this presentation. For improving the accuracy of motion prediction using optimal and advanced filters, an optimization approach for motion estimation is ongoing in the simultaneous localization and mapping (SLAM) field of research. The investigation and utilization of convex optimization for error reduction of motion estimation is an important task to avoid the trap of local minima. For object detection and behavior prediction, in order to construct an intelligent outdoor robot/vehicle that can understand the environment around itself, it is necessary for it to recognize and identify not only pedestrians but also other objects such as traffic signals, vehicles, and road characteristics. Modeling the understanding and prediction of vehicle and pedestrian behaviors is interesting for collision avoidance as well as driving behavior. To develop a system able to predict the future behavior of vehicles and pedestrians in an urban environment, it must not only recognize the scene surrounding itself but also predict their behavior based on the current situation of objects and traffic guidance, which supports decisions on future actions such as avoiding collisions and accidents and discarding unreliable samples.

REFERENCES

1. Zhang, H., A. Geiger, and R. Urtasun, Understanding high-level semantics by modeling traffic patterns. In: *International Conference on Computer Vision (ICCV)*, Sydney, NSW, Australia, 2013.
2. Hoang, V.-D., et al. 3D motion estimation based on pitch and Azimuth from respective camera and laser rangefinder sensing. In: *IEEE/RSJ International Conference on Intelligent Robots and Systems (IROS)*, Tokyo, Japan, 2013.
3. Murillo, A.C., G. Singh, J. Kosecká, and J.J. Guerrero, Localization in urban environments using a panoramic gist descriptor. *IEEE Transactions on Robotics*. **29**(1): p. 146–160, 2013.
4. Le, M.-H., et al. Vehicle localization using omnidirectional camera with GPS supporting in wide urban Area. In: *Computer Vision - ACCV 2012 Workshops*, J.-I. Park and J. Kim, Editors, Springer: Berlin Heidelberg. p. 230–241, 2013.
5. Hoang, V.-D., M.-H. Le, and K.-H. Jo, Hybrid cascade boosting machine using variant scale blocks based HOG features for pedestrian detection. *Neurocomputing*. **135**: p. 357–366, 2014.
6. Hoang, V.-D., A. Vavilin, and K.-H. Jo, Fast human detection based on parallelogram haar-like feature. In: *The 38th Annual Conference of The IEEE Industrial Electronics Society*, Montréal, Canada, 2012.
7. Chai, D., W. Forstner, and F. Lafarge, Recovering line-networks in images by junction-point processes. In: *Computer Vision and Pattern Recognition (CVPR)*, Portland, OR, 2013.
8. Murphy, L. and P. Newman, Risky planning on probabilistic Costmaps for path planning in outdoor environments. *IEEE Transactions on Robotics*. **29**(2): p. 445–457, 2013.
9. Vonasek, V., et al. Global motion planning for modular robots with local motion primitives. In: *IEEE International Conference on Robotics and Automation (ICRA)*, Karlsruhe, Germany, 2013.
10. Du Toit, N.E. and J.W. Burdick, Robot motion planning in dynamic, uncertain environments. *IEEE Transactions on Robotics*. **28**(1): p. 101–115, 2012.
11. Jaillet, L. and J.M. Porta, Path planning under kinematic constraints by rapidly exploring manifolds. *Robotics, IEEE Transactions On*. **29**(1): p. 105–117, 2013.
12. Achtelik, M.W., et al. Path planning for motion dependent state estimation on micro aerial vehicles. In: *2013 IEEE International Conference on Robotics and Automation (ICRA) IEEE*, Karlsruhe, Germany, 2013.

13. Viola, P., M.J. Jones, and D. Snow, Detecting pedestrians using patterns of motion and appearance. In: *International Conference on Computer Vision*, Nice, France, 2003.
14. Papageorgiou, C. and T. Poggio, A trainable system for object detection. *Intenational Journal Compute Vision*. **38**(1): p. 15–33, 2000.
15. Wang, X., T.X. Han, and S. Yan, An HOG-LBP human detector with partial occlusion handling. In: *International Conference on Computer Vision*, Kyoto, Japan, 2009.
16. Li, B., D. Huang, C. Wang, and K. Liu, Feature extraction using constrained maximum variance mapping. *Pattern Recognition*. **41**(11): p. 3287–3294, 2008.
17. Fraundorfer, F. and D. Scaramuzza, Visual odometry :Part II: Matching, robustness, optimization, applications. *IEEE Robotics and Automation Magazine*. **19**(2): p. 78–90, 2012.
18. Yongchun, F., L. Xi, and Z. Xuebo, Adaptive active visual servoing of nonholonomic mobile robots. *IEEE Transactions on Industrial Electronics*. **59**(1): p. 486–497, 2012.
19. Sato, T., S. Sakaino, E. Ohashi, and K. Ohnishi, Walking trajectory planning on stairs using virtual slope for biped robots. *IEEE Transactions on Industrial Electronics*. **58**(4): p. 1385–1396, 2011.
20. Scaramuzza, D. and F. Fraundorfer, Visual odometry [tutorial]. *IEEE Robotics and Automation Magazine*. **18**(4): p. 80–92, 2011.
21. Levin, A. and R. Szeliski, Visual odometry and map correlation. In: *IEEE Computer Society Conference on Computer Vision and Pattern Recognition*, IEEE, Washington DC, 2004.
22. Scaramuzza, D., A. Martinelli, and R. Siegwart, A toolbox for easily calibrating omnidirectional cameras. In: *IEEE IEEE/RSJ International Conference on Intelligent Robots and Systems*, 2006.
23. Mei, C. and P. Rives, Single view point omnidirectional camera calibration from planar grids. In: *IEEE International Conference on Robotics and Automation (ICRA)*. p. 3945–3950, Roma, Italy, 2007.
24. Geyer, C. and K. Daniilidis, Catadioptric projective geometry. *International Journal of Computer Vision*. **45**(3): p. 223–243, 2001.
25. Hartley, R. and A. Zisserman, *Multiple View Geometry in Computer Vision*, Vol. 2, Cambridge University Press, 2004.
26. Nistér, D., O. Naroditsky, and J. Bergen, Visual odometry for ground vehicle applications. *Journal of Field Robotics*. **23**(1): p. 3–20, 2006.
27. Hongdong, L. and R. Hartley, Five-point motion estimation made easy. In: *18th International Conference on Pattern Recognition*, Hong Kong, China, 2006.
28. Harris, C. and M. Stephens, A combined corner and edge detector. In: *Alvey Vision Conference*, Citeseer, 1988.
29. Lowe, D., Distinctive image features from scale-invariant Keypoints. *International Journal of Computer Vision*. **60**(2): p. 91–110, 2004.
30. Bay, H., T. Tuytelaars, and L.V. Gool, SURF: Speeded up robust features. In: *Proceedings of the 9th European Conference on Computer Vision - Volume Part I*, Springer-Verlag: Graz, Austria, p. 404–417, 2006.
31. Mikolajczyk, K. and C. Schmid, A performance evaluation of local descriptors. *IEEE Transactions on Pattern Analysis and Machine Intelligence*. **27**(10): p. 1615–1630, 2005.
32. Dalal, N. and B. Triggs, Histograms of oriented gradients for human detection. In: *Conference on Computer Vision and Pattern Recognition*, San Diego, CA, 2005.
33. Le, M.-H., K.-H. Jo, and Urban Area, *Scene Understanding through Long Range Navigation by Vision Based Mobile Robot*, University of Ulsan. p. 1–85, 2012.
34. Meng, Y. and B. Tiddeman, *Implementing the Scale Invariant Feature Transform (SIFT) Method*, School of Computer Science - University of St Andrews, Scotland, 2008.

35. Scaramuzza, D. and R. Siegwart, Appearance-guided monocular omnidirectional visual odometry for outdoor ground vehicles. *IEEE Transactions on Robotics*. **24**(5): p. 1015–1026, 2008.
36. Labrosse, F., The visual compass: Performance and limitations of an appearance-based method. *Journal of Field Robotics*. **23**(10): p. 913–941, 2006.
37. Barrow, H.G., et al. Parametric correspondence and chamfer matching: Two new techniques for image matching. In: *5th International Joint Conference on Artificial Intelligence*, Morgan Kaufmann Publishers Inc.: Cambridge, USA. p. 659–663, 1977.
38. Le, M.-H., V. Hoang, A. Vavilin, and K. Jo, One-point-plus for 5-DOF localization of vehicle-mounted omnidirectional camera in long-range motion. *International Journal of Control, Automation and Systems*. **11**(5): p. 1018–1027, 2013.
39. Hoang, V.-D. M-H. Le, and K-H. Jo, Motion estimation based on two corresponding points and angular deviation optimization. *IEEE Transactions on Industrial Electronics*. **64**(11): p. 8598–8606, 2017.
40. Hoang, V.-D. and, K-H. Jo, A simplified solution to motion estimation using an omnidirectional camera and a 2-D LRF sensor. *IEEE Transactions on Industrial Informatics*. **12**(3): p. 1064–1073, 2016.

12 Human–Robot Interaction

Kouhei Ohnishi

CONTENTS

12.1 HUMAN–ROBOT INTERACTION

Humans can identify the physical characteristic of the object in an instant just by touching it. They can easily distinguish whether the object is soft like a sponge, is rigid like an iron, has elasticity like a balloon, or is moving by itself. That sensation is an ability of the human being called "haptic sense". "Real-haptics" is a technology to reconstruct haptic sense by acquiring dynamic physical information that is transferred bi-directionally between the surrounding environment and the human. An abandonment of haptics causes difficulty in further advances in automated machines or may even result in threatening the safety and security of the process. A robot without real-motion is not a robot. The interaction between the robot and the human is deeply characterized by motion. From this point of view, the chapter focuses on force interaction between robots and humans. There are many papers and tutorials on visual and/or auditory interactions between machines and humans. Their focus is not always limited to robots; therefore they should be discussed in the category of signal processing. For that reason, this chapter excludes them from consideration. In 1920, Czech novel writer Karel Čapek published a science fiction drama entitled "RUR" [1]. The word "robot" was introduced for the first time in the drama. However, the robots in this drama are very different from those in today's industrial systems. The robots in the drama appear as if they are duplicates of humans. They can perform everything better than human beings. In the 1960s, almost 40 years after RUR, the first industrial robot was released from Unimation. Unlike in science fiction, the first industrial robot was a playback type. From the historical background, the industry was focused on mass production. Society required more flexible production lines, and the playback type robots were highly welcomed. In some application areas, so-called automation apparatuses were replaced by industrial robots because of their flexible and versatile abilities. This was the required role for robots in 20th century. However, in the 21st century, new roles for robots are arising. The required working applications for robots are diverging from the fields of agriculture, civil and construction engineering, utility service, medical and welfare applications, rescue and recovery of disaster, ocean and cosmic development, and personal support of daily service, to entertainment and so on. They need human

Table 12.1
Comparison of Conventional Robots and Upcoming New Robots

	Conventional robots	New robots to come
Target application	Production line	Non-production area which depends on human ability of motion task
Applied environment	Structured (fixed task in modeled environment)	Unstructured (variable task in unmodeled environment)
Task motion	Upper limb motion	Upper limb motion with mobility
Installation	Fixed	Movable
Required motion	Position-based motion	Compliant according to environment
Relation with humans	Replacement of humans	Cooperation with humans

interactions to a greater or lesser extent and thus, force control is essential in the implementation. Old-fashioned industrial robots do not always cover this field. In general usage, they are independent, and sometimes, humans are not allowed to approach within a certain distance. It is hoped that the coming new robots will be able to to perform some part of human motion. However, "human motion" is an implicit knowledge that can only be performed by humans due to the combination of high skill, rich experience, superior judgment, and high morality (not by industrial robots alone). The new robots, therefore, at least should have the ability to physically perform the motion done by humans. Humans can take care of the implicit knowledge part. This means that we should take as our aim not to replace the human task by the robot but to combine both the human judgment and the robot motion in a proper way. This is essential in interactions between humans and robots. To simplify the issue, Table 12.1 compares the new robots and the conventional ones. From this interpretation, the following four points are necessary in the coming new robots.

a. On-site self-mobility to reach an adequate point according to the situation
Technically, this ability could be installed by using present technology, including bi-pedal walking, wheeled platforms, and so on. This ability is essential for a robot to take part as a physical agent of the human in cooperation. The robot should be able to travel to all the places required by the human.
b. Flexible and adaptive operation in a hard-realtime manner.
Even with eyes closed, a human can recognize the characteristics of an object, whether it is stiff or soft, warm or cool, alive or mineral, and so on. This ability gives not only such recognition but also a kind of "fell" of the touched object. A human adjusts the imposed force according to this ability. It is not hard to imagine that time is an important issue in this ability. It is preferred that recognition occurs as fast as possible to avoid accidents.
c. Extending human ability.

The benefit of using robots needs to be clear. It is preferred that the efficiency or effort of the work is improved by using robots. For example, amplifying the human force by a certain amount or reducing the human force by a certain amount to produce a tiny force in the micro world can extend human ability. The ability is beneficial for applications such as high-power output motion in construction work and vivid touch sense in highly sensitive manipulation in microsurgery. The integration with controllability of applied force onto the object will be a key technology for human–robot interaction.

d. Improvement of task environment of humans.
New robots to come in the future are expected to complete tasks with humans in sites where humans have difficulty working, such as hazardous environments, construction sites, farms, the human body, and so on. The main contribution of robots to society would be this benefit. Robots cooperating with humans will relax the severe conditions of the task environment and will provide a safer environment. Humans can possibly remotely operate robots from a distant safe place. This realizes not only a human-friendly environment but also the diversity of the human task. The basis of this ability is teleoperation with haptics or simply telehaptics. Telehaptics should realize the vivid sensation of touching the distant object. Humans operate the robot as if the object were just in front of them with the feeling of intuitive touching.

Since 1970 the machine tools have not had capability of autonomous nor intelligent motions. They could not adapt themselves to the working environment. In other words they have been so stiff for long time. With the help of visual processing and industrial sensing together with highly developed information processing, the application area of automation has been widened. However, there still exist many tasks that depend on human ability and cannot be executed by robots alone. Due to the aging of skilled workers and escaping of youth from real production processes, the numbers of laborers with high ability are reducing in industry. This is causing not only a deterioration of efficiency and productivity but also an increase of danger in the production process. The new robot is expected to overcome such risks by extending the human ability. Real-haptics is one of the most important key elements in new robots. Real-haptics is a technology that transmits the information of the touching sensation in a bilateral way to produce tactile or force sensation on the real object and/or the real environment. Humans are inherently equipped with this sensation. This sensation is too natural to imagine the situation in the case of losing it. If this sensation is lost, motion becomes quite dangerous. Humans may injure themselves or may destroy surrounding objects without knowing. Good control of touching force heavily depends on this sensation. In fact, most of industrial robots work without this sensation. Real-haptics technology can provide the touch sensation to the new robots to produce well-controlled force for achieving various tasks.

The sensing of the force and transmitting the force signal are the basic functions of real-haptics. By conventional technology, humans are capable of processing auditory and visual information. One important fact to point out is that transmission of auditory and visual signals is always unilateral. However, the force signal should be always bilateral, because the so-called "law of action and reaction" governs this signal. When a human is applying force to an object, the object is always applying the same amount of the reaction force to the human. The sensation of the force comes from the reaction force. In a remotely operated situation with real-haptics application, this reaction force should come from the distant place, ideally in the same instant. However, this reactive force originates from the active force by the human. This means that there always should exist both forward and backward paths simultaneously in the force signal. Figure 12.1 shows the difference of three human sensations schematically. A first prototype of force transmission was conducted in Argonne National Laboratory (ANL) in the United States in the 1950s. Dr. R. C. Goertz at ANL published the first important paper in 1952, which emphasized bilateral characteristics for transmission of force sensation [2]. Their aim of research was to remotely manipulate radioactive elements from behind the shield. Dr. R. C. Goertz had developed two kinds of bilateral systems for the issue: one equipped with only a mechanical system and another one electrically powered. While he succeeded in transferring vivid sensation in the mechanical one, the electrical system faced difficulty in the operation. The mechanical one directly connects master and slave with linkage and gears. Although the device was successful, there is a limitation of

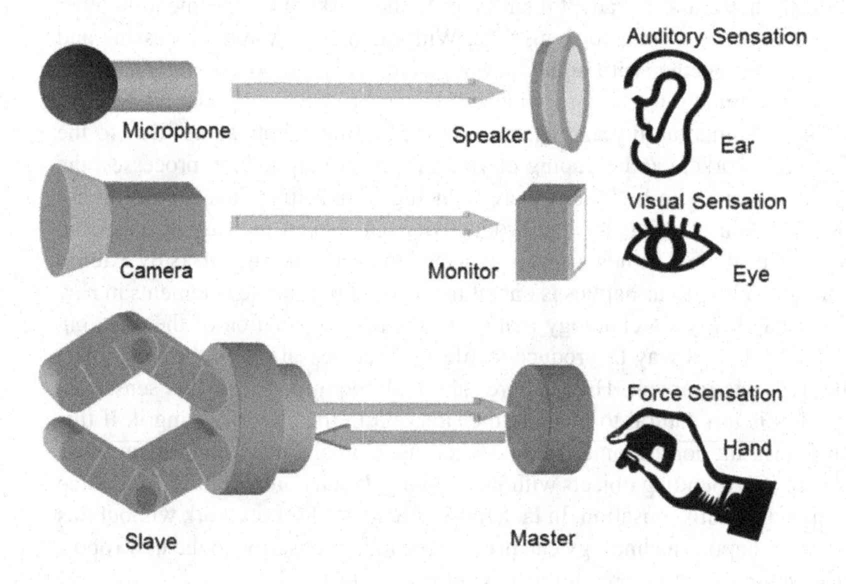

Figure 12.1 Comparisons of human sensations.

working distance due to mechanical connection. Since then, much research on haptics has been carried out to develop a successful electrical system. The difficulty of realizing haptics comes from the simultaneous satisfaction of synchronization and bilaterality of master and slave. The former and latter characteristics are expressed in the following two simple equations:

$$x_m - x_s = 0 \tag{12.1}$$

$$f_m + f_s = 0 \tag{12.2}$$

However the simultaneous realization of these equations is contradictory. Equation 12.1 means that the motion should be stiff against the disturbance; in control engineering, it is characterized by robust position control. However, Equation 12.2 requires soft motion against the disturbance to achieve pure force control; in control engineering, it is characterized by robust force control. The history of haptics has been more or less to find a compromise for this contradiction. The problem was almost perfectly overcome by the introduction of acceleration-based bilateral control (ABC) [3].

Figure 12.2 shows the schematic block diagram of ABC. In common mode or summing mode, a highly compliant or soft motion control is applied by force servoing to achieve Equation 12.2. On the contrary, in differential mode or subtracting mode, a highly stiff motion control is applied by a position regulator to achieve

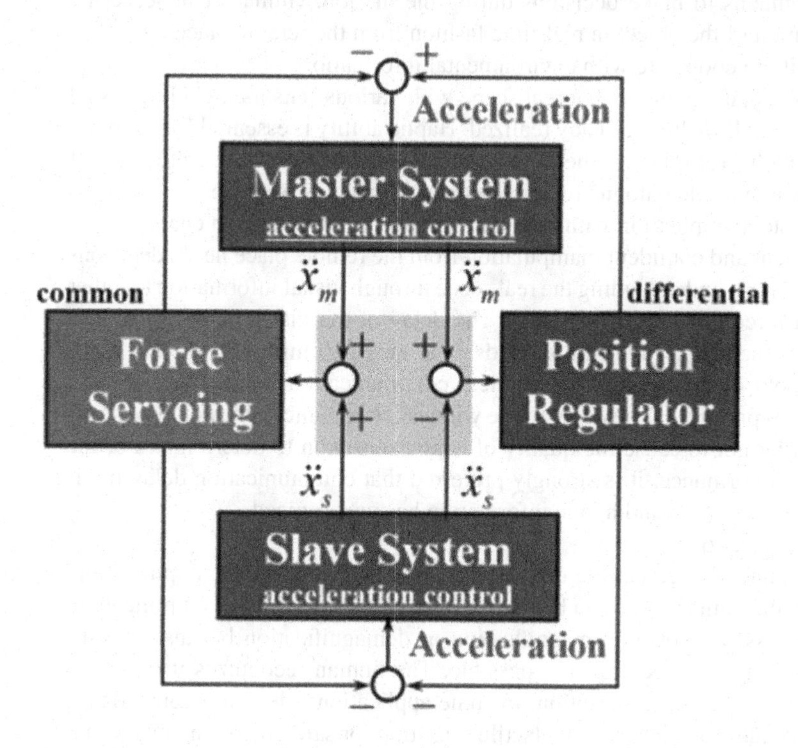

Figure 12.2 Block diagram of acceleration-based bilateral control.

Equation 12.1. These two modes are perpendicular to each other in the acceleration plane and linearly independent. In short, bilaterality is assured in common mode and synchronization is realized in differential mode in the acceleration plane. Note that the position synchronization is achieved after several sampling periods later than bilateral force transmissions. However, humans cannot feel this delay if the sampling time of the control is very short, such as several hundred micro-seconds. It is natural that the controller should be as fast as possible to realize the vivid sensation of touching. According to experiments, 100 micro-seconds (or faster) as a sampling time of the controller will be preferred; however, the performance specification is not so difficult for recent microcontrollers. Another merit of Figure 12.2 to send a real sensation of touching is that there are no force sensors required in the system. The force sensor is fragile, and its dynamic range is limited. If the force sensor is not needed, the total system will be robust, inexpensive, and simple. By introducing real-haptics, the following five features will be possible to improve quality-of-life in future society.

a. Ability to sense and transmit the touch sensing
Through touching at the slave side, the stiffness of the object, its motion, and various information on the surroundings are transmitted to the master side. The surrounding information includes vibration and surface condition for humans to make decisions during the motion. Humans can sense the situation of the object in real-time fashion from the remote place.
b. Ability to cooperate with environmental information
Robots that can move in a real scene with various sensors including visual sensation have been already realized. Haptic ability is essential for cooperation with such robots. When robots understand how to make careful contact with humans, a dialogue relation with such robots is possible.
c. Remote manipulation with wire and wireless communication channel
Efficient and confident manipulation from the remote place needs decisions by humans understanding the real scene through visual information together with force and tactile sensation. The delay of the visual signal should be short, such as several milliseconds or at most, 10 milliseconds. For more freedom of human motion, wireless communication between master and slave is preferred. The delay of the wireless communication should be small in order not to cause the quality of haptic sensation to deteriorate. For the best performance, it is strongly preferred that communication delay in visual information and haptic information be synchronized.
d. Superman effect
The human force can be extended from 1 to 10,000 times amplification. Also the human force can be reduced from 1 to 0.0001 times amplification, if necessary. Not only magnification or demagnification but also modification of the force signal is possible. The human recognizes it as a kind of illusion of force sensation. In some applications, there are demands for techniques to emphasize the tactile sensation for safe operation or easy manipulation. The concept can also be extended to virtual reality applications.

If the slave side is constructed in the computer, humans can have some effect on the virtual environment. The sensation given by the virtual environment can be constructed from measured real characteristics or modeled characteristics of the real object. This can be used in highly-realistic training or entertainment.

e. Record of the action and its playback
The master's action, including position change and force change, can be recorded. This information can be reproduced in the slave side. Like visual or auditory recording, human action is recorded by haptic devices. If such action includes expert performance, it is always reproduced by the slave machine as if it wwre operated from the master side. This is a kind of preservation of human action and human skills.

In addition to these functions, it is necessary to maintain safe operation even in dangerous situations. One possible example of application is haptic forceps to send the haptic sensation to the surgeon. Figure-12.3 shows an example of suturing (stitching) with 5 degrees-of-freedom motion. The white material is a model of the kidney; the system succeeds in inserting needle and making knots with force sensing. Force sensation in surgical application is important for safety. It can help the surgeon to avoid applying unnecessary force to the organ or surgical instruments. The string used in suturing tasks is sometimes very narrow; it is several micro-meters in so-called microsurgery tasks. It is so fragile that applying too much force results in breaking the string, making the surgeon perform the suture task again. The operation time becomes longer due to multiple repetitions of the same tasks. Real-haptics would be beneficial in such tasks by reducing the force output at the end-effector. Not only can the force applied to the string be reduced, but the surgeon can experience that the string is less fragile by amplifying the force information at the surgeon side. The experimental results on pushing motion by a linear motor are shown in Figure 12.4. The object is the sponge shown in Figure 12.4c. Figure 12.4a shows

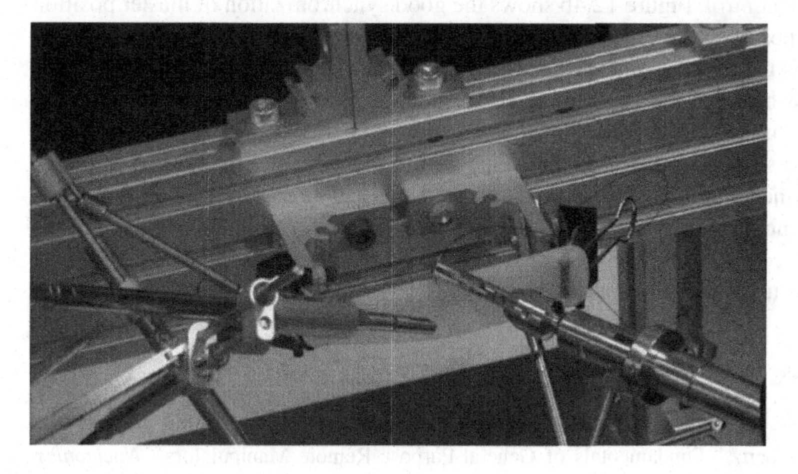

Figure 12.3 Suturing by haptic forceps.

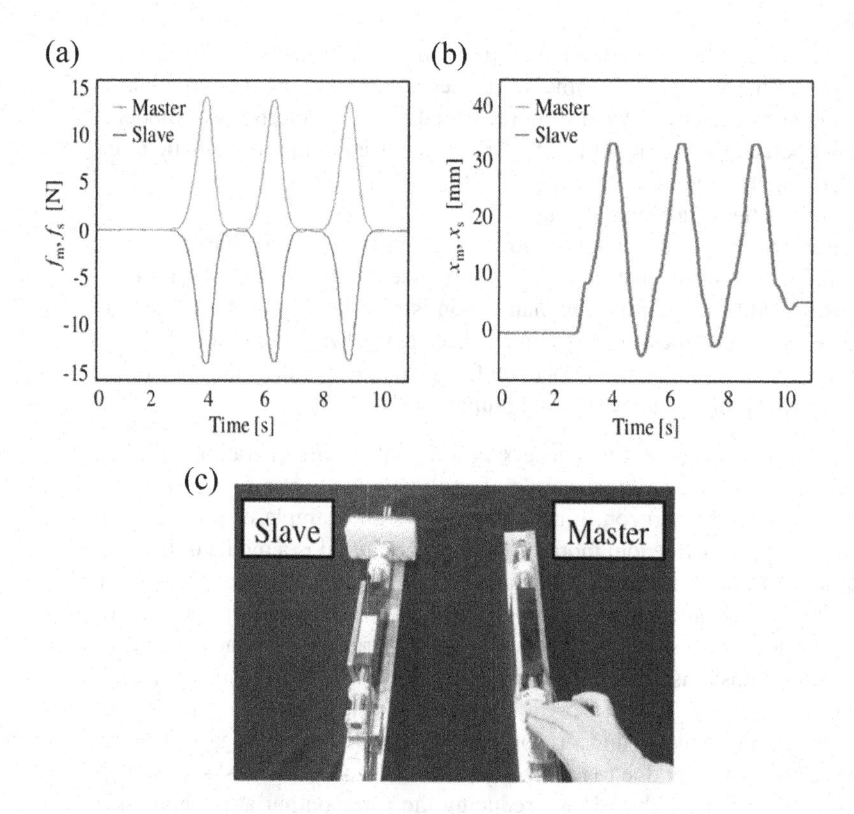

Figure 12.4 Experiment of pushing sponge.

very good coincidence of the imposed force at the master side and the reactive force at the slave side. This means that the action and reaction law is realized artificially by bilateral control. Figure 12.4b shows the good synchronization of master position and slave position. With the simultaneous satisfaction of two characteristics, an operator feels the sensation of the sponge although he is not directly touching it. The power flow between the master and the slave in the bilateral control system is quite essential to understand the real-haptics. The power factor in a motion system is convenient for such a purpose [4]. This chapter introduced a simple method to install haptics in the automatic system and suggested a proper cooperation of a robot and a human in more intuitive way. Superior motion dynamics of experienced persons can be digitalized into data with the technology. The technology will be beneficial to the upcoming super-mature society around the world.

REFERENCES

1. K. Čapek. *R.U.R.(Rossum's Universal Robots).* Penguin Classics, London, 2014.
2. R. C. Goertz. Fundamentals of General-Purpose Remote Manipulators. *Nucleonics*, 10(11):36–42, 1952.

3. Wataru Iida and Kouhei Ohnishi.Reproducibility and Operationality in Bilateral Teleoperation. In *The 8th IEEE International Workshop on Advanced Motion Control(AMC2004)*, Kawasaki, pages 217–222, March 2004.
4. T. Mizoguchi, T. Nozaki, and K. Ohnishi.A Method to Derive on Time Mechanical Power Factor. In *The 13th IEEE International Conference on Industrial Technology(ICIT2014)*, Busan, pages 73–78, February 2014.

Index

Printed in the United States
by Baker & Taylor Publisher Services